Smooth-Automorphic Forms and Smooth-Automorphic Representations

Series on Number Theory and Its Applications* ISSN 1793-3161

Published

Vol. 12 Multiple Zeta Functions, Multiple Polylogarithms and Their Special Values
by Jianqiang Zhao

Vol. 13 An Introduction to Non-Abelian Class Field Theory: Automorphic Forms of Weight 1 and 2-Dimensional Galois Representations
by Toyokazu Hiramatsu & Seiken Saito

Vol. 14 Problems and Solutions in Real Analysis (Second Edition)
by Masayoshi Hata

Vol. 15 Derived Langlands: Monomial Resolutions of Admissible Representations
by Victor Snaith

Vol. 16 Elementary Modular Iwasawa Theory
by Haruzo Hida

Vol. 17 Smooth-Automorphic Forms and Smooth-Automorphic Representations
by Harald Grobner

*For the complete list of the published titles in this series, please visit:
www.worldscientific.com/series/sntia

Series on Number Theory and Its Applications Vol. 17

Smooth-Automorphic Forms and Smooth-Automorphic Representations

Harald Grobner
University of Vienna, Austria

World Scientific

NEW JERSEY • LONDON • SINGAPORE • BEIJING • SHANGHAI • HONG KONG • TAIPEI • CHENNAI • TOKYO

Published by

World Scientific Publishing Co. Pte. Ltd.
5 Toh Tuck Link, Singapore 596224
USA office: 27 Warren Street, Suite 401-402, Hackensack, NJ 07601
UK office: 57 Shelton Street, Covent Garden, London WC2H 9HE

Library of Congress Control Number: 2023008551

British Library Cataloguing-in-Publication Data
A catalogue record for this book is available from the British Library.

Series on Number Theory and Its Applications — Vol. 17
SMOOTH-AUTOMORPHIC FORMS AND SMOOTH-AUTOMORPHIC REPRESENTATIONS

ISBN 978-981-124-616-6 (hardcover)
ISBN 978-981-124-617-3 (ebook for institutions)
ISBN 978-981-124-618-0 (ebook for individuals)

For any available supplementary material, please visit
https://www.worldscientific.com/worldscibooks/10.1142/12523#t=suppl

Desk Editors: Sanjay Varadharajan/Lai Fun Kwong

Typeset by Stallion Press
Email: enquiries@stallionpress.com

Printed in Singapore

To all who love automorphic forms. And to Café Hummel, Café-Restaurant Hold, Café Florianihof, Grande Cocktailbar, Weinstube Josefstadt, Restaurant Ufertaverne, and my dear Bar Pegaso – havens, where I found the leisure to write this book. May it not overly confuse you.

Preface

It would stand to reason to motivate our considerations by trying to answer the following question:

> **Question**: What is an automorphic representation?

However, as we shall try to convince the reader throughout this book, this is not the only (and thus, *a priori*, not necessarily the best) of all starting points – and that is not just for the obvious reason that the reader could find an answer to this question in many excellent sources (of which we only quote the fundamental [Corvallis]-volume (and here in particular [Bor-Jac79]), [Mœ-Wal95], [Sha10] or, for GL_n, [Bum98] and [Gol-Hun11]).

In 1888, Richard Dedekind wrote an influential essay on the foundations of mathematics, entitled *What are numbers and what should they be?*. In line with his thinking, we propose to rather ask ourselves:

> **Question**: What is an automorphic representation *and what should it be?*

In order to better illustrate our intentions behind this (in tendency rather grandiose) point, let G be a connected reductive group over a number field F, and let $\mathcal{A}(G)$ denote the space of automorphic forms $\phi : G(\mathbb{A}) \to \mathbb{C}$, where $\mathbb{A}$ is the ring of adèles of F. One of the defining properties of automorphic forms $\phi \in \mathcal{A}(G)$ is that they are K_∞-finite, where K_∞ is a fixed maximal compact subgroup of the real Lie group $G_\infty = G(F \otimes_\mathbb{Q} \mathbb{R})$:

$$\dim_\mathbb{C} \langle \phi(\_k), k \in K_\infty \rangle < \infty.$$

As an effect, not the full Lie group G_∞ (and hence not the whole group $G(\mathbb{A})$) acts on $\mathcal{A}(G)$ by right-translation, but only the Lie algebra $\mathfrak{g}_\infty =$

$\mathrm{Lie}(G_\infty)$, the compact Lie group K_∞ and the group $G(\mathbb{A}_f)$. In accordance with the usual theory we are led to set:

Definition 0.1. An *automorphic representation* is a subquotient $\pi = V/U$, where $\{0\} \subseteq U \subset V \subseteq \mathcal{A}(G)$ are subspaces that are stable under right-translation by $\mathfrak{g}_\infty$, K_∞ and $G(\mathbb{A}_f)$.

From this book's point of view, however, we understand this lack of an action of G_∞ as a defect, i.e., from the perspective of representation theory (and in the sense borrowed from Dedekind) this is not what an automorphic representation "should be". Indeed, we want an automorphic representation to be *a true representation of* $G(\mathbb{A})$. Therefore, we shall concern ourselves with a similar concept which we shall call "*smooth*-automorphic representation", in order to distinguish it from the classical one.

As a very first observation and as indicated by the above, the functions building up a "smooth-automorphic representation" *cannot all be* K_∞*-finite*, since this is not compatible with right-translation by G_∞. On the other hand, clearly the concept of smooth-automorphic representations should also not be too far away from the concept of classical automorphic representations, in order to be a useful addition to the theory. Indeed, in this regard we expect smooth-automorphic representations to satisfy certain properties, which we shall formulate in the course of this introduction as "wishes".

Wish 1. A smooth-automorphic representation Π should be a representation of $G(\mathbb{A})$ and should be "smooth", in the sense that all its elements are smooth vectors (together with some sort of topological compatibility).

Wish 2. For an admissible smooth-automorphic representation Π, the subspace

$$\Pi_{(K_\infty)} := \{\varphi \in \Pi \mid \dim_{\mathbb{C}}\langle \varphi(\_k) : k \in K_\infty \rangle < \infty\}$$

of K_∞-finite vectors should be an admissible automorphic representation in the classical sense of Def. 0.1. Vice versa, any admissible automorphic representation π should give rise to an admissible smooth-automorphic representation by taking an appropriate "completion" $\overline{\pi} = \Pi$.

To give an example of what we expect a smooth-automorphic representation to look like, let us consider the space $\mathcal{A}_{cusp}(G) \subseteq \mathcal{A}(G)$ of cuspidal automorphic forms. It is stable under the actions of $\mathfrak{g}_\infty$, K_∞ and $G(\mathbb{A}_f)$. Recall the following:

Definition 0.2. A *cuspidal* automorphic representation is a subquotient $\pi = V/U$ with $\{0\} \subseteq U \subset V \subseteq \mathcal{A}_{cusp}(G)$.

Fact. Up to isomorphism, any cuspidal representation is actually a *subrepresentation* $\pi = V$. So, if π is *irreducible*, then $V = \langle\phi\rangle_{(\mathfrak{g}_\infty, K_\infty, G(\mathbb{A}_f))}$ is spanned as a $(\mathfrak{g}_\infty, K_\infty, G(\mathbb{A}_f))$-module by a single cuspidal automorphic form $\phi \in \mathcal{A}_{cusp}(G)$.

In fact, more can be said about ϕ under a suitable technical assumption that we now formulate. Recall that there is a certain subgroup $A_G^{\mathbb{R}}$ of the center of G_∞, isomorphic to $\mathbb{R}^m_{>0}$ for some $m = m(G)$, with the property that $A_G^{\mathbb{R}}G(F)\backslash G(\mathbb{A})$ has finite invariant volume. Our assumption is that $A_G^{\mathbb{R}}$ acts trivially on π. If this is satisfied, then one can prove that ϕ lies in the space $L^2(A_G^{\mathbb{R}}G(F)\backslash G(\mathbb{A}))$ of measurable, left $A_G^{\mathbb{R}}G(F)$-invariant functions $f : G(\mathbb{A}) \to \mathbb{C}$, which are square-integrable on $A_G^{\mathbb{R}}G(F)\backslash G(\mathbb{A})$.

Identifying two functions in $L^2(A_G^{\mathbb{R}}G(F)\backslash G(\mathbb{A}))$, if they only differ on a zero-set, turns this space into a Hilbert space with the usual inner product, on which $G(\mathbb{A})$ acts unitarily by right-translation. Denote by $\widehat{\pi}$ the closure (or, equivalently, the completion) of $\langle\phi\rangle_{G(\mathbb{A})}$ inside $L^2(A_G^{\mathbb{R}}G(F)\backslash G(\mathbb{A}))$. This is a Hilbert-subspace, which is stable under the action of $G(\mathbb{A})$. It is well known (but not self-evident) that $\pi \subset \widehat{\pi}$ is a dense subspace. Let us call $\widehat{\pi}$ the *L^2-completion* or the *Hilbert space completion* of π. It turns out that

$$\pi = \widehat{\pi}^\infty_{(K_\infty)},$$

in other words, the elements of π are precisely the globally smooth, K_∞-finite vectors in $\widehat{\pi}$. Thus, for π cuspidal and irreducible, the space of globally smooth vectors $\widehat{\pi}^\infty$ just looks like the object we set out to find.

But, as they say, the devil is in the details, and indeed there are a few problems with the strategy we just outlined.

The first and most obvious drawback is that this strategy *a priori* only works for subrepresentations $\pi = V$ of $\mathcal{A}(G)$ spanned by square-integrable functions. A second, also quite obvious problem arises when looking more closely at the above construction and comparing it with the properties we expect of it.

Wish 3. Suppose we had found a general strategy to complete a given admissible automorphic representation π to an admissible smooth-automorphic representation $\Pi = \overline{\pi}$ (as desired in "wish 2"). Then, an *irreducible* π should give rise to an *irreducible* $\overline{\pi} = \Pi$ and vice versa. Moreover, for π cuspidal and irreducible the following triangle should commute

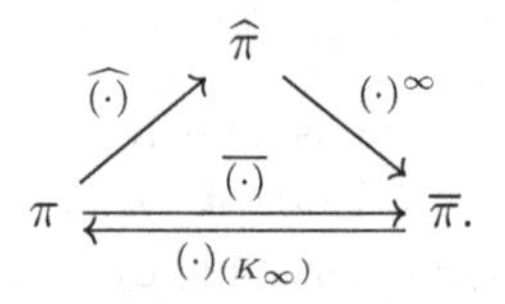

For the above special case where π is cuspidal and irreducible, this wish produces conflicting consequences, because all in a sudden, we would have found a $G(\mathbb{A})$-stable, proper subspace $\Pi = \overline{\pi} = \widehat{\pi}^\infty \subset \widehat{\pi}$, naively contradicting the fact that $\widehat{\pi}$ is irreducible. Less naively, this amounts to *adjusting our notion of irreducibility* in a way that also takes into account the topology on $\widehat{\pi}$ and one on $\overline{\pi}$. In other words, a working theory of smooth-automorphic representations will *need to deal with questions of topologies* on its representation spaces (and so to resolve the above naive contradiction by revealing $\overline{\pi}$ as a non-closed subspace of $\widehat{\pi}$). Consequently, a certain level of background knowledge from functional analysis will be entirely indispensable, and we shall provide a concise while at the same time comprehensive reminder of the necessary notions and concepts at the beginning of Parts I and II of this book.

As a conclusion to this introduction, let us formulate two more "wishes" that shall also guide our exposition:

Wish 4. Some sort of "decomposition theorem" should hold, linking local representation theory with global, automorphic representation theory: Recall that we have for an irreducible automorphic representation

$$\pi \underset{(\mathfrak{g}_\infty, K_\infty, G(\mathbb{A}_f))}{\cong} \bigotimes_{\mathsf{v}}{}' \pi_{\mathsf{v}},$$

where π_{v} are irreducible admissible smooth representations of $G(F_{\mathsf{v}})$ for each non-archimedean place $\mathsf{v} < \infty$ and irreducible admissible $(\mathfrak{g}_{\mathsf{v}}, K_{\mathsf{v}})$-modules for each archimedean place $\mathsf{v} \mid \infty$, which are furthermore unique up to isomorphism. Accordingly, for an irreducible smooth-automorphic representation we expect an analogous result

$$\Pi \underset{G(\mathbb{A})}{\cong} \overline{\bigotimes}{}' \Pi_{\mathsf{v}},$$

but now with the benefit that the Π_{v} are irreducbile admissible smooth $G(F_{\mathsf{v}})$-representation at all places v. In turn, this decomposition should be compatible with the decomposition of $\pi = \Pi_{(K_\infty)}$, i.e., there should be isomorphisms $\pi_{\mathsf{v}} \underset{G(F_{\mathsf{v}})}{\cong} \Pi_{\mathsf{v}}$ for $\mathsf{v} < \infty$ and $\pi_{\mathsf{v}} \underset{(\mathfrak{g}_{\mathsf{v}}, K_{\mathsf{v}})}{\cong} (\Pi_{\mathsf{v}})_{(K_{\mathsf{v}})}$ for $\mathsf{v} \mid \infty$.

As indicated by the latter "wish", local representation theory shall cover an important part of this book, as a certain depth of knowledge thereof will be required for our global considerations.

In fact, after having recalled some necessary background from functional analysis, we will give an introduction to the representation theory of local groups. Unlike most references dealing with this important subject, we try to present the theory in a uniform way, i.e., treating archimedean as well as non-archimedean local groups together. We express our hope to have served the reader, who has a (in a certain way bizarre) passion for conceptualism, by following this (not always totally easy) path to local representation theory.

Our last "wish" is purely global again. It subsumes a version of "Langlands's subquotient theorem" for smooth-automorphic representations.

Wish 5. Some sort of notion of *parabolic support*, respectively, *cuspidal support* of a smooth-automorphic representation should exist, which extends the usual notion of parabolic support and cuspidal support of automorphic forms to smooth-automorphic forms.

Finally, it should be noted that the concept of extending automorphic forms to include those that are not necessarily K_∞-finite has its roots in the late 1980s and early 1990s: Just to name three pioneers of the theory, we would like to mention Casselman ([...]*one plausible, and perhaps useful, extension of the notion of automorphic form would be to include functions* [...] *which are $Z(\mathfrak{g})$-finite but not necessarily K-finite.*[1]), Wallach (see in particular [Wal94], §6.1, which, to the author's knowledge, is the first place, where a notion of a smooth-automorphic form was given), and Bernstein.

In that regard, this book is intended to be just another contribution to keep the theory moving forward by consolidating and bringing together in one place relevant facts from the literature, while presenting several results and insights in the global theory, that are either genuinely new, or never appeared with a (detailed) proof in the literature.

It is our hope that this book will be a fruitful source for those, who do want to devote themselves to the (in the author's opinion beautiful) theory of smooth-automorphic representations.

[1] [Cas89II], Rem. 3.6.

Standing safety instructions

(1) The letters $\mathbb{Z}$, $\mathbb{Q}$, $\mathbb{R}$, $\mathbb{C}$ have their usual (non-constructivistic) meaning. Similarly, if we put a lower > 0 (or ≥ 0), e.g., $\mathbb{R}_{>0}$ (or $\mathbb{Z}_{\geq 0}$), we mean the respective subset of elements greater (or equal to) 0. The letter $\mathbb{N}$ denotes the natural numbers. For our purposes it is irrelevant, whether or not we start counting with 0 (and hence view $\mathbb{N}$ as a semiring) or with 1. Even the reader, who is inclined to follow Aristoteles's point of view[2] that a number shall indicate a multiple – and hence is forced to set the smallest element of $\mathbb{N}$ to be at least 2 – may adhere to her/his philosophical principles throughout this book.

(2) We assume the Axiom of Choice without hestitation.

(3) If A and B are two sets, then $A \subseteq B$ means, that A is a subset of B, which potentially could be equal to it. If, however, we write $A \subset B$, then this latter possibility is excluded.

(4) All vector spaces are assumed to be over $\mathbb{C}$ as a ground field, unless otherwise stated. In this regard, a linear map between two vector spaces is assumed to be $\mathbb{C}$-linear. If V is a vector space, we will denote by $\mathrm{Aut}_{\mathbb{C}} V$ the space of automorphisms of V, considered as usual as a group by composition (and as it stands, of course, without any topology).

(5) If $\mathscr{V}$ is a subset of a vector space, then $\langle \mathscr{V} \rangle$ denotes the $\mathbb{C}$-linear span of $\mathscr{V}$.

(6) If X is a subset of a topological space $\mathfrak{X}$, then $\mathrm{Cl}_{\mathfrak{X}}(X)$ denotes the topological closure of X in $\mathfrak{X}$.

(7) The symbol S^1 denotes the unit circle around 0 in $\mathbb{C}$. It is a compact, Hausdorff topological subgroup of $(\mathbb{C}, \cdot)$.

(8) Whenever we post an exercise, it is by no means clear that this shall be a simple finger practice, to be carried out for the purpose of exercising. It shall rather motivate the reader to consult further literature, deepen her/his knowledge and maybe finally compile a proof of the given claim, based on what s/he learned during her/his research.

(9) We appreciate a lot the excellent work that so many other mathematicians have done in order to push the field, while at the same time we love short, concise presentations. Hence, we prefer to give references to the literature, instead of repeating the arguments and proofs ourselves, whenever there is a convenient source available. Actually, we are convinced that this is the only way to comprise all the topics mentioned in our book in the aimed generality without exceeding a readable length.

---

[2] Cf. [Aristoteles, Physics], IV 12, 220b, 27.

Acknowledgements

It is my pleasure to express my gratitude to my friends and colleagues Raphaël Beuzart-Plessis, Erez Lapid, Alberto Mínguez, Marko Tadic, and Sonja Žunar. Without their valuable hints and explanations, this book could not have come into its present form. In particular Sonja's contribution can hardly be overrated.
My thanks also go to Giancarlo Castellano, who provided me with his detailed transcript of my seminar lectures on the subject.
Finally, I am indebted to the Austrian Science Fund (FWF), which has been supporting my scientific work for many years, most recently with the START-Prize (Y966) and the Stand-alone research project (P32333).

Acknowledgements

Contents

Part 1
Local Groups

We are now going to review the basics of the representation theory of reductive groups over local fields. We shall start off with a "crash course" in functional analysis, which is needed throughout over archimedean ground fields (and close to never over non-archimedean ground fields). In the following we will introduce several more or less well-behaved (respectively, more or less well-known) classes of representations, culminating in the notion of "interesting" representations (in the sense that these are the local representations of our interest) and their classification by means of Langlands triples.

Chapter 1

Basic Notions and Concepts from Functional Analysis ("Local")

1.1 Seminormed spaces

We start with the following.

Definition 1.1. A *seminorm* on a vector space V is a function $p : V \to \mathbb{R}_{\geq 0}$ with the following two properties:

(1) $p(\lambda v) = |\lambda|\, p(v) \quad \forall \lambda \in \mathbb{C}, v \in V$;
(2) $p(v + w) \leq p(v) + p(w) \quad \forall v, w \in V$.

Note that if p is a seminorm, then the first property implies that $p(0) = 0$. A seminorm is called a *norm* if $p(v) = 0$ implies $v = 0$. Not every seminorm is a norm, as the following example shows.

Example 1.2. Let $V = \mathbb{C}^2$. Then p defined by $p((x, y)) := |x - y|$ is a seminorm on V, but not a norm.

Definition 1.3. A *seminormed (vector) space* is a topological vector space whose topology is the initial topology with respect to a family of seminorms

$$\mathcal{P} = \{p_i : p_i \text{ is a seminorm on } V, i \in I\},$$

I being any (previously fixed) index set. That is, V carries the coarsest topology for which all p_i's are continuous. For us, *all seminormed spaces are assumed to be Hausdorff*, or in other words, *we assume throughout that the family $\mathcal{P}$ is point-separating.*

Remark 1.4. Our notion of a seminormed space is equivalent to the perhaps more common notion of a (complex) Hausdorff *locally convex vector space*, cf. [Bou03], Chp. II, §4, Sect. 1, corollary to Prop. 1.

Definition 1.5. Let V be a seminormed space with a family $\mathcal{P}$ of seminorms defining its locally convex Hausdorff topology, as above.

(1) A subset $\mathcal{P}_0 \subseteq \mathcal{P}$ is called a *subbasis* of seminorms, if $\mathcal{P}_0$ generates the same locally convex topology on V as $\mathcal{P}$ does.
(2) V is called a *Fréchet* space, if V is *complete* and if there exists a *countable* subbasis $\mathcal{P}_0$ of $\mathcal{P}$.
(3) V is called a *Banach* space, if it is Fréchet and if additionally the subbasis $\mathcal{P}_0$ can be chosen to contain precisely one element.[1]
(4) V is called a *Hilbert* space, if it is Banach and the unique element p of $\mathcal{P}_0$ can be written as $p(v) = \sqrt{\langle v, v \rangle}$ for some non-degenerate, positive-definite Hermitian form $\langle \cdot, \cdot \rangle : V \times V \to \mathbb{C}$.

Remark 1.6. In view of Rem. 1.4 our notion of a Fréchet space is equivalent to the notion of a complete metrizable (complex) locally convex vector space. This is a direct consequence of [Jar81], Sect. 2.8, Thm. 1. Hence, the Banach–Schauder theorem is in force:

Theorem 1.7. *A continuous, linear bijection $\varphi : V \to W$ of Fréchet spaces is bicontinuous.*

See cf. [Jar81], Sect. 5.4, Cor. 3. Finally, we recall that a Fréchet space is either finite-dimensional or of uncountable dimension. See [Jar81], Thm. 1 and Prop. 2 in Sect. 5.1

Obviously, $V = \mathbb{C}^n$, equipped with the standard Hermitian form $\langle (v_1, ..., v_n), (w_1, ..., w_n) \rangle := \sum_{i=1}^n v_i \overline{w_i}$, becomes a Hilbert space. The following well-known lemma shows that this is actually the canonical model of a finite-dimensional seminormed space (see [Bou03], Chp. I, §2, Sect. 3, Thm. 2 and Cor. 2 for a proof):

Theorem 1.8. *Let V be a finite-dimensional Hausdorff topological vector space. Then V is isomorphic to the Hilbert space $\mathbb{C}^n$. Consequently, on a finite-dimensional seminormed space V any two families $\mathcal{P}$, $\mathcal{P}'$ of seminorms on V are equivalent, i.e., generate the same locally convex topology, and any linear map $f : V \to W$ into any seminormed space W is continuous.*

---

[1] Since we assume seminormed spaces to be Hausdorff, the unique element of $\mathcal{P}_0$ must be a norm, and so a Banach space is alternatively described as a *complete, normed vector space.*

We shall need two more very basic notions from functional analysis:

Definition 1.9. A seminormed space V is called

(1) *barreled*, if every seminorm p, whose unit ball $p_{\leq 1} := \{v \in V | p(v) \leq 1\}$ is closed, is continuous.
(2) *bornological*, if every bounded seminorm p (i.e., a seminorm, which maps bounded sets onto bounded sets), is continuous.

The following lemma is standard (cf. [Bou03], Chp. III, §2, Prop. 1 & 2 and §4, Sect. 1, corollary to Prop. 2):

Lemma 1.10. *A seminormed space V is bornological, if and only if every bounded linear map $f : V \to W$, W any seminormed space, is continuous. A Fréchet space is bornological and barreled.*

1.2 Three constructions with seminormed spaces

We shall now present a few constructions involving seminormed spaces which will be needed for our local, representation-theoretic purposes. Roughly, we will need three types of constructions:

(I) forming subspaces and quotients,
(II) forming direct sums,
(III) forming tensor products.

1.2.1 *Subspaces and -quotients*

Here, the situations turns out to be quite well-behaved. We record the following easy lemma, whose proof is left as an exercise.

Lemma 1.11. *Let $W \subseteq V$ be a closed subspace of a seminormed space V. If V is Fréchet, Banach, respectively, Hilbert, then so is W with the subspace topology and so is V/W with the quotient topology.*

Of course, if W were not closed in V, then the quotient V/W would not be Hausdorff, i.e., not a seminormed space according to our definitions, nor could W be complete (when equipped with the subspace topology).

1.2.2 Direct sums

Let I be any index set and suppose we are given a seminormed space V_i for each $i \in I$. Consider the algebraic direct sum of the V_i's,

$$\bigoplus_{i\in I} V_i = \left\{ (v_i)_{i\in I} \in \prod_{i\in I} V_i \text{ such that } v_i = 0 \text{ for all but finitely many } i \right\}.$$

Then, for any choice of continuous seminorms $p_i : V_i \to \mathbb{R}_{\geq 0}$, we obtain a seminorm p on $\bigoplus_{i\in I} V_i$ via:

$$p((v_i)_{i\in I}) := \sum_{i\in I} p_i(v_i).$$

Let $\mathcal{P}$ denote the family of all such seminorms p. Then $\bigoplus_{i\in I} V_i$, equipped with the initial topology with respect to $\mathcal{P}$, is again a seminormed space (cf. [Bou03], II, §4, Sect. 5, Cor. 2 for the Hausdorffness of V) which we call the *locally convex sum* of the V_i's and denote the so-topologized version of $\bigoplus_{i\in I} V_i$ in symbols by $\boldsymbol{\bigoplus}_{i\in I} V_i$.

Lemma 1.12. *Let I be any index set and suppose we are given a seminormed space V_i for each $i \in I$.*

(1) If all V_i are complete, then so is $\boldsymbol{\bigoplus}_{i\in I} V_i$.
(2) If, moreover, I is a finite set, then:

(a) $\boldsymbol{\bigoplus}_{i\in I} V_i \cong \prod_{i\in I} V_i$.
(b) (Hence,) if all V_i are Fréchet, then so is $\boldsymbol{\bigoplus}_{i\in I} V_i$.

See [Jar81], Sect. 6.6, Prop. 7 for completeness of $\boldsymbol{\bigoplus}_{i\in I} V_i$. The rest is left as an exercise.

Warning 1.13. Using the above notations, suppose that a subbasis $\mathcal{P}_i$ is given on each V_i, and let

$$\mathcal{P}_0 := \left\{ p : \bigoplus_{i\in I} V_i \to \mathbb{R}_{\geq 0} \;\middle|\; p((v_i)_{i\in I}) = \sum_{i\in I} p_i(v_i) \text{ for some } p_i \in \mathcal{P}_i \right\}.$$

Then $\mathcal{P}_0$ is *not* a subbasis for the topology on $\boldsymbol{\bigoplus}_{i\in I} V_i$ in general. Indeed, consider a sequence $(V_n)_{n\in\mathbb{N}}$ of seminormed spaces, where each member V_n is a copy of $\mathbb{C}$. Then, on each V_n we have a subbasis $\mathcal{P}_n$ consisting of a single seminorm p_n (which is even a norm), namely the complex absolute value. Defining $\mathcal{P}_0$ as above, and taking the initial topology with respect to $\mathcal{P}_0$, the vectors

$$\begin{aligned} u_1 &= \left(\tfrac{1}{2}, 0, 0, 0, \dots\right), \\ u_2 &= \left(\tfrac{1}{2}, \tfrac{1}{4}, 0, 0, \dots\right), \\ u_3 &= \left(\tfrac{1}{2}, \tfrac{1}{4}, \tfrac{1}{8}, 0, \dots\right), \\ &\vdots \end{aligned}$$

in the so topologized version of $\bigoplus_{n\in\mathbb{N}} V_n$ form a Cauchy sequence which does not converge. This, however, obviously contradicts Lem. 1.12.

Hilbert spaces: The next natural question is how to define direct sums in the category of Hilbert spaces. So, take a family of Hilbert spaces H_i with Hermitian form $\langle\cdot,\cdot\rangle_i$ and associated norm $\|\cdot\|_i$, indexed again by an arbitrary index set I. Then we can look at tuples $(h_i)_{i\in I} \in \prod_{i\in I} H_i$ such that the *generalized sum*

$$\sum_{i\in I} \|h_i\|_i^2$$

converges.[2] The space of such tuples is then equipped with the Hermitian form

$$\langle (h_i)_{i\in I}, (g_i)_{i\in I}\rangle := \sum_{i\in I} \langle h_i, g_i\rangle_i$$

and becomes a Hilbert space, cf. [Bou03], V, §2, Prop. 1.(c), denoted $\widehat{\bigoplus}_{i\in I} H_i$ and called the *direct Hilbert sum* of the H_i's.

Warning 1.14. Note that, in general, $\widehat{\bigoplus}_{i\in I} H_i$ is larger than the algebraic direct sum $\bigoplus_{i\in I} H_i$, but always contains the latter as a dense subspace (see again [Bou03], V, §2, Prop. 1.(c)). If I is a finite set, then the direct Hilbert sum $\widehat{\bigoplus}_{i\in I} H_i$ is isomorphic (as a *Banach* space) to the locally convex sum $\bigoplus_{i\in I} H_i$ as defined further above (see [Bou03], V, §2, Sect. 2). However, this is not true in general, if I is infinite.

[2] A generalized sum is to be understood as a special case of a net, and so the notions of convergence, or limit, of a generalized sum are as for general nets.
In more detail, consider the family $\mathcal{I}$ of all finite subsets of I, ordered by inclusion; this is a directed set. For each $I_0 \in \mathcal{I}$ and each $(h_i)_{i\in I}$ one can form the finite sum $\sum_{i\in I_0} \|h_i\|_i^2$, which is simply a nonnegative real number. This yields a map from the directed set $\mathcal{I}$ to $\mathbb{R}$, so, a net. The generalized sum $\sum_{i\in I} \|h_i\|_i^2$ is defined to be precisely this net.

1.2.3 *Tensor products*

There are several inequivalent extensions of the notion of (algebraic) tensor product to seminormed spaces. For the moment, as it will be enough for the purposes of *local* representation theory, we shall merely define *projective* tensor products for seminormed spaces and review tensor products of Hilbert spaces. Further notions from functional analysis, like the inductive tensor product, will be expanded upon later, when we consider global representation theory, cf. 8.10. We refer to [AGro66], I, §1, n°'s 1–3, [BorWal00], IX, Sect. 6.1, [WarI72], App. 2.2, [Tre70], §43 and [Bou03], V, §3 for details and further reading concerning this section.

Let V and W be seminormed spaces with subbases $\mathcal{P}_0$ and $\mathcal{Q}_0$, respectively, and let $V \otimes W$ denote the algebraic tensor product of V and W. For $p \in \mathcal{P}_0$ and $q \in \mathcal{Q}_0$ we define a seminorm $p \otimes q$ on $V \otimes W$ via:

$$\begin{aligned} p \otimes q : V \otimes W &\to \mathbb{R}_{\geq 0} \\ u &\mapsto \inf\left\{ \sum_j p(e_j)q(f_j) \right\}, \end{aligned}$$

where the infimum is taken over all representations of $u \in V \otimes W$ as a sum $u = \sum_j e_j \otimes f_j$ of elementary tensors. Then $V \otimes W$ equipped with the initial topology with respect to the family of seminorms

$$\mathcal{P}_0 \otimes \mathcal{Q}_0 := \{p \otimes q : p \in \mathcal{P}_0, q \in \mathcal{Q}_0\}$$

is again a seminormed space (cf. [Tre70], Prop. 43.3 and the explanation on p. 438, *ibidem*, for this family being point-separating), called the *projective tensor product* of V and W and denoted $V \otimes_{\mathsf{pr}} W$. It is the finest locally convex topology such that the natural bilinear map $V \times W \to V \otimes W$, $(v, w) \mapsto v \otimes w$ is continuous.

Unlike our previous constructions with seminormed spaces, the projective tensor product of complete spaces need not be complete. For this reason, one also has to look at the *completed tensor product* of V and W,

$$V \mathbin{\overline{\otimes}_{\mathsf{pr}}} W := \overline{V \otimes_{\mathsf{pr}} W},$$

i.e., one has to consider the topological completion of $V \otimes_{\mathsf{pr}} W$ (which, as we just recall in passing, is again a seminormed space). The following assertions hold:

Lemma 1.15. *(1) For seminormed spaces V and W, the (linear extension of the) map*

$$V \otimes_{\mathsf{pr}} W \to W \otimes_{\mathsf{pr}} V,$$
$$v \otimes w \mapsto w \otimes v$$

is an isomorphism of seminormed spaces.

(2) For seminormed spaces V_1, V_2 and W,

$$(V_1 \oplus V_2) \otimes_{\mathsf{pr}} W \cong (V_1 \otimes_{\mathsf{pr}} W) \oplus (V_2 \otimes_{\mathsf{pr}} W).$$

(3) If V_i, W_i, $i = 1, 2$ are seminormed spaces and $\phi : V_1 \to V_2$, $\psi : W_1 \to W_2$ are continuous linear maps, then the linear map

$$\phi \otimes_{\mathsf{pr}} \psi : V_1 \otimes_{\mathsf{pr}} W_1 \to V_2 \otimes_{\mathsf{pr}} W_2$$

is continuous.

Moreover, the assertions above continue to hold if one replaces $\otimes_{\mathsf{pr}}$ by $\overline{\otimes}_{\mathsf{pr}}$ throughout.

(4) If V and W are Fréchet, respectively, Banach, then so is $V \overline{\otimes}_{\mathsf{pr}} W$.

(5) If W is finite-dimensional (and V is arbitrary), then $V \overline{\otimes}_{\mathsf{pr}} W = V \otimes_{\mathsf{pr}} W$ and

$$V \otimes_{\mathsf{pr}} W \cong \bigoplus_{j=1}^{\dim_{\mathbb{C}} W} V.$$

Hilbert spaces: For Hilbert spaces H_1, H_2, we shall denote the Hilbert space completion of the (algebraic) tensor product $H_1 \otimes H_2$ with respect to the Hermitian form defined via

$$\langle h_1 \otimes h_2, h_1' \otimes h_2' \rangle := \langle h_1, h_1' \rangle_1 \langle h_2, h_2' \rangle_2$$

(and extended by use of sesqui-linearity) by $H_1 \widehat{\otimes} H_2$. (Exercise: Convince yourself that this form indeed conserves Hausdorffness: Cf. [Bou03], V, §3, Prop. 1)

Lemma 1.16. *(1) For Hilbert spaces H_1 and H_2, the (linear extension of the) map*

$$H_1 \widehat{\otimes} H_2 \to H_2 \widehat{\otimes} H_1,$$
$$h_1 \otimes h_2 \mapsto h_2 \otimes h_1$$

is an isomorphism of Hilbert spaces.

(2) For Hilbert spaces H_1, H_2 and H_3, there is an an isomorphism of Hilbert spaces

$$(H_1 \widehat{\oplus} H_2) \widehat{\otimes} H_3 \cong (H_1 \widehat{\otimes} H_3) \widehat{\oplus} (H_2 \widehat{\otimes} H_3).$$

(3) If H_i, H'_i, $i = 1, 2$ *are Hilbert spaces and* $\phi : H_1 \to H_2$, $\psi : H'_1 \to H'_2$ *are continuous linear maps, then the linear map*

$$\phi \mathbin{\widehat{\otimes}} \psi : H_1 \mathbin{\widehat{\otimes}} H'_1 \to H_2 \mathbin{\widehat{\otimes}} H'_2$$

is continuous.

(4) If H' *is finite-dimensional (and* H *is an arbitrary Hilbert space), then* $H \mathbin{\widehat{\otimes}} H' = H \otimes H'$ *as sets and*

$$H \mathbin{\widehat{\otimes}} H' \cong \widehat{\bigoplus_{j=1}^{\dim_{\mathbb{C}} H'}} H$$

as Hilbert spaces.

We point out that the underlying Banach space of $H_1 \mathbin{\widehat{\otimes}} H_2$ and the completed tensor product $H_1 \mathbin{\overline{\otimes}_{\mathsf{pr}}} H_2$ of the Banach spaces H_1 and H_2 are not isomorphic (unless one of the two is finite-dimensional, see the last items of Lem. 1.15 and Lem. 1.16): Indeed, the completed projective tensor product of two infinite-dimensional Hilbert spaces $H_1 \mathbin{\overline{\otimes}_{\mathsf{pr}}} H_2$ is never reflexive, as explained in [AGro66], Rem. 2 on p. 49.

Chapter 2

Representations of Local Groups – The Very Basics

We are now ready to begin with our account of *local representation theory.*

2.1 Local fields and local groups

From now onwards and until the end of §7, F denotes a *local field*. For us, this means a non-discrete, locally compact, topological field of characteristic 0. More concretely, local fields split into two subcategories with very different flavors, cf. [Wei74], Chp. 1, §3, Thm. 5:

- If a local field F is connected (as a topological space), then we say it is *archimedean*. One can show that, up to isomorphism of topological fields, the only archimedean local fields are $\mathbb{R}$ and $\mathbb{C}$ equipped with the respective standard (Euclidean) topologies. Indeed, the reader, who is not familiar with the notion of an archimedean local field may simply take this as its definition.
- If F is not connected we will call it *non-archimedean*. It can be shown that then F is totally disconnected. More precisely, there is a countable neighbourhood base around 0 consisting of compact open (additive) subgroups. Unlike the case of archimedean local fields, where (up to isomorphism of topological fields) there are only two examples, $\mathbb{R}$ and $\mathbb{C}$, the situation for non-archimedean local fields is much richer. And still, there is a prototypical example of a non-archimedean local field: the field $\mathbb{Q}_p$ of p-adic numbers for any prime $p \in \mathbb{Z}_{>0}$, equipped with the topology induced by the p-adic absolute value $|\cdot|_p$. Indeed, any non-archimedean local field is isomorphic to a finite-degree extension of some $\mathbb{Q}_p$ and the reader, yet unfamiliar with non-archimedean local fields, may take this as their very definition.

In particular, the topology on F is always induced by an absolute value: There is even a canonical choice for this (see §9 for more details, which are still irrelevant here), which we shall denote by $|\cdot|_F$ of simply by $|\cdot|$, if no confusion is possible.

The dichotomy between the archimedean and the non-archimedean case is reflected in the study of linear algebraic groups over local fields. For our purpose and for now, it will be enough to recall just the most basic facts and definitions concerning linear algebraic groups (over fields of characteristic 0), see [Bor-Wal00], §0, 3.0.1, for a concise and sufficiently general summary: A group $\mathbf{G}$[1] is called *linear algebraic* if it is a Zariski-closed subgroup of some general linear group $\mathbf{GL}_n(\Omega)$, where Ω is an algebraically closed field of characteristic zero. It is furthermore called *reductive*,[2] if its unipotent radical $\mathbf{R_u}(\mathbf{G})$, i.e., its maximal Zariski-connected unipotent normal subgroup, is the trivial group. Obviously, the direct product of a finite number of Zariski-connected reductive linear algebraic groups is again a Zariski-connected reductive linear algebraic group. Suppose now that Ω is an algebraic closure of a local field F. Then, $\mathbf{G}$ is said to be defined over F, if the ideal of polynomials vanishing on $\mathbf{G}$ is generated by polynomials with coefficients in F. In the following and until the end of §7, $\mathbf{G}$ will denote a Zariski-connected reductive linear algebraic group defined over F, and we will set $G := \mathbf{G}(F) = \mathbf{G} \cap \mathbf{GL}_n(F)$. Note that

- G is a *real Lie group* with finitely many connected components, if F is archimedean (cf. [Pla-Rap94], Thm. 3.5 & Thm. 3.6), and
- G is a *t.d. group*, i.e., a locally compact, topological group, which has a countable basis of neighborhoods of the identity and is totally disconnected (the latter property leading to the very notion "*t.d.*"), if F is non-archimedean (cf. [Pla-Rap94], §3.1 & §3.3).

For us, such groups G will be collectively known as *local groups*. Inside our local group G we also fix a choice of a maximal compact subgroup

[1] At this point we would like to thank *Spurius Carvilius Ruga* (fl. 230 BC) for introducing the letter "G" into the (latin) alphabet.

[2] Just as a short remark for the reader, interested also in etymology, the choice of the notion "reductive" is based on the following observation: It turns out that the condition $\mathbf{R_u}(\mathbf{G}) = \{id\}$ is equivalent to every finite-dimensional representation of $\mathbf{G}$ being completely reducible. Since we will not need this fact, and only mention it for linguistic reasons, we limit ourselves to referring to [JMil17], Thm. 22.42 for details and a proof.

$K \subseteq G$ (their existence being guaranteed by [Pla-Rap94], Prop. 3.10,[3] in the archimedean, and by [Pla-Rap94], Prop. 3.16, in the non-archimedean case). It is an important technicality to note that a maximal compact subgroup of a non-archimedean local group is open, see [Pey87], Cor. 1 to Thm. 2.

Later on we will have to assume additional properties on K. As this is not necessary, yet, and needs further objects and notation to be explained, we will only specify these assumptions, when they become necessary. See §4.3.1 and §7.2.

When F is archimedean, we recall that – contrasting the non-archimedean case, cf. [Pla-Rap94], Thm. 3.13 – all maximal compact subgroups are G-conjugate to one another. Indeed, suppose first that $F \cong \mathbb{C}$. Then, applying Weil's restriction of scalars from $\mathbb{C}$ to $\mathbb{R}$, one obtains a connected reductive group $\mathrm{Res}_{\mathbb{C}/\mathbb{R}}(\mathbf{G})$ over $F = \mathbb{R}$ with the property that $G \cong \mathrm{Res}_{\mathbb{C}/\mathbb{R}}(\mathbf{G})(\mathbb{R})$, cf. [Spr79], §3.3. Hence, one may reduce the proof to considering the case $F \cong \mathbb{R}$, in which case the result can be found in [Pla-Rap94], Prop. 3.10.(b).

Moreover, if F is archimedean, we will use the standard notation $\mathfrak{g} = \mathrm{Lie}(G)$ for the real Lie algebra of G and we will write $\mathcal{U}(\mathfrak{g})$ for the universal enveloping algebra of the complexification $\mathfrak{g}_{\mathbb{C}} := \mathfrak{g} \oplus i\mathfrak{g} \cong \mathfrak{g} \otimes_{\mathbb{R}} \mathbb{C}$. Analogous notation will be used for subgroups of G without further mention, e.g., $\mathfrak{k} = \mathrm{Lie}(K)$.

Remark 2.1 (For real representation-theorists; and as a justification of our use of references for archimedean representation theory in the following). For readers having a background in the representation theory of real reductive groups, it might be worthwhile to note that when F is archimedean, a local group G is a *real reductive Lie group* in the terminology of all the following sources: [Bor-Wal00], 0.3.1, [Kna02], VII.2.(i)–(v), [Kna-Vog95], Def. 4.29.(i)–(iv), [Spr79], 5.1, [Vog81], Chp. 0.1.1.(a)–(f) and [Wal89], 2.1.1. It is moreover of "inner type" in the sense of [Wal89], 2.2.8, respectively, in "Harish-Chandra class" in the sense of [Kna-Vog95], Def. 4.29.(v) and [Kna02], VII.2.(vi) and "of class $\mathcal{H}$" in

[3] As a purely technical remark, we recall that we may (and henceforth will) assume that a local archimedean group G is closed under taking the transpose of matrices, cf. [Pla-Rap94], Thm. 3.7. The reader, who is not inclined to go through the details of the referenced proofs, may safely ignore this technicality.

the sense of [Var77], II.1.1.

This is well-known, but for the sake of completeness, we sketch the argument: Firstly, as above, applying Weil's restriction of scalars $\mathrm{Res}_{\mathbb{C}/\mathbb{R}}$, one may again reduce the proof to considering $F \cong \mathbb{R}$, cf. [Spr79], §3.3. In this case, it follows from a combination of [Var77], Thms. II.1.13, II.1.14 and II.1.16 and [Kna02], Prop. 7.19.(b) that being a "real reductive group in Harish-Chandra class" in the sense of [Kna02], VII.2.(i)–(vi) is equivalent to being "of class $\mathcal{H}$" in the sense of [Var77], II.1.1. It now follows from [Var77], Prop. II.1.2 that any local archimedean group G (in our terminology here) is a real reductive group in Harish-Chandra class in the sense of [Kna02], VII.2.(i)–(vi), respectively, "of class $\mathcal{H}$" in the sense of [Var77], II.1.1. It is therefore a fortiori a real reductive group in the terminology of [Bor-Wal00], 0.3.1, [Kna-Vog95], Def. 4.29, [Spr79], 5.1, and [Wal89], 2.1.1, which is moreover of "inner type" in the sense of [Wal89], 2.2.8, respectively, in "Harish-Chandra class" in the sense of [Kna-Vog95], Def. 4.29.(v). Furthermore, it satisfies [Vog81], Chp. 0.1.1.(a)–(e). That it also satisfies [Vog81], Chp. 0.1.1.(f) ("all Cartan subgroups are abelian") is left to the reader, see [Vog81], Ex. 0.1.4.(c).

Knowing this, one may alternatively see [Var77], Thm. II.1.12 for a proof that all maximal compact subgroups of a local archimedean group G are conjugate.

2.2 Local representations

To account for the topological considerations that become necessary in the archimedean case, one refines the familiar definition of an (abstract) group representation as follows:

Definition 2.2. A *representation* of the local group G is a pair (π, V), where V is a vector space and π is a homomorphism $\pi : G \to \mathrm{Aut}_{\mathbb{C}} V$ of groups. If the ground field F is archimedean, then we additionally require V to be a Fréchet space and the map

$$\begin{aligned} G \times V &\to V \\ (g, v) &\mapsto \pi(g)v \end{aligned} \tag{2.3}$$

to be continuous.

Before we proceed with further notions, we shall briefly focus on the additional two conditions (V being Fréchet and $G \times V \to V$ being continuous), which we imposed in the archimedean case, respectively, on why we imposed them:

Remark 2.4 (Fréchet-ness of V). Nothing can be reached in archimedean representation theory without real and/or complex analysis. Hence, we shall (not only) assume that our vector space V is a topological vector space, (but also that it is) complete (to make limits exist) and Hausdorff (to make limits unique). Moreover, we shall assume that its topology is locally convex, in order make sense of integrals of vector-valued functions (cf. (4.7) for a major example). So, in summary, and according to our definitions, it is very natural to impose the condition that V is a complete seminormed space.

And indeed, in many regards this is a working starting-point for local, archimedean representation theory – a starting-point, which, as we already reveal, we will in fact be forced to take up again in the global part of this book. Our remaining, additional assumption that V is metrizable, i.e., all together Fréchet, cf. Rem. 1.6, does not only turn out to simplify basic "local" life quite substantially (for instance, in light of Thm. 1.7), but is also indispensable in our presentation of the Langlands classification and all applications connected to it.

Remark 2.5 (Continuity of $G \times V \to V$). In view of the last remark, some condition of continuity regarding our action on V turns out to be a very natural, even necessary, assumption. However, there are several other natural ways to define a notion of continuity in our scenario and it is worthwhile to compare them with our choice.
Let us firstly suppose that we have in fact been furnished by a group homomorphism

$$\begin{aligned} \pi : G &\to \mathrm{Aut}^{\mathrm{ct}}_{\mathbb{C}} V, \\ g &\mapsto \pi(g) \end{aligned}$$

into the subgroup of bicontinuous linear automorphisms in $\mathrm{Aut}_{\mathbb{C}} V$.
Then, one way to define a notion of continuity for representations would be to demand the continuity of the so-called *orbit maps*

$$\begin{aligned} c_v : G &\to V, \\ g &\mapsto \pi(g)v \end{aligned}$$

for all $v \in V$ (traditionally called "strong continuity"). Another way could be to ask for the continuity of the maps

$$G \to \mathbb{C},$$
$$g \mapsto v'(\pi(g)v)$$

for all $v \in V$ and all v' in the dual space $V' = \mathrm{Hom}_{\mathrm{ct}}(V, \mathbb{C}) = \{f : V \to \mathbb{C}$ linear and continuous$\}$ (called "weak continuity").
Obviously, our chosen condition of continuity of $G \times V \to V$, cf. 2.25, implies "strong continuity", which in turn implies "weak continuity".
If V is a Banach space, then there is even another way to define continuity (maybe, naively speaking, even the most natural one) by requiring that the group homomorphism

$$\pi : G \to \mathrm{Aut}^{\mathrm{ct}}_{\mathbb{C}} V,$$
$$g \mapsto \pi(g)$$

is itself continuous, once we equip $\mathrm{Aut}^{\mathrm{ct}}_{\mathbb{C}} V$ with the operator norm (with respect to V')), whence this notion of continuity is called "continuity in the operator norm". It implies our chosen condition of continuity of $G \times V \to V$. The following proposition compares the above mentioned notions of continuity.

Proposition 2.6. *If V is a Banach space, then our notion of continuity, "strong continuity" and "weak continuity" are equivalent, whereas "continuity in the operator norm" is strictly stronger than our notion of continuity. For general V, i.e., a Fréchet space, still our notion of continuity is equivalent to "strong continuity".*

Proof. The non-trivial part of the first statement is Prop. 4.2.2.1 in [WarI72], as combined with the following observation: Consider the action of the multiplicative group $G = S^1$ by translation on $C(G; \mathbb{C})$, equipped with the supremum norm. It is not "continuous in the operator norm", but, certainly continuous according to our notion of continuity. For the second assertion we recall our Lem. 1.10 together with the [BouII04], VIII, §2, Prop. 1. □

Definition 2.7. Let (π_i, V_i), $i = 1, 2$, be representations of G. We say that π_1 is *isomorphic*, or, likewise, *equivalent*, to π_2, (in symbols: $\pi_1 \cong \pi_2$; or $\pi_1 \underset{G}{\cong} \pi_2$, if we want to stress the group G; or $V_1 \cong V_2$, if the actions are clear), if there is a linear bijection $f : V_1 \to V_2$, which is G-*equivariant*, i.e., $\pi_2(g)f(v_1) = f(\pi_1(g)v_1)$ for all $g \in G$, $v_1 \in V_1$. If the ground field F is archimedean, then we also assume that f is bicontinuous.

Remark 2.8. As for archimedean ground fields F we restricted our attention to Fréchet spaces, the condition of bicontinuity of f above can be relaxed to continuity: See Thm. 1.7.

Let us now see how the concept of a representation goes with the standard constructions with vector spaces, namely taking subspaces, quotients, direct sums and tensor products. In the case of an archimedean ground field, the reader is advised to recall our constructions with seminormed spaces from §1.2.

Definition 2.9. Let (π_i, V_i), $i = 1, 2$, be representations of G. We say π_2 is a *subrepresentation* (respectively, *subquotient*) of π_1, if $V_2 \subseteq V_1$ and $\pi_2(g) = \pi_1(g)|_{V_2}$ for all $g \in G$ (respectively, if there are subrepresentations $U_2 \subseteq U_1 \subseteq V_1$ such that $V_2 = U_1/U_2$ and, writing $v_2 \in V_2$ as $v_2 = u_1 + U_2$, $\pi_2(g)v_2 = \pi_1(g)u_1 + U_2$ for all $g \in G$). If in the latter $U_1 = V_1$, we omit the "sub" and call π_2 a *quotient* of π_1.

We point out that in the archimedean case, when we write $V_2 \subseteq V_1$, as in Def. 2.9, then this means an inclusion of *Fréchet spaces*, i.e., that V_2 is not only a vector-subspace of V_1, but also that its original Fréchet space topology coincides with the subspace topology from V_1. In light of Lem. 1.11, the subrepresentations of a given representation (π, V) of an archimedean local group G are hence nothing else but the closed G-invariant vector-subspaces of V equipped with the restricted action of G. Similarly, if G is a non-archimedean local group, then the subrepresentations of a given representation (π, V) are exactly the G-invariant vector-subspaces of V equipped with the restricted action of G. Obviously, regardless of the ground field, V and $\{0\}$ are always subrepresentations of a given representation (π, V). This leads us to the fundamental notion of irreducibility:

Definition 2.10. A representation (π, V) of G is called *irreducible*, if $V \neq \{0\}$ and if it admits no proper subrepresentation (i.e., a subrepresentation different from V and $\{0\}$).

We now turn to direct sums and tensor products. We will do this in form of valuable exercises.

Exercise 2.11. Let (π_i, V_i), $i = 1, 2$, be representations of archimedean local groups G_i. Show that

$$\pi_1 \oplus \pi_2 : (G_1 \times G_2) \times (V_1 \oplus V_2) \to V_1 \oplus V_2,$$

$$((g_1, g_2); (v_1, v_2)) \mapsto (\pi_1(g_1)v_1; \pi_2(g_2)v_2)$$

is a representation of $G_1 \times G_2$. It is called the *direct sum of* π_1 *and* π_2. Furthermore, show that if $G_1 = G_2 =: G$, the above construction leads a representation of G defined as

$$G \times (V_1 \oplus V_2) \to V_1 \oplus V_2,$$

$$(g; (v_1, v_2)) \mapsto (\pi_1(g)v_1; \pi_2(g)v_2)$$

and still denoted by $\pi_1 \oplus \pi_2$.

Remark 2.12. Exercise 2.11 has an obvious analogue for representations (π_i, V_i), $i = 1, 2$, of non-archimedean local groups G_i:

$$\pi_1 \oplus \pi_2 : (G_1 \times G_2) \times (V_1 \oplus V_2) \to V_1 \oplus V_2,$$

$$((g_1, g_2); (v_1, v_2)) \mapsto (\pi_1(g_1)v_1; \pi_2(g_2)v_2)$$

is a representation of $G_1 \times G_2$. In the case $G_1 = G_2 =: G$, restriction to the diagonal leads again a representation of G:

$$\pi_1 \oplus \pi_2 : G \times (V_1 \oplus V_2) \to V_1 \oplus V_2,$$

$$(g; (v_1, v_2)) \mapsto (\pi_1(g)v_1; \pi_2(g)v_2).$$

Exercise 2.13. Let (π_i, V_i), $i = 1, 2$, be representations of archimedean local groups G_i. Show that the map $(G_1 \times G_2) \times (V_1 \otimes_{\mathsf{pr}} V_2) \to (V_1 \otimes_{\mathsf{pr}} V_2)$, defined by the (linear extension of the) assignment $((g_1, g_2); v_1 \otimes v_2) \mapsto \pi_1(g_1)v_1 \otimes \pi_2(g_2)v_2$ is continuous and extends to a representation

$$\pi_1 \overline{\otimes}_{\mathsf{pr}} \pi_2 : (G_1 \times G_2) \times (V_1 \overline{\otimes}_{\mathsf{pr}} V_2) \to (V_1 \overline{\otimes}_{\mathsf{pr}} V_2)$$

of $G_1 \times G_2$. Moreover, in the case $G_1 = G_2 =: G$, show that restriction to the diagonal $G \hookrightarrow G \times G$ provides a representation of G. In each case, it is called the *(tensor) product of* π_1 *and* π_2. *Hint*: Use Lem. 1.15 and [WarI72], Prop. 4.1.2.4, as well as the paragraph below it.

Remark 2.14. Again, Exc. 4.17 has an obvious analogue for representations (π_i, V_i), $i = 1, 2$, of non-archimedean local groups G_i:

$$\pi_1 \otimes \pi_2 : (G_1 \times G_2) \times (V_1 \otimes V_2) \to V_1 \otimes V_2,$$

defined by the (linear extension of the) assignment $((g_1, g_2); v_1 \otimes v_2) \mapsto \pi_1(g_1)v_1 \otimes \pi_2(g_2)v_2$ is a representation of G. Moreover, in the case $G_1 = G_2 =: G$, show that restriction to the diagonal $G \hookrightarrow G \times G$ provides a representation of G.

2.3 Smooth representations

We continue with the fundamental notion of *smoothness*.

Definition 2.15. Let (π, V) be a representation of G. A vector $v \in V$ is called *smooth*, if its orbit map

$$c_v : G \to V,$$
$$g \mapsto \pi(g)v$$

is smooth. The space of all smooth vectors of V is denoted V^∞.

We remark/recall that for F archimedean, smoothness of c_v just comes with the usual definition: It means that all partial derivatives of c_v (with respect to local coordinates on the real manifold underlying G) of all orders exist (in the respective seminorms on V) and are continuous.

If the ground field F is non-archimedean, then a smooth function is per definition a *locally constant*[4] one: This means there exists some open compact subgroup $C \subseteq G$, such that that for every $g \in G$

$$c_v(gc) = c_v(g) \quad \forall c \in C,$$

i.e., that c_v is right-invariant under an open compact subgroup C of G.

Lemma 2.16. *The subspace V^∞ of V is stable under the restricted action of G, i.e., $\pi(g)v \in V^\infty$ for all $g \in G$ and $v \in V^\infty$:*

Proof. Indeed, if F is non-archimedean, then $c_{\pi(g)v}$ is right-invariant under the open compact subgroup gCg^{-1} of G, whereas if F is archimedean, then $c_{\pi(g)v} = c_v \circ R_g$, where $R_g : G \xrightarrow{\sim} G$, $h \mapsto hg$, denotes the smooth map of right-translation by g, whence $c_{\pi(g)v}$ is smooth as it is the composition of smooth maps. □

Hence, if the ground field F is non-archimedean, then (π, V^∞) is again a representation of G.

However, if F is archimedean, then $V^\infty \subseteq V$ is dense in the subspace topology, cf. [WarI72], Prop. 4.4.1.1. In particular, if $V^\infty \neq V$, then the subspace topology on V^∞ is not complete, hence not Fréchet. It is nonetheless possible to define a Fréchet topology on V^∞, and we shall now outline how this is done. We start by topologizing the space $C^\infty(G, V)$ of smooth functions from G to V (smoothness being defined as above). For any $f \in C^\infty(G, V)$

[4] In our opinion, this should have better been called *uniformly* locally constant. However, the literature decided differently (and we decided to follow the literature in this question).

and any X in the Lie algebra $\mathfrak{g}$ of G, one gets a new smooth function $X \cdot f : G \to V$, given by

$$(X \cdot f)(g) := \left.\frac{\mathrm{d}}{\mathrm{d}t}\right|_{t=0} f(g \exp(tX)).$$

Extending this action by use of partial differentiation, this holds true, if we replace $X \in \mathfrak{g}$ by a general element D of the universal enveloping algebra $\mathcal{U}(\mathfrak{g})$ of $\mathfrak{g}_{\mathbb{C}}$. More precisely, if $D = X_1 \cdot ... \cdot X_k$, then

$$(D \cdot f)(g) := \left.\frac{\mathrm{d}}{\mathrm{d}t_1 \cdot ... \cdot \mathrm{d}t_k}\right|_{t_1=...=t_k=0} f(g \exp(t_1 X_1)... \exp(t_k X_k)), \quad (2.17)$$

and this action is extended linearly to all $D \in \mathcal{U}(\mathfrak{g})$. For each $D \in \mathcal{U}(\mathfrak{g})$, each continuous seminorm p_α on V and each compact subset $C \subseteq G$, we get continuous seminorms $p = p_{D,\alpha,C}$ on $C^\infty(G, V)$ via

$$p(f) := \sup_{c \in C} \{p_\alpha(Df(c))\}.$$

Equipping $C^\infty(G, V)$ with the locally convex topology defined by them, makes $C^\infty(G, V)$ into a seminormed space. Note that since the topology on V has a countable subbasis of seminorms, the same holds for $C^\infty(G, V)$, because the real manifold underlying G is paracompact and hence countable at infinity, i.e., the countable union of compact subsets. It turns out that $C^\infty(G, V)$ is complete, hence, in summary, a Fréchet space. We now consider the linear map

$$V^\infty \to C^\infty(G, V),$$

$$v \mapsto c_v.$$

One easily verifies, see [WarI72], p. 253, that its image is a closed and hence complete subspace of $C^\infty(G, V)$. Therefore, pulling back the subspace-topology, inherited from $C^\infty(G, V)$, to V^∞ we obtain a Fréchet topology on V^∞, as desired. It is useful to note that this topology coincides with the locally convex topology given by the seminorms $p_{\alpha,D}(v) := p_\alpha(\pi(D)v)$, where p_α runs through the continuous seminorms on V and D through the elements of the universal enveloping algebra $\mathcal{U}(\mathfrak{g})$. See [Cas89I], Lem. 1.2 for this simple fact. Of course, here $\pi(D)$ denotes the differentiated linear action of $\mathfrak{g}$ on V, as extended linearly to all of $\mathcal{U}(\mathfrak{g})$.

Finally, one checks as in [WarI72], p. 253, that $G \times V^\infty \to V^\infty$, $(g, v) \mapsto \pi(g)v$ is continuous, whence defines a representation of G on V^∞.

Regardless of the nature of our ground field F, this representation will be denoted (π^∞, V^∞). We obtain

Lemma 2.18. *Let (π, V) be a representation of G. Then,*

(1) $(\pi^\infty)^\infty \cong \pi^\infty$.
(2) If F is archimedean, the identity map $V^\infty \hookrightarrow V$ is continuous.

Proof. The first assertion is obvious, if F is non-archimedean and to be found in [WarI72], p. 254, if F is archimedean. The second assertion follows from our explicit description of the locally convex topology on V^∞ using the seminorms $p_{\alpha,D}$ and the fact that $1 \in F \subset \mathcal{U}(\mathfrak{g})$. See also [WarI72], p. 253. □

Definition 2.19. A representation (π, V) of G is *smooth*, if the identity map $V^\infty \hookrightarrow V$ defines an isomorphism $\pi^\infty \cong \pi$.

The following lemma simplifies the situation in our current setup:

Lemma 2.20. *A representation (π, V) of G is smooth if and only if every vector $v \in V$ is smooth.*

Proof. For F non-archimedean, this is obvious by our definitions. For F archimedean it follows from Thm. 1.7 and Lem. 2.18.(2). □

Corollary 2.21. *Any subrepresentation and any quotient of a smooth representation is smooth.*

Proof. Given the previous lemma, only the case of a quotient V/U of a smooth representation (π, V) of an archimedean local group G needs a short argument: Indeed, we have to show that every vector $v+U \in V/U$ is smooth. By assumption $c_v : G \to V$ is smooth for all $v \in V$. Furthermore, the natural projection $p_U : V \twoheadrightarrow V/U$ is continuous, hence bounded, and so p_U maps smooth curves to smooth curves by [Kri-Mic97], Cor. 2.11. Therefore, the composition $c_{v+U} = p_U \circ c_v$ is smooth by [Kri-Mic97], Cor. 3.14. □

Exercise 2.22. Let (π_i, V_i), $i = 1, 2$, be representations of an archimedean local group G. Show that

$$(\pi_1 \oplus \pi_2)^\infty \cong \pi_1^\infty \oplus \pi_2^\infty$$

and conclude that the direct sum of two smooth representations is smooth. *Hint:* In order to establish an equality of vector spaces $(\pi_1 \oplus \pi_2)^\infty =$

$\pi_1^\infty \oplus \pi_2^\infty$ use that the natural projections $V_1 \oplus V_2 \twoheadrightarrow V_i$, respectively, injections $V_i \hookrightarrow V_1 \oplus V_2$, are continuous, hence bounded and so smooth by [KriMic97], Cor. 2.11. To see equality of topologies, use [WarI72], App. 2.2, Ex. 2 and our Lem. 1.15.(2) for the completed projective tensor product.

Exercise 2.23. Let G be an archimedean local group. Show that the tensor product representation $\pi_1 \overline{\otimes}_{\mathsf{pr}} \pi_2$ of smooth G-representations (π_1, V_1) and (π_2, V_2) is smooth.
Hint: Try to understand [WarI72], Prop. 4.4.1.10 and its proof.

For later use, we urge the reader to provide the short argument for the analogous assertions of Exc. 2.22 and 2.23 in the non-archimedean case, too.

On the negative side, we mention that there are important representations, which are not smooth: Consider, for instance, $V = L^2(G)$, acted upon by right-translation $(\pi(g)f)(h) = f(hg)$ for a prominent example.[5]

The following lemma is a useful reformulation of smoothness, which is in turn special for the non-archimedean case.

Lemma 2.24. *Let F be non-archimedean and let (π, V) be a representation of G. Then, the following are equivalent:*

(1) π is smooth.
(2) The spaces of C-invariant vectors, $V^C := \{v \in V | \pi(c)v = v \ \forall c \in C\}$, C running through the open compact subgroups of G, cover all of V, i.e.,

$$V = \bigcup_{\substack{C \text{ compact open} \\ \text{subgroups of } G}} V^C,$$

(3) The map

$$\begin{aligned} G \times V &\to V \\ (g, v) &\mapsto \pi(g)v \end{aligned} \tag{2.25}$$

is continuous, if V is equipped with the discrete topology.

This result, whose proof we leave as an exercise to the reader, leads to the notable scholium that the fundamental notion of smoothness by force introduces a condition of continuity of $G \times V \to V$ also in the non-archimedean

[5]The prominency of this example will become clearer in Chp. 5.

case. In a sense, this tells us that there has never been a chance to completely escape questions of continuity even when working only over non-archimedean ground fields. This fact, however, opens a window of opportunity: It will allow us to somewhat resolve the still existing asymmetry in our terminology – namely in what it needs to be a smooth representation in the archimedean and in the non-archimedean case – which shall be reconciled in the very concept of an *admissible representation*, to be introduced right below.

Before we introduce this fundamental notion, let us ask ourselves what can be said about irreducibility when passing to smooth vectors V^∞. As it turns out, this will be just another instance of where an already built-in notion of continuity of $G \times V \to V$ into the concept of being sheerly a representation, i.e., when F is archimedean, becomes a striking advantage:

Proposition 2.26. *Let (π, V) be a representation of G. If F is archimedean, then π is irreducible, if and only if π^∞ is irreducible. Both implications may fail, if F is non-archimedean.*

Proof. For F archimedean, this is shown in [WarI72], p. 254. So, let F be non-archimedean. For the failure of the implication "π^∞ irreducible $\Rightarrow \pi$ irreducible", we need to find a non-smooth representation, whose subrepresentation of smooth vectors is irreducible. As a prominent family of examples, start from any irreducible, smooth representation (π, V) and consider its linear dual $\mathrm{Hom}(V, \mathbb{C})$ acted upon by left translation:

$$\begin{aligned} \ell : G &\to \mathrm{Aut}_{\mathbb{C}}(\mathrm{Hom}(V, \mathbb{C})), \\ g &\mapsto \big(f \mapsto f\big(\pi(g)^{-1}\_\big)\big). \end{aligned}$$

The resulting representation $(\ell, \mathrm{Hom}(V, \mathbb{C}))$ is not smooth in general, however, its space of smooth vectors is always irreducible, cf. [Ren10], Cor. VI.2.2. Hence, whenever it is non-smooth, $(\ell, \mathrm{Hom}(V, \mathbb{C}))$ is not irreducible, although $(\ell^\infty, \mathrm{Hom}(V, \mathbb{C})^\infty)$ is. (We will reconsider this example in §4.1, as it leads to the important concept of the contragredient representation of (π, V).)

For the failure of the other implication "π irreducible $\Rightarrow \pi^\infty$ irreducible", we need to find an irreducible representation, which, by irreducibility, has no non-zero smooth vectors. To this end, consider the complex p-adic numbers $\mathbb{C}_p$, i.e., the topological completion of the algebraic closure of $\mathbb{Q}_p$.[6] It is an algebraically closed field of characteristic 0, [Was82], Prop. 5.2, which has the same cardinality as $\mathbb{C}$, the field of complex numbers. Hence, (recall that we use the Lemma of Zorn) there must be an isomorphism of fields $\mathbb{C}_p \xrightarrow{\sim} \mathbb{C}$ (Use [SLan02], Thm. VIII.1.1 for a proof). In summary, we obtain an *injective* group homomorphism $\chi : \mathbb{Q}_p^\times \hookrightarrow \mathbb{C}^\times$. Putting $F = \mathbb{Q}_p$, $\mathbf{G} = \mathbf{GL}_1$ and $V = \mathbb{C}$, such a homomorphism defines an irreducible representation $\pi : G \to \mathrm{Aut}_{\mathbb{C}}(V)$, $\pi(g)v := \chi(g) \cdot v$, for which $V^\infty = \{0\}$: Indeed, if $0 \neq v \in V$ were smooth, then there would have to be an open compact subgroup $C \subseteq G$ such that $\pi(c)v = v$ for all $c \in C$. However, v being non-zero by assumption, this simply means that $\chi(c) = 1$ for all $c \in C$. The character χ being injective now leads to a contradiction. Hence, $V^\infty = \{0\}$ is trivial and so π^∞ is not irreducible. □

Remark 2.27. For non-archimedean local fields F, the situation is not as bad as the previous proposition seems to imply: We shall see later on, cf. Thm. 3.35, that the functor $V \mapsto V^\infty$ in fact preserves a "topologized version" of irreducibility, analogously to the archimedean case, for a very important class of representations.

2.4 Admissible representations

The attentive reader will have noticed that our definition of a *representation* (as well as the attached fundamental notions of being *irreducible* or *smooth*) extends verbatim from G to all of its non-trivial closed subgroups H – simply by replacing the letter G by H and then applying literally the same definition. Having made this observation, we may henceforth speak of representations of our fixed maximal compact subgroup K[7], which are likewise irreducible or smooth.

Definition 2.28. We call a representation (π, V) of G *pro-admissible* (with respect to K), if for every irreducible finite-dimensional representation (ρ, W) of K,

$$\dim_{\mathbb{C}} \mathrm{Hom}_K(W, V) < \infty,$$

[6] For existence and uniqueness (up to isomorphism of fields) of the latter, recall that we use the Lemma of Zorn.

[7] ...although K does not necessarily come under the purview of local groups (If F is archimedean, it does not, if the underlying linear algebraic $\mathbb{R}$-group $\mathbf{K}$ of the real Lie group $K = \mathbf{K}(\mathbb{R})$ is not connected; and it actually never does, if F is non-archimedean).

where $\mathrm{Hom}_K(W, V)$ denotes the space of K-equivariant, linear maps $f : W \to V$.

Observe that we could avoid any reference to continuity of $f \in \mathrm{Hom}_K(W, V)$ (for F archimedean), as any linear map from a finite-dimensional semi-normed space is continuous, cf. Thm. 1.8. This shows that pro-admissibility is a purely algebraic notion, a fact, which will become important in §3.2.

Definition 2.29. A representation (π, V) of G is called *admissible*, if

(1) (π, V) is pro-admissible, if F is archimedean,
(2) (π, V) is smooth and pro-admissible, if F is non-archimedean.

It is easy to show that the notion of admissibility is independent of the choice of the maximal compact subgroup K of G. Indeed, for F archimedean, it is a simple consequence of the fact that any two maximal compact subgroups are conjugate by an element in G, and we leave it a as a valuable exercise to the reader to fill in the details. For F non-archimedean, it is a corollary to the even more instructive Exc. 2.36 below, which will provide us an alternative, but equivalent definition for a representation being admissible for all non-archimedean local groups.

The following lemma is clear by the very definitions (and a simple combination with Cor. 2.21, respectively, Lem. 2.18, when F is non-archimedean):

Lemma 2.30. *Any subrepresentation of a (pro-)admissible representation is (pro-)admissible. The representation on the space of smooth vectors of any pro-admissible representation is admissible.*

Let (π, V) be a representation of G. For an irreducible finite-dimensional representation (ρ, W) of K, let us denote by $V(\rho)$ the *ρ-isotypic component* of V, i.e., the image of the natural map

$$\begin{aligned} \mathrm{Hom}_K(W, V) \otimes W &\to V, \\ f \otimes w &\mapsto f(w). \end{aligned} \tag{2.31}$$

As one immediately verifies, one obtains an isomorphism of vector space

$$\mathrm{Hom}_K(W, V) \otimes W \cong V(\rho),$$

i.e., the natural map is injective. Indeed, if $f \otimes w$ is mapped onto $0 \in V$, then either $f \equiv 0$ (in which case $f \otimes w = 0$ in $\mathrm{Hom}_K(W, V) \otimes W$) or, $f \not\equiv 0$. But in the latter case, f is injective, as (ρ, W) is assumed to be irreducible. So, $f(w) = 0$ in V implies $w = 0$ in W, whence $f \otimes w = 0$

in $\mathrm{Hom}_K(W,V) \otimes W$. Therefore, $\mathrm{Hom}_K(W,V)$ is finite-dimensional if and only if $V(\rho)$ is, i.e., (π,V) is pro-admissible, if and only if all its ρ-isotypic components are finite-dimensional.

Let now F be archimedean and let dk denote the (unique) Haar measure on K, which gives K total volume $\mathrm{vol}_{dk}(K) = \int_K 1 \; dk = 1$. Moreover, let again (π,V) be a representation of G and (ρ,W) an irreducible finite-dimensional representation of K. For $v \in V$, consider the integral

$$E_\rho(v) := \dim_{\mathbb{C}}(W) \cdot \int_K \mathrm{tr}(\rho(k)^{-1}) \; \pi(k)v \; dk. \tag{2.32}$$

One easily checks that the integrand is continuous as a function $K \to V$, hence $E_\rho(v)$ is well-defined (i.e., convergent) for all $v \in V$. The resulting map

$$\begin{aligned} E_\rho : & V \to V \\ & v \mapsto E_\rho(v), \end{aligned} \tag{2.33}$$

turns out to be a continuous projection onto $V(\rho)$, cf. [HCh66], p. 8. We also refer to [Cas22], Prop. 5.6 for a detailed proof of this fact.

Proposition 2.34. *Let (π,V) be an admissible representation of a local group G and let $U \subseteq V$ be a subrepresentation. Then the representation on the quotient V/U is admissible.*

Proof. Let first F be non-archimedean. By our above observation (equivalence of finite-dimensionality of Hom-spaces and isotypic components) and Cor. 2.21, we are left to prove that the ρ-isotypic components $(V/U)(\rho)$ are finite-dimensional for all irreducible finite-dimensional representation (ρ,W) of K.

So let (ρ,W) be such a representation. If (ρ,W) is not smooth, then obviously $\mathrm{Hom}_K(W,V/U) = \{0\}$. Indeed, if $0 \neq f \in \mathrm{Hom}_K(W,V/U)$, then f must be injective by the irreducibility of (ρ,W). Hence, $f(W)$ were a K-subrepresentation of the restriction $(\pi|_K, V/U)$, which were isomorphic to W by injectivity of f, on the one hand, but V/U being a smooth G-representation, cf. Cor. 2.21, implies that also its restriction to a K-representation is smooth, whence so were the K-subrepresentation $f(W) \cong W$, a contradiction. Therefore, $(V/U)(\rho) = \{0\}$ for (ρ,W) not smooth and so $(V/U)(\rho)$ is trivially finite-dimensional for all such representations (ρ,W).

So, let (ρ, W) be a smooth irreducible finite-dimensional representation of K and let us denote by p_U the canonical projection $p_U : V \twoheadrightarrow V/U$. Obviously, p_U is a linear G-equivariant map between smooth representations, so [Bus-Hen06], Prop. 2.3.(2), implies that $(V/U)(\rho) = p_U(V(\rho))$. Therefore, as $\dim_{\mathbb{C}}(p_U(V(\rho))) \leq \dim_{\mathbb{C}}(V(\rho)) < \infty$, the isotypic component $(V/U)(\rho)$ is finite-dimensional for all smooth irreducible finite-dimensional representations (ρ, W) of K. This shows the claim in the non-archimedean case.

Let now F be archimedean. By the definition of admissibility and our above observation, we again have to show that the ρ-isotypic components $(V/U)(\rho)$ are finite-dimensional for all irreducible finite-dimensional representations (ρ, W) of K. Recall that now the canonical linear projection $p_U : V \twoheadrightarrow V/U$ is continuous, hence commutes with integration over K. So, using the projectors E_ρ from (2.33), we see that for any $v + U \in V/U$,

$$
\begin{aligned}
E_\rho(v+U) &= \dim_{\mathbb{C}}(W) \cdot \int_K \mathrm{tr}(\rho(k)^{-1})\ (\pi(k)v + U)\ \mathrm{d}k \\
&= \dim_{\mathbb{C}}(W) \cdot \int_K \mathrm{tr}(\rho(k)^{-1})\ p_U(\pi(k)v)\ \mathrm{d}k \\
&= p_U\left(\dim_{\mathbb{C}}(W) \cdot \int_K \mathrm{tr}(\rho(k)^{-1})\ \pi(k)v\ \mathrm{d}k\right) \\
&= p_U(E_\rho(v)) \\
&= E_\rho(v) + U.
\end{aligned}
$$

Hence, we obtain identifications of vector spaces

$$
\begin{aligned}
(V/U)(\rho) &= E_\rho(V/U) \\
&= \{E_\rho(v) + U,\ v \in V\} \\
&= (E_\rho(V) + U)/U \\
&\cong E_\rho(V)/(E_\rho(V) \cap U) \\
&= V(\rho)/U(\rho), \qquad (2.35)
\end{aligned}
$$

which shows that $(V/U)(\rho)$ is finite-dimensional for all irreducible finite-dimensional representations (ρ, W) of K. This shows the claim in the archimedean case and hence the proposition. □

Exercise 2.36. Show that a smooth representation (π, V) of a non-archimedean local group G is admissible if and only if $\dim_{\mathbb{C}} V^C < \infty$ for all open compact subgroups C of G.

Hint: If (π, V) is admissible, then it is smooth by definition, whence our Lem. 2.24 together with [Bum98], Prop. 4.2.2, implies that $\dim_{\mathbb{C}} V^C < \infty$ for all compact open subgroups C of G. Conversely, if $V = V^\infty$ satisfies $\dim_{\mathbb{C}} V^C < \infty$ for all compact open subgroups C of G, then by the proof of [Bum98], Prop. 4.2.2, $V(\rho)$ is finite-dimensional for all irreducible finite-dimensional representations (ρ, W) of K, which have open kernel $\ker \rho$ in K. Complete the proof by showing that for an irreducible finite-dimensional representation (ρ, W) of K, which has non-open kernel, necessarily $V(\rho) = \{0\}$, hence such isotypic components are in particular finite-dimensional. Hint (in the hint): Show that none of such representations is smooth and then reconsult the non-archimedean part of the proof of Prop. 2.34.

As a scholium, which we could take from this exercise, we note that for smooth representations (π, V) of a non-archimedean local group G there is some sort of (informal) duality-principle: Either one fixes a maximal compact subgroup K and demands finite-dimensionality of all $V(\rho)$, where $\rho = \rho_K$ runs through all irreducible finite-dimensional representations of K; or one fixes a "type" of a minimal representation – the trivial one – and demands finite-dimensionality of all $V^C = V(\mathbf{1}_C)$, where C runs through all compact, open subgroups C of G.

It is worth noting that the equivalent condition for being admissible, established in Exc. 2.36, is actually the more common definition of an admissible representations of a non-archimedean local group, which the reader will find in close-to-all text-books on the topic (at least in those, which the author of the present book knows). For the sake of making our approach as uniform as possible in the archimedean and in the non-archimedean case, however, we have chosen not to take it as our definition, but rather to give preference to working with ρ-isotypic components.

Indeed, our definition of admissibility finally lets us overcome the built-in asymmetry that we imposed a continuity condition in our notion of a representation *only* in the archimedean case: Recalling from Lem. 2.24, that smoothness for non-archimedean representations amounts to continuity of $G \times V \to V$ with respect to a topology (the discrete one) on V that is natural, given the very nature of the topology on G itself (being totally disconnected), and, likewise, recalling that the notion of archimedean representation includes continuity of $G \times V \to V$, again with respect to a topology on V (namely a Fréchet one) which is natural, given the very nature of the topology on G (being a smooth paracompact manifold), our notion of admissibility finally serves as some sort of complement to bring these two ends together again, as it is illustrated in the following diagram:

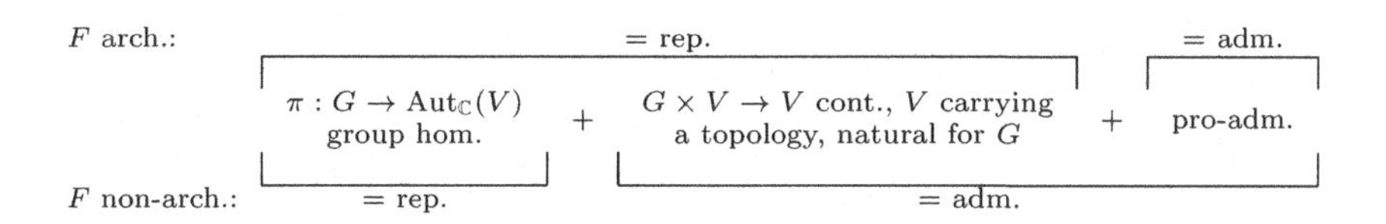

One could summarize this observation in the following dictum, which expresses in a uniform manner what it means to be an admissible representation of a local group:

An admissible representation of a local group G is a pair (π, V), where V is a vector space and $\pi : G \to \mathrm{Aut}_{\mathbb{C}} V$ is a homomorphism of groups, such that the natural map

$$\begin{aligned} G \times V &\to V \\ (g, v) &\mapsto \pi(g)v \end{aligned}$$

becomes continuous, if V is given a certain topology, which is associated to the topology on G in a natural way, and such that all finite-dimensional irreducible representations of one (and hence any) maximal compact subgroup K of G appear in V with finite multiplicity.

Chapter 3

Langlands Classification: Step 1 – What to Classify?

We resume our notation $G = \mathbf{G}(F)$, K, and (π, V) for a representation of G from the previous chapter. The class of *irreducible* representations will play a prominent role for the rest of Part I, and we shall therefore now try to investigate them on a deeper level.

The first, most naive idea would certainly be to try to classify all of them up to equivalence of G-representations – a hopeless plan, given the current status of the available techniques, but even more, an unnecessarily overambitious undertaking regarding our final aims in the global part of the theory in Part II: Not every irreducible representation may locally constitute to an automorphic or smooth-automorphic representation. In this respect, we shall (and will) explain in the next two sections on which special subfamily of irreducible representations we shall concentrate our attention – leading finally to our notion of *interesting* representation[1] – whose description up to equivalence of G-representations shall then be the purpose of what one usually refers to as the Langlands classification.

3.1 The relationship of smoothness and admissibility for irreducible representations – a detour to $(\mathfrak{g}, K)$-modules

So far, we have introduced two particular notions *smoothness* and, *admissibility*, which were tuned in a way, such that, when combined, they reestablish the symmetry in our definitions in the case of an archimedean and a non-archimedean ground field. It is hence instructive in a first step to compare the classes of *irreducible admissible* representations, on the one hand, and *irreducible smooth* representations, on the other. In the

[1] In the sense that it is a local representation *of our interest.*

non-archimedean case, we have the following beautiful result (cf. [Ren10], Thm. VI.2.2 for a full proof):

Theorem 3.1 (Jacquet). *Let F be non-archimedean. If (π, V) is irreducible and smooth, then it is admissible.*

Irreducible, non-admissible (i.e, by the last theorem, irreducible non-smooth) representations should be thought of being exotic. Indeed, by irreducibility, such representations obviously have to satisfy $V^{\infty} = \{0\}$ (see Lem. 2.26 for a family of examples of such representations). For this reason, and also in foresight of our global goals, we shall exclude them from our considerations.

We turn now to the archimedean case. Then the analogue of Thm. 3.1 does not hold true. However, it is quite a non-trivial job to find a counterexample: The first example of an irreducible representation, which is not admissible, was found by Enflo, Read and Soergel only in the 80's, cf. [Soe88]. From Soergel's counterexample one may also construct an irreducible smooth representation, which is not admissible (This is a non-trivial exercise for the very interested reader: Use Lem. 2.18 and Prop. 2.26 in order to identify the space of smooth vectors of Soergel's example as an irreducible and smooth representation. Then apply [Bor-Wal00], Prop. III.7.11 and reread Soergel's argument, replacing continuous induction by smooth induction, and so obtain a counterexample to Thm. 3.1 in the archimedean case.)

In summary, the situation looks as follows:

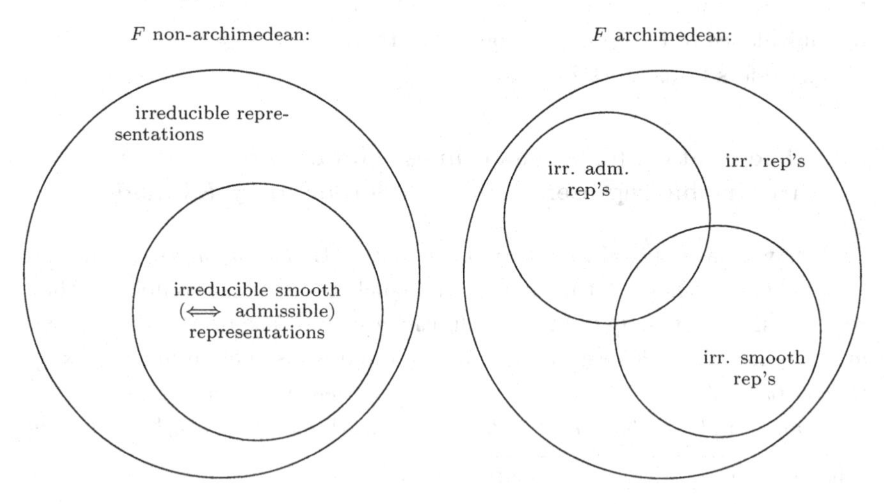

It is still tempting, however – and, as it will turn out in §3.2 below, also very instructive for singling out those irreducible representation, which shall finally be the object of our interest in the archimedean case – to try to repair this "defect" that there are irreducible smooth representations, which are not admissible. To this end one has to look deeper into the basic structure of archimedean representations. The remedying concept is that of a $(\mathfrak{g}, K)$-*module*:

Definition 3.2. A $(\mathfrak{g}, K)$-*module* is a triple (π_0, Π_0, V_0), where V_0 is a complex vector space, $\Pi_0 : K \to \mathrm{Aut}_{\mathbb{C}}(V_0)$ is a group homomorphism and $\pi_0 : \mathcal{U}(\mathfrak{g}) \to \mathrm{Aut}_{\mathbb{C}}(V_0)$ an algebra homomorphism such that

(1) $\dim_{\mathbb{C}}(\langle \Pi_0(K)v_0 \rangle) < \infty$, for all $v_0 \in V_0$, i.e., every vector is K-*finite*, and such that the map

$$K \times \langle \Pi_0(K)v_0 \rangle \to \langle \Pi_0(K)v_0 \rangle$$
$$(k, v) \mapsto \Pi_0(k)v \tag{3.3}$$

with the canonical locally convex topology on the finite-dimensional space $\langle \Pi_0(K)v_0 \rangle$, cf. Thm. 1.8, is continuous.

(2) For all $X \in \mathcal{U}(\mathfrak{g})$, $k \in K$ and $v_0 \in V_0$,

$$\Pi_0(k)\pi_0(X)v_0 = \pi_0(Ad(k)X)\Pi_0(k)v_0.$$

(3) For all $v_0 \in V_0$ and $Y \in \mathfrak{k}$,

$$\frac{\mathrm{d}}{\mathrm{d}t}\bigg|_{t=0} (\Pi_0(\exp(tY)))v_0 = \pi_0(Y)v_0,$$

i.e., $\pi_0|_{\mathfrak{k}}$ equals the differentiated action of $\mathfrak{k}$ defined by Π_0.

Note that condition (2) is only needed when K is disconnected. Moreover, because of condition (3) we will notationwise identify Π_0 with π_0 in what follows and simply denote a $(\mathfrak{g}, K)$-module by a pair (π_0, V_0).

Remark 3.4. As mentioned earlier, it can be shown that there exists a linear algebraic group $\mathbf{K}$ over $\mathbb{R}$ such that $K = \mathbf{K}(\mathbb{R})$. In this perspective, one may prove that V_0 being a $(\mathfrak{g}, K)$-module is equivalent to V_0 being the union of finite-dimensional, representations of the linear algebraic group $\mathbf{K}/\mathbb{C}$. Hence, the notion of a $(\mathfrak{g}, K)$-module is truly algebraic.

A *morphism* of $(\mathfrak{g}, K)$-modules (π_0, V_0) and (ψ_0, W_0) is a linear map $\varphi_0 : V_0 \to W_0$, which is equivariant under the action of $\mathcal{U}(\mathfrak{g})$ and K. An *isomorphism* of $(\mathfrak{g}, K)$-modules is a bijective morphism of $(\mathfrak{g}, K)$-modules. In such a case we write $(\pi_0, V_0) \cong (\psi_0, W_0)$ or simply $\pi_0 \cong_{(\mathfrak{g},K)} \psi_0$.

A $(\mathfrak{g}, K)$-*submodule* of a $(\mathfrak{g}, K)$-module (π_0, V_0) is a subspace $W_0 \subseteq V_0$, which is stable under the action of $\mathcal{U}(\mathfrak{g})$ and K.

Finally, a $(\mathfrak{g}, K)$-module (π_0, V_0) is called *irreducible*, if $V_0 \neq \{0\}$ and if V_0 admits no proper $(\mathfrak{g}, K)$-submodules.

Now, let (π, V) be a representation of G (we remind the reader that we are still considering the case of an archimedean ground field). We will write

$$V_{(K)} := \{v \in V | \dim_{\mathbb{C}}(\langle \pi(K)v \rangle) < \infty\} \tag{3.5}$$

for the subspace of K-finite vectors of V. It depends on the choice of K. The following result is an easy consequence of our definitions and builds a bridge between G-representations and $(\mathfrak{g}, K)$-modules:

Lemma 3.6. *Let F be archimedean and let (π, V) be a representation of G. Then, $V_0 := V_{(K)} \cap V^\infty$ is stable under $\mathcal{U}(\mathfrak{g})$ and K and hence defines a $(\mathfrak{g}, K)$-module $(\pi_0, V_0) := (\pi^\infty_{(K)}, V^\infty_{(K)})$.*

Remark 3.7. It is straight forward to check that V^∞ is stable under $\mathcal{U}(\mathfrak{g})$ (see [WarI72], Prop. 4.4.1.2; the very argument may be extrapolated from [Kna86], p. 53). Now, by its definition, our choice of a Fréchet-topology on V^∞ is precisely the coarsest locally convex topology with respect to which the action of $\mathcal{U}(\mathfrak{g})$ is continuous, see also [Cas89I], Lem. 1.2.

The following lemma is almost equally obvious, but it tells us that passing over to K-finite vectors, one does not loose that much information:

Lemma 3.8. *Let F be archimedean and let (π, V) be a representation of G. Then, $V_{(K)}$ is dense in V.*

Proof. See, e.g., [Var77], Lem. II.7.10. □

Observing that Lem. 3.8 remains valid, when G is replaced by K, one directly obtains the following corollary:

Corollary 3.9. *Let (π, V) be an irreducible representation of K. Then V is finite-dimensional.*

Hence, we could have safely removed the condition of being finite-dimensional, whenever we spoke of irreducible representations of K ever since, e.g., in the definition of an admissible G-representation. For sake of completeness, we remark that in the non-archimedean case the assertion of Cor. 3.9 remains valid for irreducible, smooth representations of K: They are all finite-dimensional, cf. [Ren10], Thm. IV.1.1.(i).

Let us now see, why the concept of underlying $(\mathfrak{g}, K)$-modules serves a some sort of remedy for the lacking implication that irreducibility and smoothness imply admissibility. To this end, and also for later use, we record the following definition and a deep result of Harish-Chandra:

Definition 3.10. Let $\mathcal{Z}(\mathfrak{g})$ be the center of $\mathcal{U}(\mathfrak{g})$ and let (π_0, V_0) be a $(\mathfrak{g}, K)$-module. It is called...

(1) ... *$\mathcal{Z}(\mathfrak{g})$-finite*, if the annihilating ideal $\mathcal{J} \lhd \mathcal{Z}(\mathfrak{g})$

$$\mathcal{J} := \{Z \in \mathcal{Z}(\mathfrak{g}) \mid \pi_0(Z)v_0 = 0 \text{ for all } \ v_0 \in V_0\}$$

is of finite codimension, i.e., $\dim_{\mathbb{C}}(\mathcal{Z}(\mathfrak{g})/\mathcal{J}) < \infty$.

(2) ... *admissible*, if the space $\mathrm{Hom}_K(W, V_0)$ of K-equivariant linear maps $f : W \to V_0$ is finite-dimensional for all irreducible (finite-dimensional) representations (ρ, W) of K.

(3) ... *finitely generated*, if there is a finite-dimensional subspace $W \subseteq V_0$, such that $\pi_0(K)W \subseteq W$ (i.e., W is K-stable) and $\pi_0(\mathcal{U}(\mathfrak{g}))W = V_0$.

(4) ... *of finite length*, if every strictly ascending chain of $(\mathfrak{g}, K)$-submodules is finite.

We make the following two observations: Firstly, if (π, V) is a representation of G, and (ρ, W) an irreducible finite-dimensional representation of K, then

$$\mathrm{Hom}_K(W, V) = \mathrm{Hom}_K(W, V_{(K)}). \tag{3.11}$$

Indeed, for any $f \in \mathrm{Hom}_K(W, V)$, $k \in K$ and $w \in W$, $\pi(k)f(w) = f(\rho(k)w) \in f(W)$, so $\dim_{\mathbb{C}}(\langle \pi(K)f(w)\rangle) \leq \dim_{\mathbb{C}} f(W) \leq \dim_{\mathbb{C}} W < \infty$. Therefore, every $f(w)$ is K-finite, whence $f(W) \subseteq V_{(K)}$ and so (3.11). As a consequence, a smooth G-representation (π, V) is admissible, if and only if its underlying $(\mathfrak{g}, K)$-module $(\pi_{(K)}, V_{(K)})$ is.

For the second observation let (π_0, V_0) be any $(\mathfrak{g}, K)$-module and let (ρ, W) be any irreducible (finite-dimensional) representation of K. We define the *ρ-isotypic component* $V_0(\rho)$ of (π_0, V_0) as the image of the natural map

$$\begin{aligned} \mathrm{Hom}_K(W, V_0) \otimes W &\to V_0, \\ f \otimes w &\mapsto f(w). \end{aligned}$$

One may argue exactly as in §2.4 that this map is injective, i.e., that one has an isomorphism of vector spaces $\mathrm{Hom}_K(W, V_0) \otimes W \cong V_0(\rho)$, whence the space $\mathrm{Hom}_K(W, V_0)$ is finite-dimensional, if and only if the ρ-isotypic component $V_0(\rho)$ is finite-dimensional. Now assume that (π_0, V_0) is the underlying $(\mathfrak{g}, K)$-module of a smooth G-representation (π, V) and recall the continuous projections $E_\rho : V \twoheadrightarrow V(\rho)$ from (2.33). Then (3.11) implies that E_ρ restricts to a projection

$$E_\rho : V_{(K)} \twoheadrightarrow V_{(K)}(\rho) = V(\rho). \tag{3.12}$$

The following lemma is obvious from the definitions...

Lemma 3.13. *Let (π, V) be an admissible representation of G. Then (π_0, V_0) is an admissible $(\mathfrak{g}, K)$-module.*

... whereas the next result is a fundamental theorem of Harish-Chandra:

Theorem 3.14 (Harish-Chandra). *Let (π_0, V_0) be a $(\mathfrak{g}, K)$-module. Then any two of the following conditions imply the others:*

(1) (π_0, V_0) is $\mathcal{Z}(\mathfrak{g})$-finite
(2) (π_0, V_0) is admissible
(3) (π_0, V_0) is finitely generated
(4) (π_0, V_0) is of finite length

Moreover, condition (4) implies all the others.

Proof. See [Vog81], Cor. 5.4.16 for a detailed proof. □

We hence obtain the following corollary, which provides the announced analogue of Thm. 3.1 in the archimedean case.

Corollary 3.15. *Let (π_0, V_0) be an irreducible $(\mathfrak{g}, K)$-module. Then it is admissible.*

3.2 Infinitesimal equivalence – advantages and disadvantages

The analogue of Jacquet's theorem, Thm. 3.1, for $(\mathfrak{g}, K)$-modules, Cor. 3.15, is not their only nice property. Indeed, it is the following result from which $(\mathfrak{g}, K)$-modules gain their structural importance:

Theorem 3.16 (Harish-Chandra, [HCh53]; see also [Var77], Thm. II.7.14). *Let F be archimedean, and let (π, V) be an admissible representation of G. Then:*

(1) $V_{(K)} \subseteq V^{\infty}$, i.e., every K-finite vector is smooth, or, equivalently $V_0 = V_{(K)}$.
(2) The assignments $W \mapsto W_{(K)}$ and $S \mapsto \overline{S}$ set up pairwise inverse bijections between G-subrepresentations $W \subseteq V$ on the one hand and $(\mathfrak{g}, K)$-submodules $S \subseteq V_{(K)}$ on the other.[2] In particular, (π, V) is irreducible if and only if $(\pi_{(K)}, V_{(K)})$ is irreducible as $(\mathfrak{g}, K)$-module.

Remark 3.17. Usually, the second part of the theorem is only stated for Banach spaces V, because it is derived from the fact – shown by Harish-Chandra in our reference [HCh53] – that every K-finite vector is analytic (a notion, which we did not define and which is stronger than being smooth). The notion of being analytic makes no sense in our more general setup of Fréchet spaces, but it may be adopted to a notion called "weakly analytic": A vector $v \in V$ is called weakly analytic, if for every $v' \in V'$, the function $G \to \mathbb{C}$, $g \mapsto v'(\pi(g)v)$ is real analytic. By [HCh53], for every K-finite vector v the latter map satisfies an elliptic differential equation with real analytic coefficient functions, implying that every $v \in V_{(K)}$ is weakly analytic, and one may therefrom derive the second part of Thm. 3.16 analogously as for Banach spaces, see [Var77], Thm. II.7.14.

Remark 3.18. In light of Soergel's counterexample one cannot remove the assumption of admissibility in Thm. 3.16. Indeed, suppose we are given an irreducible smooth representation (π, V) of G. Then, if the above theorem remained true in this generality, (π_0, V_0) would be an irreducible and hence, by Cor. 3.15, also an admissible $(\mathfrak{g}, K)$-module. In other words, we had $\dim_{\mathbb{C}}(\mathrm{Hom}_K(W, V_0)) < \infty$ for all irreducible (finite-dimensional) representations (ρ, W) of K. However, applying (3.11) to the smooth representation (π, V), we would get $\dim_{\mathbb{C}}(\mathrm{Hom}_K(W, V)) = \dim_{\mathbb{C}}(\mathrm{Hom}_K(W, V_0)) < \infty$, whence (π, V) would be admissible, too. But this is impossible by the above mentioned counterexample of Soergel.

By Thm. 3.16, the internal structure of an admissible representation may be understood in purely algebraic terms, that is, forgetting all topological intricacies, by looking at its underlying $(\mathfrak{g}, K)$-module (where we recall that the notion of admissibility, due to the automatic continuity of linear maps

[2]Here, $\overline{S}$ denotes the topological closure of S in V.

on finite-dimensional seminormed spaces, Thm. 1.8, was algebraic in nature itself). This motivates the following general definition.

Definition 3.19. Let F be archimedean. Two G-representations (π_1, V_1) and (π_2, V_2) are *infinitesimally equivalent*, in symbols $\pi_1 \underset{(\mathfrak{g},K)}{\cong} \pi_2$, if the underlying $(\mathfrak{g}, K)$-modules are isomorphic, i.e., $\pi^\infty_{1_{(K)}} \cong \pi^\infty_{2_{(K)}}$.

Exercise 3.20. Show that – although the space of K-finite vectors $V_{(K)}$ in a G-representation (π, V) depends on the choice of K – the notion of infinitesimal equivalence is independent of it. *Hint:* Use that a maximal compact subgroup K of an archimedean local group G is unique up to conjugation.

Certainly, regarding the reduction to an object of algebraic nature, it is one of the biggest advantages of the notion of infinitesimal equivalence, that it allows a much simpler analysis of archimedean representations. Indeed, building on work of Harish-Chandra, in [Lan89], Langlands was able to give a parameterization of all irreducible admissible representations of G on Banach spaces up to infinitesimal equivalence and established in this way one of the most fundamental results in the theory. When combined with the work of Knapp–Zuckerman, [Kna-ZucI82, Kna-ZucII82] (which treats *connected semisimple* Lie groups, i.e., in our terminology here, connected archimedean local groups with finite center) and the (unpublished) thesis of Mirković, [Mir86] (which, according to Knapp and Vogan, cf. [Kna-Vog95], p. 913, allows the treatment of general archimedean local groups), Langlands's parameterization becomes a proper classification. In fact, we will devote Chapter 5 below to a certain presentation of the Langlands classification.

However, on the side of disadvantages, the problem with Langlands classification is that infinitesimal equivalence is (unfortunately) much weaker than equivalence: In fact, an infinitesimal equivalence class of G-representations typically (!) contains *infinitely many* equivalence classes of G-representations, as it shall be illustrated by the following paradigmatic example.

Example 3.21. Let G be the special unitary group of signature $(1, 1)$, i.e.,

$$G = \mathrm{SU}(1,1) = \left\{ \begin{pmatrix} a & b \\ \overline{b} & \overline{a} \end{pmatrix} : a, b \in \mathbb{C}, |a|^2 - |b|^2 = 1 \right\},$$

viewed as the group of real points $G = \mathbf{G}(\mathbb{R})$ of an algebraic group $\mathbf{G}$ over the local archimedean field $F = \mathbb{R}$. As our choice of a maximal compact subgroup, we fix

$$K = \left\{ k_\vartheta = \begin{pmatrix} e^{i\vartheta} & 0 \\ 0 & e^{-i\vartheta} \end{pmatrix}, \vartheta \in \mathbb{R} \right\}.$$

Consider the action of G on the open unit disc $D \subset \mathbb{C}$ by linear fractional transformations,

$$\begin{pmatrix} a & b \\ \bar{b} & \bar{a} \end{pmatrix} \cdot z := \frac{az+b}{\bar{b}z+\bar{a}}.$$

This action is transitive, and the stabilizer of $0 \in D$ is seen to be precisely K, so the quotient G/K is isomorphic to D. It follows that G also acts by left translation on the space of smooth functions on D. That is to say, G acts on

$$C^\infty(D) := \{f : D \to \mathbb{C}, f \text{ is smooth in the sense of real manifolds}\},$$

via the map

$$\begin{aligned} \pi_{C^\infty} : G &\to \operatorname{Aut}_{\mathbb{C}}(C^\infty(D)), \\ g &\mapsto (f \mapsto f(g^{-1}\_\_)). \end{aligned}$$

We now construct an infinite family of pairwise inequivalent representations on subspaces of $C^\infty(D)$, which, however, will turn out to be all infinitesimally equivalent. To this end, for each $1 \le p < \infty$, let

$$H^p(D) := \{f : D \to \mathbb{C} \text{ holomorphic}, f \text{ has } L^p \text{ boundary values}\} \subset C^\infty(D).$$

Each of these spaces is seen to be stable under π_{C^∞}, so we can restrict π_{C^∞} to $H^p(D)$ and obtain a group homomorphism $G \to \operatorname{Aut}_{\mathbb{C}}(H^p(D))$, which we denote by π_{H^p}. Moreover, $H^p(D)$ can be topologized by embedding it into $L^p(S^1)$, and so inherits a Banach space structure (for $p = 2$ this is even a Hilbert space structure). It is left to the reader to check that $(\pi_H^p, H^p(D))$ is a representation of G, i.e., that the map

$$\begin{aligned} G \times H^p(D) &\to H^p(D), \\ (g, f) &\mapsto \pi_{H^p}(g)f \end{aligned}$$

is continuous.

To check infinitesimal equivalence, we determine the space of K-finite vectors $H^p(D)_{(K)}$ inside $H^p(D)$. For every function of the form $f_n : z \mapsto z^n$ (this is an element of $H^p(D)$), a direct computation yields

$$\pi_{H^p}(k_\vartheta)f_n = e^{-2ni\vartheta} f_n,$$

so a general $f \in H^p(D)$ is K-finite if and only if its Taylor expansion $f = \sum_{n=0}^{\infty} c_n f_n$ around $0 \in D$ has only finitely many terms, i.e., is a polynomial. In other words, for each $1 \leq p < \infty$,

$$H^p(D)_{(K)} = \mathbb{C}[z],$$

independent of p, which implies that all representations $(\pi_{H^p}, H^p(D))$ are infinitesimally equivalent, as claimed (The very careful reader may want to combine the observation made in Rem. 3.18, which shows that all these representations are admissible, and then apply Thm. 3.16.(1), showing that $(H^p(D))^{\infty}_{(K)} = H^p(D)_{(K)}$, in order to conclude this). However, it may also be shown that they are all pairwise inequivalent (Exercise!).

On a deeper level, this example shall make apparent, that the notion of infinitesimal equivalence – although it stems in view of Jacquet's theorem from our wish to reestablish the symmetry between archimedean and non-archimedean representation theory – is not (yet) the right setup for our purposes, as we really intend to work with G-representations and proper equivalence classes.

Resolving this intriguing problem is the theme of the theory of "globalizations". For us, fortunately, Casselman and Wallach have provided a particularly suitable solution. Roughly speaking, they prove that Langlands's classification (of irreducible admissible representations up to infinitesimal equivalence) is equivalent to a classification of irreducible admissible smooth representations (but now up to equivalence of G-representations), if our representation is nicely compatible with the seminorms on the given Fréchet-space. The general formulation of Casselman and Wallach's result makes use of the following notion:

Definition 3.22. Let F be archimedean. A smooth representation (π, V) of G is called of *moderate growth* if, for every continuous seminorm p on V, there exist a real number $d = d_p$ and a continuous seminorm $q = q_p$ on V such that

$$p(\pi(g)v) \leq \|g\|_{\infty}^{d_p} q_p(v) \qquad \text{for all } g \in G, v \in V,$$

where we set

$$\|g\|_{\infty} := \max_{1 \leq i,j \leq n} \{|g_{ij}|, |(g^{-1})_{ij}|\}.$$

Remark 3.23. Recall that we defined $\mathbf{G}$ as an F-subgroup of some $\mathbf{GL}_n(\Omega)$, cf. §2.1, giving rise to coordinates g_{ij} and $(g^{-1})_{ij}$ in F.

The following lemma provides us with many interesting examples:

Lemma 3.24. *Let (π, V) be a G-representation on a Banach space V. Then (π^∞, V^∞) is of moderate growth. If (π, V) is a smooth G-representation of moderate growth and $W \subseteq V$ is a G-subrepresentation, then W and V/W are of moderate growth.*

Proof. The first assertion is a simple consequence of Lem. 2.18 and [Wal92], Lem. 11.5.1. For the second assertion, use our Cor. 2.21 to identify W and V/W as smooth representations of G and then apply [Wal92], Lem. 11.5.2. □

For our purposes the right notion of a globalization turns out to be provided by *Casselman–Wallach representations*:

Definition 3.25. A smooth admissible representation (π, V) of moderate growth, whose underlying $(\mathfrak{g}, K)$-module $(\pi_{(K)}, V_{(K)})$ (cf. Lem. 3.6) is finitely generated, is henceforth called a *Casselman–Wallach representation.*

Let us denote by $CW(G)$ the category with objects all Casselman–Wallach representations of G and with morphisms the continuous linear G-equivariant maps, whose images are topological summands, and by $HC(G)$ the category with objects all admissible finitely generated $(\mathfrak{g}, K)$-modules and with morphisms the morphisms of $(\mathfrak{g}, K)$-modules. Then, for all representations $(\pi, V) \in CW(G)$ the underlying $(\mathfrak{g}, K)$-module of K-finite vectors $(\pi_{(K)}, V_{(K)})$ obviously lies in $HC(G)$ (cf. Lem. 3.13). Moreover, for any continuous linear G-equivariant map $f : V \to W$ between two Casselman–Wallach representations, obviously $f(V_{(K)}) \subseteq W_{(K)}$ and so $f_{(K)} := f|_{V_{(K)}}$ defines a morphism of the underlying $(\mathfrak{g}, K)$-modules $V_{(K)}$ and $W_{(K)}$. In fancier language, this shows that taking K-finite vectors defines a functor $CW(G) \to HC(G)$, given on the level of objects by $(\pi, V) \mapsto (\pi_{(K)}, V_{(K)})$ and on the level of morphisms by $f \mapsto f_{(K)}$. We can now state Casselman and Wallach's result:

Theorem 3.26 (Casselman–Wallach). *The functor $CW(G) \to HC(G)$, $(\pi, V) \mapsto (\pi_{(K)}, V_{(K)})$, $f \mapsto f_{(K)}$ is an equivalence of categories. In particular, two Casselman–Wallach representations (π, V) and (ψ, W) are infinitesimally equivalent if and only if they are equivalent as G-representations, and any admissible, finitely generated $(\mathfrak{g}, K)$-module (π_0, V_0) is the space of K-finite vectors in a Casselman–Wallach representation, denoted $(\overline{\pi_0}^{\mathsf{CW}}, \overline{V_0}^{\mathsf{CW}})$, which is unique up to equivalence of*

G-representations. The resulting functor $HC(G) \to CW(G)$, $(\pi_0, V_0) \mapsto (\overline{\pi_0}^{\mathsf{cw}}, \overline{V_0}^{\mathsf{cw}})$, $f_0 \mapsto \overline{f_0}^{\mathsf{cw}}$ satisfies

$$((\overline{\pi_0}^{\mathsf{cw}})_{(K)}, (\overline{V_0}^{\mathsf{cw}})_{(K)}) \underset{(\mathfrak{g},K)}{\cong} (\pi_0, V_0) \quad \text{and} \quad (\overline{\pi_{(K)}}^{\mathsf{cw}}, \overline{V_{(K)}}^{\mathsf{cw}}) \underset{G}{\cong} (\pi, V).$$

Both functors map irreducibles onto irreducibles.

Proof. This is the contents of cf. [Wal92], Thm. 11.6.7 & Cor. 11.6.8. For sake of completeness, however, we sketch the argument for the assertion on irreducibility: Firstly, given an irreducible Casselman–Wallach representation (π, V), the irreducibility of $(\pi_{(K)}, V_{(K)})$ follows from Thm. 3.16. Conversely, let (π_0, V_0) be an irreducible (and hence admissible, cf. Cor. 4.4) $(\mathfrak{g}, K)$-module and let $(\overline{\pi_0}^{\mathsf{cw}}, \overline{V_0}^{\mathsf{cw}})$ be Casselman–Wallach representation, whose space of K-finite vectors is isomorphic to (π_0, V_0). Then, $(\overline{\pi_0}^{\mathsf{cw}}, \overline{V_0}^{\mathsf{cw}})$ is in particular admissible and its space of K-finite vectors is irreducible. Hence, again by Thm. 3.16 $(\overline{\pi_0}^{\mathsf{cw}}, \overline{V_0}^{\mathsf{cw}})$ is irreducible as a representation of G. □

A Casselman–Wallach representation $(\overline{\pi_0}^{\mathsf{cw}}, \overline{V_0}^{\mathsf{cw}})$ from Thm. 3.26 is called a *Casselman–Wallach completion* of the $(\mathfrak{g}, K)$-module (π_0, V_0). As it is clear from Thm. 3.26, its G-equivalence class $[(\overline{\pi_0}^{\mathsf{cw}}, \overline{V_0}^{\mathsf{cw}})]$ is unique, hence we will by abuse of language also speak of $(\overline{\pi_0}^{\mathsf{cw}}, \overline{V_0}^{\mathsf{cw}})$ as *the* Casselman–Wallach completion of (π_0, V_0). The following lemma shows that the category of Casselman–Wallach representations is well-behaved under taking subrepresentations and quotients:

Lemma 3.27. *Let (π, V) be a Casselman–Wallach representation of G and $U \subseteq V$ a subrepresentation. Then the representations on U and V/U are also Casselman–Wallach representations of G.*

Proof. We use Cor. 2.21 (U is smooth), Lem. 2.30 (U is admissible), Lem. 3.24 (U is of moderate growth), in oder to identify the representation on U as a Casselman–Wallach representation – $U_{(K)}$ being obviously finitely generated. In order to show that also V/U is a Casselman–Wallach representation, we first focus on the $(\mathfrak{g}, K)$-module $V_{(K)}/U_{(K)}$: Firstly, it is obviously finitely generated, since $V_{(K)}$ is finitely generated by assumption. Secondly, recall that V/U is a smooth and admissible G-representation by Cor. 2.21 and Prop. 2.34, whence its underlying $(\mathfrak{g}, K)$-module is $(V/U)_{(K)}$, which is admissible by Lem. 3.13. Now, our (3.12) and (2.33), as combined with [Wal89], Lem. 3.3.3, provide a K-equivariant linear bijection

$$(V/U)_{(K)} \cong \bigoplus_{[(\rho,W)]} E_\rho(V/U),$$

the direct algebraic sum being over all equivalence classes $[(\rho, W)]$ of irreducible finite-dimensional K-representations (ρ, W). Hence, recalling (2.35),

$$\begin{aligned}(V/U)_{(K)} &\cong \bigoplus_{[(\rho,W)]} V(\rho)/U(\rho) \\ &\cong \bigoplus_{[(\rho,W)]} V(\rho)/ \bigoplus_{[(\rho,W)]} U(\rho) \\ &\cong V_{(K)}/U_{(K)},\end{aligned}$$

where we again used [Wal89], Lem. 3.3.3, for the last K-equivariant linear bijection. Therefore, $V_{(K)}/U_{(K)}$ is an admissible $(\mathfrak{g}, K)$-module and so

$$0 \to U_{(K)} \hookrightarrow V_{(K)} \twoheadrightarrow V_{(K)}/U_{(K)} \to 0$$

is an exact sequence in $HC(G)$. Hence, taking Casselman–Wallach completions, Thm. 3.26 and [Wal92], §11.6.8, imply that we obtain an exact sequence in $CW(G)$

$$0 \to U \hookrightarrow V \twoheadrightarrow \overline{(V_{(K)}/U_{(K)})}^{\mathrm{cw}} \to 0.$$

Hence, $\overline{(V_{(K)}/U_{(K)})}^{\mathrm{cw}}$ being Fréchet, the Banach–Schauder theorem, Thm. 1.7, implies that the resulting G-equivariant, continuous linear bijection $V/U \to \overline{(V_{(K)}/U_{(K)})}^{\mathrm{cw}}$ is an isomorphism of G-representations $V/U \cong \overline{(V_{(K)}/U_{(K)})}^{\mathrm{cw}}$. The latter being a Casselman–Wallach representation, hence so is the isomorphic representation V/U. □

We remark that as a side-effect of the above proof we have found that the K-equivariant linear bijection

$$(V/U)_{(K)} \cong V_{(K)}/U_{(K)} \tag{3.28}$$

is in fact an isomorphism of $(\mathfrak{g}, K)$-modules (since $(V/U)_{(K)} \cong (\overline{(V_{(K)}/U_{(K)})}^{\mathrm{cw}})_{(K)} \cong V_{(K)}/U_{(K)}$, where the last isomorphism is a consequence of Thm. 3.26).

Let us now change – or better: enlarge – our point of view: As implied by Lem. 2.18, Lem. 2.30, Thm. 3.16 and Lem. 3.24, the representation on the space of smooth vectors in an irreducible admissible representation *on a Hilbert space* is a Casselman–Wallach representation, and hence the representations on the spaces of smooth vectors in any two such representations are infinitesimally equivalent if and only if they are equivalent as G-representations. The next theorem in fact provides a "converse" to this construction.

Theorem 3.29 (Casselman). *Let (π, V) be a Casselman–Wallach representation. Then there exists an admissible representation $(\widehat{\pi}, \widehat{V})$ on a Hilbert space $\widehat{V}$, such that*

$$(\pi, V) \cong (\widehat{\pi}^\infty, \widehat{V}^\infty).$$

If (π, V) is irreducible, then so is $(\widehat{\pi}, \widehat{V})$.

Proof. Let (π, V) be a Casselman–Wallach representation. Then its underlying $(\mathfrak{g}, K)$-module $(\pi_{(K)}, V_{(K)})$ is admissible (cf. Lem. 3.13) and finitely generated (by assumption), hence, by [Wal89], Cor. 4.2.4 and §4.2.5, *ibidem*, there exists an admissible representation $(\widehat{\pi}, \widehat{V})$ on a Hilbert space $\widehat{V}$, such that $\widehat{\pi}_{(K)} \cong \pi_{(K)}$ as $(\mathfrak{g}, K)$-modules. In particular, $(\widehat{\pi}_{(K)}, \widehat{V}_{(K)})$ is finitely generated. Combining this simple observation with Lem. 2.18, Lem. 2.30 and Lem. 3.24, we see that $(\widehat{\pi}^\infty, \widehat{V}^\infty)$ is a Casselman–Wallach representation. But as $(\widehat{\pi}^\infty, \widehat{V}^\infty)$ is infinitesimally equivalent with our given Casselman–Wallach representation (π, V), they must be equivalent by Thm. 3.26. This shows the first assertion. In order to show the second assertion, suppose that $(\widehat{\pi}, \widehat{V})$ is not irreducible. As $(\widehat{\pi}, \widehat{V})$ is admissible, Thm. 3.16 then implies that its underlying $(\mathfrak{g}, K)$-module $(\widehat{\pi}_{(K)}, \widehat{V}_{(K)})$ is not irreducible. But $(\widehat{\pi}_{(K)}, \widehat{V}_{(K)})$ is isomorphic to $(\pi_{(K)}, V_{(K)})$, which, as the underlying $(\mathfrak{g}, K)$-module of an irreducible admissible representation is irreducible (by Thm. 3.16 again). This is a contradiction, so $(\widehat{\pi}, \widehat{V})$ is irreducible itself. □

Warning 3.30. For given (π, V), the equivalence class of the Hilbert space representation $[(\widehat{\pi}, \widehat{V})]$ is not unique.

The theorem and our observation in the paragraph before, tells us that the following two approaches to a classification of representations of G are equivalent:

- Classify irreducible Casselman–Wallach representations up to equivalence.
- Classify the representations on the spaces of smooth vectors inside irreducible admissible representations on Hilbert spaces up to equivalence.

3.3 Interesting representations

After this archimedean intermezzo, we now revert to the case of an arbitrary local ground field F. According to our observation above, we can finally isolate a family of representations (π, V) of G for all local fields F, for which we seek to classify the respective spaces of smooth vectors (π^∞, V^∞) up to equivalence of G-representations. In this sense, we adopt the following terminology:

Definition 3.31. A representation (π, V) of G is called *interesting*, if it is

(1) irreducible, admissible and V is a Hilbert space, if F is archimedean,
(2) irreducible and admissible, if F is non-archimedean.

Moreover for technical reasons we shall assume that an interesting representation (π, V) always has a *central character*, i.e., that there exists a character (i.e., a continuous group homomorphism) $\omega_\pi : Z_G \to \mathbb{C}^*$ of the center Z_G of G, such that $\pi(z)v = \omega_\pi(z)v$ for all $v \in V$, $z \in Z_G$. We remark that this is automatic, if F is non-archimedean, cf. [Hen07], Prop. 1.4

3.4 Unitary representations

We conclude this chapter by introducing an important family of representations, the *unitary representations.* In this section, F is an arbitrary local field.

Definition 3.32. (1) A representation (π, V) of G is called *unitary*, if V comes equipped with a G-invariant, non-degenerate, positive-definite Hermitian form $\langle \cdot, \cdot \rangle$, i.e., if

$$\langle \pi(g)v, \pi(g)w \rangle = \langle v, w \rangle \qquad \text{for all } g \in G, v, w \in V.$$

If F is archimedean, we additionally assume that $\langle \cdot, \cdot \rangle$ generates the topology on V, i.e., that $(V, \langle \cdot, \cdot \rangle)$ is a Hilbert space. We denote by $[\widehat{\pi}]$ the *unitary equivalence class* of π, i.e. the family of unitary representations which are isomorphic to (π, V).

(2) Let (π, V) a unitary representation of G. If V is a Hilbert space and $G \times V \to V$, $(g, v) \mapsto \pi(g)v$ is continuous, then (π, V) is said to be *topologically irreducible*, if $V \neq \{0\}$ and if there is no G-invariant proper, closed subspace of V.

Obviously, Def. 3.32.(2) only defines a new notion, if F is non-archimedean (and agrees with our normal notion of irreducibility if F is archimedean). The following result by Harish-Chandra (as accompanied by work by Gel'fand and Bernstein) tells us why this additional notion of topological irreducibility is useful, because it nicely relates to the other notions introduced so far:

Theorem 3.33. *If (π, V) is a topologically irreducible unitary representation, then it is pro-admissible.*[3] *Moreover, V^∞ is dense in V.*

Proof. For F archimedean, the first assertion is shown in [Wal89], Thm. 3.4.10, while the second is a far more general fact, see [WarI72], Prop. 4.4.1.1. For F non-archimedean we refer to [Car79], Cor. 2.3, for density of V^∞ in V and to Thm. 2.8, *ibidem* for pro-admissibility of (π, V). For a detailed proof of the latter, see also [Gar09], Thm. 8. □

In particular (by definition of admissibility), irreducible unitary representations are automatically admissible in the archimedean setting. In fact, we obtain

Corollary 3.34. *Let F be archimedean. Then all irreducible unitary representations are interesting.*

Proof. This clear from the definitions, once we recall [Wal89], Lem. 1.2.1 (which establishes Schur's lemma in this case, and hence the existence of a central character). □

Contextual Remark. Historically, irreducible unitary representations had actually been studied first, i.e., before a proper notion of an admissible representation was established. In fact, the systematic study of irreducible admissible representations of semisimple Lie groups originates in the work of Harish-Chandra, cf. [HCh53, HCh54] (... *following an idea of Chevalley* as Harish-Chandra writes (cf. [HCh53], p. 185)).

In the non-archimedean case, the analogue of Cor. 3.34 holds true for the space of smooth vectors of a topologically irreducible unitary representation (π, V). We recall this as a consequence of the following.

Theorem 3.35. *Let F be non-archimedean. Then the assignment $(\pi, V) \mapsto (\pi^\infty, V^\infty)$ sets up a functorial bijection between the topological equivalence classes (i.e., the isomorphisms are all assumed to be (bi-)continuous) of topologically irreducible unitary representations of G and the equivalence classes of irreducible unitary admissible representations of G. Its inverse is given by taking the Hilbert space closure.*

[3] Recall that our fixed choice of a maximal compact subgroup K, with respect to which we defined pro-admissibility, was arbitrary. Hence, the more precise assertion of the first claim of Thm. 3.33 is to say that a topologically irreducible unitary representation is pro-admissible with respect to any choice of a maximal compact subgroup.

Proof. This is explained in [Car79], §2.8, to which we refer. See also [Hen07], Thm. 1.5, for a more explicit statement (while we also invite the reader to consider the elegant proof of Thm. 1 in [Ber74], which may be taken as the most original source of the theorem's assertion). □

Corollary 3.36. *Let F be non-archimedean. Then (π^∞, V^∞) is interesting for all topologically irreducible unitary representations (π, V).*

Another important property of topologically irreducible unitary representations of local groups is that they factor along decompositions $G = G_1 \times G_1$ into direct factors of local groups G_i. More precisely, let (π_i, V_i) be topologically irreducible unitary representations of local groups G_i (defined over a local field F). Then, the map $(G_1 \times G_2) \times (V_1 \otimes V_2) \to (V_1 \otimes V_2)$, defined by the (linear extension of the) assignment $((g_1, g_2); v_1 \otimes v_2) \mapsto \pi_1(g_1)v_1 \otimes \pi_2(g_2)v_2$ is continuous, if $V_1 \otimes V_2$ is given the subspace topology from $V_1 \widehat{\otimes} V_2$, and in fact extends continuously to a unitary representation

$$\pi_1 \widehat{\otimes} \pi_2 : (G_1 \times G_2) \times (V_1 \widehat{\otimes} V_2) \to (V_1 \widehat{\otimes} V_2)$$

of $G_1 \times G_2$ on the Hilbert space $V_1 \widehat{\otimes} V_2$ (Exercise!). The above mentioned principle of factorization of topologically irreducible unitary representations of $G = G_1 \times G_2$ is now provided by

Theorem 3.37 ("All local groups are tame"). *Let $G = G_1 \times G_2$ be the direct product of two local groups over a local field F and let (π_i, V_i) be topologically irreducible unitary representations of G_i, $i = 1, 2$. Then the unitary product representation $\pi_1 \widehat{\otimes} \pi_2$ of $G_1 \times G_2$ on the Hilbert space $V_1 \widehat{\otimes} V_2$ is topologically irreducible. Conversely, given a topologically irreducible unitary representation (π, V) of G, there are topologically irreducible unitary representations (π_i, V_i) of G_i, $i = 1, 2$, which are unique up to topological equivalence, such that π is topologically equivalent to $\pi_1 \widehat{\otimes} \pi_2$.*

Proof. This is a classical result, due to Dixmier, Harish-Chandra and Bernstein. For a comprehensive and concise modern treatment, we refer to [Dei10], Thm. 7.5.29 as combined with our Thm. 3.33. □

Replacing G by K is Def. 3.32, we obtain an obvious notion of unitary representations and topologically irreducibility for our maximal compact subgroup K. The following theorem then provides a converse of Cor. 3.34 for compact Lie groups:

Theorem 3.38 (see [Wal89], Lem. 1.4.8). *Let F be archimedean, and let (ρ, W) be a representation of K on a Hilbert space. Then, without loss of generality, (ρ, W) is unitary.*

Analogously, the following theorem provides a converse of Cor. 3.36 in the non-archimedean case:

Theorem 3.39 (see [Ren10], Thm. IV.1.1.(iv)). *Let F be non-archimedean, and let (ρ, W) be a representation of K. Then, without loss of generality, (ρ^∞, W^∞) is unitary.*

From Thm. 3.38 and Thm. 3.39 we obtain what ones calls the *unitarian trick*: whenever we have an interesting representation (π, V) of G, the restriction $(\pi|_K, V)$ is a representation of the maximal compact subgroup $K \subseteq G$ which, by the theorem, may be assumed unitary.

We end our short summary on unitary representations with a fundamental result of Harish-Chandra, which ties up with Thm. 3.26 and Thm. 3.29 and puts the special class of irreducible unitary representations of an archimedean local group G into the context of §3.3:

Theorem 3.40 (Harish-Chandra, cf. [Wal89], Thm. 3.4.11). *Let F be archimedean, and let (π_1, V_1) and (π_2, V_2) be irreducible unitary representations of G. Then π_1 is equivalent to π_2 if and only if they are infinitesimally equivalent.*

By the above theorem, the unitary equivalent class $[\widehat{\pi}]$ of an irreducible unitary representation π of an archimedean local group G hence coincides with the infinitesimal equivalence class of π among unitary representations.

Chapter 4

Langlands Classification: Step 2

In the last chapter, we introduced the concepts of interesting and unitary representations and explained how they relate to the other notions encountered so far. In this chapter, we shall discuss tempered representations and parabolic induction and thus come another step closer to the Langlands classification.

4.1 The contragredient representation

Classically, the *contragredient representation* of a complex representation (π, V) is a representation on a suitable linear dual vector space, i.e., a suitable subspace of $\mathrm{Hom}(V, \mathbb{C})$, which is derived from π in a natural way. We shall now see how this definition can be adjusted to fit our needs. *For the rest of this section, we assume our representation (π, V) to be interesting.*

If the ground field F is non-archimedean, then to construct a representation of G on $\mathrm{Hom}(V, \mathbb{C})$ simply means to give a group homomorphism from G to $\mathrm{Aut}_{\mathbb{C}}(\mathrm{Hom}(V, \mathbb{C}))$, which is accomplished in the familiar way:

$$\begin{aligned} l : G &\to \mathrm{Aut}_{\mathbb{C}}(\mathrm{Hom}(V, \mathbb{C})), \\ g &\mapsto \big(f \mapsto f\big(\pi(g^{-1})\_\_\big)\big). \end{aligned}$$

Observe that, although (π, V) is interesting, and hence $(\pi^\infty, V^\infty) = (\pi, V)$, the resulting representation $(l, \mathrm{Hom}(V, \mathbb{C}))$, however, is not smooth (hence not admissible) in general, which is why we restrict to its space of smooth vectors:

Definition 4.1. (Let F be non-archimedean, (π, V) be an interesting representation.) The *contragredient representation* of (π, V) is $(\pi^\vee, V^\vee) := (l^\infty, \mathrm{Hom}(V, \mathbb{C})^\infty)$.

The following result confirms that this notion behaves as expected.

Proposition 4.2 (see [Cas95], Prop. 2.1.10 & Cor. 2.1.13). $(\pi^\vee, V^\vee)$ *is again irreducible and admissible, hence interesting.*

Let us now turn to the more complicated case of an archimedean ground field F. The "dual space" of interest here would be the space $\mathrm{Hom}_{ct}(V^\infty, \mathbb{C})$ of continuous linear functionals on V^∞. This dual becomes a seminormed space with the *strong topology*: Explicitly, a family of seminorms, defining its locally convex topology, is given by $\mathcal{P} = \{p_B, B \subseteq V^\infty \text{ bounded}\}$, where

$$p_B(f) := \sup_{v \in B} |f(v)|, \qquad f \in \mathrm{Hom}_{ct}(V, \mathbb{C}).$$

The resulting topology is again complete, but (unless V^∞ is Banach) is not Fréchet in general, so it cannot be a representation according to our definitions.

Indeed, instead of considering linear functionals on V^∞, we will first restrict our attention to the space of K-finite vectors inside V^∞, i.e. $V^\infty_{(K)}$. By Thm. 3.16, this is none other than $V_{(K)}$. The next step is then to construct a space $V^\vee_{(K)}$ whose elements have a property reminiscent of K-finiteness. To achieve this, note first of all that since V is admissible, for each irreducible representation (ρ, W) of K (which is automatically finite-dimensional by Cor. 3.9), the attached ρ-isotypic component $V(\rho)$ of V is finite-dimensional, cf. §2.4. Upon replacing V by $V_{(K)}$, we analogously obtained the ρ-isotypic component $V_{(K)}(\rho)$ of $V_{(K)}$. Moreover, recall that we are given an identification $V_{(K)}(\rho) = V(\rho)$ (see (3.11)) and hence the natural map defined by summation becomes a K-equivariant linear bijection

$$\bigoplus_{[(\rho, W)]} V(\rho) \cong V_{(K)} \tag{4.3}$$

cf. [Wal89], Lem. 3.3.3. Here, the direct algebraic sum ranges over all equivalence classes of irreducible representations (ρ, W) of K . We now set

$$V^\vee_{(K)} := \{f \in \mathrm{Hom}(V_{(K)}, \mathbb{C}), f(V(\rho)) = \{0\} \text{ for all but finitely many } [(\rho, W)]\}$$

We get linear actions of both $\mathfrak{g}$ and K on $V^\vee_{(K)}$, both of which we shall denote by $\pi^\vee_{(K)}$: Explicitly, for $f \in V^\vee_{(K)}$, $X \in \mathfrak{g}$ and $k \in K$, we have

$$\pi^\vee_{(K)}(X)f := f(\pi_{(K)}(-X)\_\_)$$

and

$$\pi^\vee_{(K)}(k)f := f(\pi(k^{-1})\_\_).$$

We obtain

Lemma 4.4. $(\pi^\vee_{(K)}, V^\vee_{(K)})$ *is an irreducible admissible* $(\mathfrak{g}, K)$*-module.*

Proof. It is straight forward to check that $(\pi^\vee_{(K)}, V^\vee_{(K)})$ satisfies the conditions of Def. 3.2. We sketch the argument for irreducibility: Suppose we are given a proper $(\mathfrak{g}, K)$-submodule $\widetilde{V}$ of $V^\vee_{(K)}$. By [Wal89], Lem. 3.3.3, the natural map defined by summation defines a K-equivariant linear bijection

$$\bigoplus_{[(\rho,W)]} V^\vee(\rho) \cong V^\vee_{(K)}.$$

Observe that $V^\vee(\rho) = \mathrm{Hom}(V(\rho), \mathbb{C})$ and abbreviate $\widetilde{V}_\rho := \widetilde{V} \cap V^\vee(\rho)$. Then by the above decomposition of $V^\vee_{(K)}$, we obtain the analogous decomposition of $\widetilde{V}$:

$$\bigoplus_{[(\rho,W)]} \widetilde{V}_\rho \cong \widetilde{V}. \tag{4.5}$$

Properness implies that there must be an irreducible representation (ρ_0, W_0) of K, such that $\widetilde{V}_{\rho_0} \subset V^\vee(\rho_0)$. Hence, there is a non-zero $v_{\rho_0} \in V(\rho_0)$ such that $f_{\rho_0}(v_{\rho_0}) = 0$ for all $f_{\rho_0} \in \widetilde{V}_{\rho_0}$. Now, let $\widetilde{f} \in \widetilde{V}$ be arbitrary. Then, by (4.5), $\widetilde{f} = \oplus_{[(\rho,W)]}\widetilde{f}_\rho$ for unique $\widetilde{f}_\rho \in \widetilde{V}_\rho$. Hence, by (4.3), $\widetilde{f}(v_{\rho_0}) = \widetilde{f}_{\rho_0}(v_{\rho_0})$, which vanishes, because $\widetilde{f}_{\rho_0} \in \widetilde{V}_{\rho_0}$ and the choice of v_{ρ_0}. Therefore,

$$v_{\rho_0} \in \{v \in V_{(K)} | \widetilde{f}(v) = 0 \text{ for all } \widetilde{f} \in \widetilde{V}\},$$

and so the latter is a non-zero subspace of $V_{(K)}$. One easily checks that it is also a $(\mathfrak{g}, K)$-module, which, by $\widetilde{V} \subset V^\vee_{(K)}$, is properly contained in $V_{(K)}$. This is a contradiction to the irreducibility of $(\pi_{(K)}, V_{(K)})$. Admissibility is now a consequence of Cor. 3.15. (See also [Vog81], Cor. 8.5.3.) □

Having constructed an irreducible admissible $(\mathfrak{g}, K)$-module $(\pi^\vee_{(K)}, V^\vee_{(K)})$, we may take its Casselman–Wallach completion $(\overline{\pi^\vee_{(K)}}^{\mathsf{CW}}, \overline{V^\vee_{(K)}}^{\mathsf{CW}})$ to obtain an irreducible smooth admissible representation of G, which has $(\pi^\vee_{(K)}, V^\vee_{(K)})$ as its underlying $(\mathfrak{g}, K)$-module, cf. Thm. 3.26. The next result, which is a consequence of a theorem of Dixmier–Malliavin, cf. [Dix-Mal78], Thm. 3.3, shows how to interpret the Casselman–Wallach completion as a subspace of the continuous dual $(V^\infty)' = \mathrm{Hom}_{\mathrm{ct}}(V^\infty, \mathbb{C})$:

Theorem 4.6. $\overline{V^\vee_{(K)}}^{\mathsf{CW}}$ *is isomorphic to the* $\mathbb{C}$*-linear span of*

$$\left\{ v'\left(\int_G f(g)\pi^\infty(g)\_\_ \ \mathrm{d}g \right) : v' \in (V^\infty)', f \in C^\infty_c(G, \mathbb{C}) \right\} \tag{4.7}$$

inside $(V^\infty)'$. *(Here,* $\mathrm{d}g$ *is some fixed but arbitrary choice of a Haar measure on* G.*)*

We can finally put all pieces together:

Definition 4.8. (Let F be archimedean, (π, V) be interesting.) The *contragredient representation* $((\pi^\infty)^\vee, (V^\infty)^\vee)$ of (π^∞, V^∞) is the Casselman–Wallach completion of $(\pi^\vee_{(K)}, V^\vee_{(K)})$, identified with a linear subspace of $\mathrm{Hom}_{ct}(V^\infty, \mathbb{C})$ as per Thm. 4.6.

Again, taking the contragredient representation does not lead us outside the class of interesting representations:

Proposition 4.9. *The contragredient $((\pi^\infty)^\vee, (V^\infty)^\vee)$ is the space of smooth vectors of an interesting representation.*

Proof. Realizing $(V^\infty)^\vee$ inside $(V^\infty)'$ as per Thm. 4.6, shows that $(V^\infty)^\vee$ has a central character. By Thm. 3.29, $((\pi^\infty)^\vee, (V^\infty)^\vee)$ is isomorphic to the representation on the space of smooth vectors in an irreducible admissible representation $(\widehat{\pi}, \widehat{V})$ on a Hilbert space. Density of $(V^\infty)^\vee \cong \widehat{V}^\infty$ in the latter (cf. [WarI72], Prop. 4.4.1.1) and the continuity of the map $G \times \widehat{V} \to \widehat{V}$ implies that the central character extends continuously to $(\widehat{\pi}, \widehat{V})$, so $(\widehat{\pi}, \widehat{V})$ is interesting and hence the last assertion follows. □

As a final remark, regardless of the nature of the ground field F, taking the double contragredient just yields the original representation (up to isomorphy):

Proposition 4.10. *There is a (even canonical, i.e., uniquely determined, up to multiplication by non-zero complex numbers) isomorphism*

$$(((\pi^\infty)^\vee)^\vee, ((V^\infty)^\vee)^\vee) \cong (\pi^\infty, V^\infty).$$

Proof. Firstly, by Prop. 4.9 and Prop. 4.2 we notice that taking the contragredient of the contragredient is a well-defined operation. Now, in the archimedean case, the existence of the claimed isomorphism follows easily from Def. 4.8, recalling the finite-dimensionality of the spaces $V^\vee(\rho) = \mathrm{Hom}(V(\rho), \mathbb{C})$ (here we use admissibility of (π, V) as a crucial ingredient) and Thm. 3.26. In the non-archimedean case we refer to [Cas95], Prop. 2.1.10, or [Ren10] Prop. III.1.7, or [Bum98] pp. 427–428. Again, admissibility is crucial. The uniqueness of the respective isomorphism is an easy consequence of Schur's Lemma, cf. [Wal89], Lem. 3.3.2 and [Ren10], Prop. III.1.8. □

4.2 Square-integrable and tempered representations

With the contragredient at hand, we are now able to introduce the notion of L^r-representations for $r \geq 1$. We start with the following:

Definition 4.11. Let (π, V) be interesting, $v \in V^\infty$, $v^\vee \in (V^\infty)^\vee$, and set

$$\begin{aligned} c_{v,v^\vee} &: G \to \mathbb{C}, \\ & g \mapsto v^\vee(\pi(g)v). \end{aligned}$$

A map of this form (for some (π, V), v and $v^\vee$) is called a *matrix coefficient* of G. In the archimedean case, it is called K-finite, if both $v \in V^\infty$ and $v^\vee \in (V^\infty)^\vee$ are K-finite, i.e., by Thm. 3.16.(i), $v \in V_{(K)}$ and by Prop. 4.9, $v^\vee \in V^\vee_{(K)}$.

Remark 4.12. (1) The name "matrix coefficient" is explained as follows. Let W be a finite-dimensional complex representation of a group G, and set $n = \dim_{\mathbb{C}} W$. Moreover let (e_i) be a basis of W, and let $(e_i^\vee)$ denote the dual basis. Then for each $g \in G$,

$$(e_i^\vee(\pi(g)e_j))_{i,j=1,\ldots,n}$$

is precisely the matrix representing $\pi(g)$ with respect to the basis (e_i).

(2) Regarding the space $C(G)$ of continuous functions $f : G \to \mathbb{C}$ as being acted upon by G by right-translation, then for fixed $v^\vee$, the assignment

$$\begin{aligned} V &\to C(G), \\ v &\mapsto c_{v,v^\vee} \end{aligned} \tag{4.13}$$

obviously defines a G-equivariant linear map.

Let now $r \in [1, \infty)$. In view of (4.13), a natural attempt of defining L^r-representations, is to impose the condition that all matrix coefficients $c_{v,v^\vee} \in C(G)$ actually lie in the subspace of continuous L^r-functions. More precisely, one would require that

$$\int_{G/Z} |c_{v,v^\vee}(g)|^r \, dg < \infty \qquad \text{for all } v \in V^\infty \text{ and } v^\vee \in (V^\infty)^\vee, \tag{4.14}$$

where $Z = Z_G$ denotes the center of G (and where we assume $c_{v,v^\vee}$ to be K-finite, if F is archimedean). Of course, for fixed v and $v^\vee$, the integral is only defined if

$$|c_{v,v^\vee}(z)|^r = 1 \qquad \text{for all } z \in Z,$$

which, if we take into account the existence of a central character ω_π, makes us impose the condition that $|\omega_\pi(z)| = 1$ for all $z \in Z$. Obviously, this is the case if and only if $\pi|_Z$ is unitary. Summing up, we obtain:

Definition 4.15. Let (π, V) be an interesting G-representation.

(1) We call (π, V) a *pro-*L^r representation if $\pi|_Z$ is unitary and condition (4.14) above is satisfied.
(2) We furthermore call (π, V) an L^r*-representation*, if it is pro-L^r and unitary (as a whole).

Exercise 4.16. This one is somewhat for later: Let $G^{(1)} := \ker H_G \subseteq G$ be the kernel of the Harish-Chandra height function as defined in Def. 4.26 below. Show that an interesting representation (π, V) with unitary central character is pro-L^r, if and only if the following integral is finite

$$\int_{G^{(1)}} |c_{v,v^\vee}(g)|^r \, dg\,.$$

Hint: The composition $G^{(1)} \to G \to G/Z$ has compact kernel and finite cokernel.

The following exercise is more difficult:

Exercise 4.17. Let $G = G_1 \times ... \times G_k$ be an arbitrary, but finite product of local groups over a local field F. Show that a representation (π, V) is an L^r-representation of G, if and only if there exist L^r-representations (π_i, V_i) of G_i such that

$$\pi \cong \begin{cases} \pi_1 \,\widehat{\otimes}\, ... \,\widehat{\otimes}\, \pi_k & F \text{ archimedean,} \\ \pi_1 \otimes ... \otimes \pi_k & F \text{ non-archimedean.} \end{cases}$$

In the archimedean case, one furthermore has $\pi^\infty \cong \pi_1^\infty \,\overline{\otimes}_{\mathsf{pr}} ... \,\overline{\otimes}_{\mathsf{pr}}\, \pi_k^\infty$.
Hint: In the non-archimedean case, for a rigorous proof use Thm. 3.35,Thm. 3.37, Cor. 3.36, Prop. 4.2 and line (2) in the proof of [Bor-Wal00], Cor. X.6.2. In the more involved archimedean case, combine the following sources for a proof: Thm. 3.37, Cor. 3.34, Thm. 3.16, Lem. 4.4, Thm. 3.38, [Dei10], Lem. 7.5.12 and [Wal89], Lem. 3.3.3. Finally, in order to derive that $\pi^\infty \cong \pi_1^\infty \,\overline{\otimes}_{\mathsf{pr}} ... \,\overline{\otimes}_{\mathsf{pr}}\, \pi_k^\infty$, use that π^∞ as well as all π_i^∞ are irreducible Casselman–Wallach representations (by Lem. 2.18, Lem. 2.30, Thm. 3.16, Lem. 3.24 and Prop. 2.26) and combine this fact with Thm. 3.26 and [Vog08], Lem. 9.9.(3).

We resume our discussion with the following crucial observation:

Proposition 4.18. *If (π, V) is L^r, then it is $L^{r'}$ for all $r' > r$.*

Proof. For F non-archimedean, see [Sil82], Prop. 2.5., respectively, Cor. 2.6 therein. For F archimedean, combine [Bor97], Lem. 2.23 with the boundedness of $c_{v,v^\vee}$: In fact, for boundedness, using Thm. 3.26 and the Fréchet-Riesz representation theorem (cf. [Lar73], Thm. 13.4.2), identify the K-finite $v^\vee$ with $\langle \cdot, v_0 \rangle$ for a unique $v_0 \in V$. Hence, the Cauchy–Schwarz inequality together with unitarity of π gives $|c_{v,v^\vee}(g)| = |\langle \pi(g)v, v_0 \rangle| \leq \|\pi(g)v\| \cdot \|v_0\| = \|v\| \cdot \|v_0\|$ for all $g \in G$. □

At this stage, for the convenience of the reader, we recall that the notion of an L^2-representation is synonymous to a *square-integrable representation* or simply to a *discrete series representation*. In the context of these representations the prefix "pro-" turns out to be almost redundant, as the following result shows:

Proposition 4.19. *If (π, V) is pro-L^2, then there exists an L^2-representation π' such that $\pi^\infty \cong (\pi')^\infty$.*

Proof. For F non-archimedean, this is shown in [Cas95], Prop. 2.5.4. For F archimedean, one first shows, combining [Bor-Wal00], Lem. IV.1.9 and our Ex. 4.16 above, that π is infinitesimally equivalent to an L^2-representation π' and then applies Thm. 3.16 (irreducibility of the underlying $(\mathfrak{g}, K)$-modules), Lem. 3.24 (smoothness of π^∞ and π'^∞), Lem. 2.30 (admissibility of π^∞ and π'^∞), Lem. 2.18 (moderate growth of π^∞ and π'^∞) and Thm. 3.26 (isomorphy of π^∞ and π'^∞). □

Finally, we consider the following classes of representations, which are in a certain sense "almost" square-integrable:

Definition 4.20. Let (π, V) be an interesting G-representation.

(1) We call (π, V) *pro-tempered*, if it is pro-$L^{2+\varepsilon}$ for all $\varepsilon > 0$.
(2) We call (π, V) *tempered*, if it is pro-tempered and unitary.

Corollary 4.21. *If (π, V) is square-integrable, then it is tempered.*

Proof. This follows readily from Prop. 4.18. □

Exercise 4.22. For those readers, who plan to go through the proof of the Langlands classification in the non-archimedean case in great detail, the following exercise will be useful later: Suppose F is non-archimedean and G a local group over F. Then the family T of all unitary equivalence classes $[\widehat{\tau}]$ of tempered representations τ of G equals the family T' of all

equivalence classes $[\tau']$ of irreducible admissible representations τ' of G, which satisfy Def. VII.2.1 in [Ren10], i.e., are tempered in the sense of the latter reference. (The definition of [Ren10] is more common in non-archimedean representation theory.)
Hint: This result is folklore, but somewhat non-trivial. Lacking a good reference, we suggest to go through the following steps to compile your own proof: In order to show $T \subseteq T'$, use [Sil82], Cor. 2.6, [Sil79], Lem. 4.5.3 and Thm. 4.5.1 to see that a $\tau \in [\hat{\tau}] \in T$ satisfies the usual growth condition of its matrix coefficients, whence $[\hat{\tau}] \in T'$ by [Wld03], Prop. III.2.2, [Ren10], Rem. VII.2.1.3 and Cor. VII.2.6. To obtain $T' \subseteq T$, just go into the opposite direction: Use [Ren10], Cor. VII.2.6 again to see that any $\tau' \in [\tau'] \in T'$ is unitary and then [Ren10], Rem. VII.2.1.3 to see that each such τ' satisfies (ii) in [Wld03], Prop. III.2.2, whence, it is pro-tempered by [Sil82], Cor. 2.6 and so $[\tau'] \in T$.

4.3 Parabolic induction

4.3.1 *Parabolic subgroups and attached data*

The next and final topic for this chapter is parabolic induction. To this end, we need some preparations from the theory of algebraic groups.

Recall that an F-subgroup of $\mathbf{G}$ is called *parabolic*, if the quotient $\mathbf{G}/\mathbf{P}$ is a complete (or, equivalently, projective) variety. For a parabolic subgroup $\mathbf{P}$, we have a *Levi decomposition* $\mathbf{P} = \mathbf{L} \cdot \mathbf{N}$, where $\mathbf{L}$ is a connected reductive linear algebraic F-group (unique up to conjugation) and $\mathbf{N} = \mathbf{R_u}(\mathbf{P})/F$ is the unipotent radical of $\mathbf{P}$, cf. [Bor91], Prop. 20.5. Furthermore one denotes by $\mathbf{A_P}$ a maximal F-split central torus of $\mathbf{L}$: More precisely, this means that $\mathbf{A_P}$ is a linear algebraic F-subgroup of the center $\mathbf{Z_L}$ of $\mathbf{L}$ and that $\mathbf{A_P}$ is isomorphic over F to a finite direct product $\mathbf{GL}_1 \times \cdots \times \mathbf{GL}_1$.

From now on, we assume to have fixed a minimal (in the sense of inclusion) parabolic subgroup $\mathbf{P}_0$ *with Levi decomposition* $\mathbf{P}_0 = \mathbf{L}_0 \cdot \mathbf{N}_0$ *and a maximal* F*-split central torus* $\mathbf{A}_{\mathbf{P}_0}$ *as above.*

A parabolic subgroup $\mathbf{P}$ will be called *standard* if $\mathbf{P} \supseteq \mathbf{P}_0$. We may and will assume that $\mathbf{A_P} \subseteq \mathbf{A}_{\mathbf{P}_0}$, $\mathbf{L} \supseteq \mathbf{L}_0$ and $\mathbf{N}_0 \subseteq \mathbf{N}$. We will consider $\mathbf{G}$ as a standard parabolic subgroup of itself, in particular we assume to have fixed a maximal F-split central torus $\mathbf{A_G}$ of $\mathbf{G}$, such that $\mathbf{A_G} \subseteq \mathbf{A}_{\mathbf{P}_0}$.

Let now be $\mathbf{P} = \mathbf{LN}$ a general standard parabolic F-subgroup of $\mathbf{G}$. Denote by $\mathbf{X}_F(\mathbf{L})$ the group of F-rational characters $\mathbf{L} \to \mathbb{C}^\times$ (Exercise: Show that $\mathbf{X}_F(\mathbf{L})$ is a finitely generated, free abelian group), and set

$$\mathfrak{a}_{\mathbf{P}} := \mathrm{Hom}(\mathbf{X}_F(\mathbf{L}), \mathbb{R}),$$

the space of all group-homomorphisms $\mathbf{X}_F(\mathbf{L}) \to \mathbb{R}$. It is a finite-dimensional $\mathbb{R}$-vector space. We denote its dual by

$$\check{\mathfrak{a}}_{\mathbf{P}} := \mathbf{X}_F(\mathbf{L}) \otimes_{\mathbb{Z}} \mathbb{R} \cong \mathbf{X}_F(\mathbf{A_P}) \otimes_{\mathbb{Z}} \mathbb{R} \tag{4.23}$$

(the latter isomorphy of $\mathbb{R}$-vector spaces being a consequence of the fact that $\mathbf{X}_F(\mathbf{L})$ is a subgroup of finite index in $\mathbf{X}_F(\mathbf{A_P})$ (which, in obvious notation, denotes the group of F-rational characters $\mathbf{A_P} \to \mathbb{C}^\times$)) and further set

$$\check{\mathfrak{a}}_{\mathbf{P},\mathbb{C}} := \check{\mathfrak{a}}_{\mathbf{P}} \otimes_{\mathbb{R}} \mathbb{C}$$

for the complexification of $\check{\mathfrak{a}}_{\mathbf{P}}$. The natural pairing between $\check{\mathfrak{a}}_{\mathbf{P}}$ and $\mathfrak{a}_{\mathbf{P}}$ is denoted $\langle \cdot, \cdot \rangle$.

The inclusion $\mathbf{A_P} \subseteq \mathbf{A}_{\mathbf{P}_0}$, respectively, the restriction of characters from $\mathbf{L}$ to $\mathbf{L}_0$, defines natural injections $\mathfrak{a}_{\mathbf{P}} \hookrightarrow \mathfrak{a}_{\mathbf{P}_0}$ and $\check{\mathfrak{a}}_{\mathbf{P}} \hookrightarrow \check{\mathfrak{a}}_{\mathbf{P}_0}$, where the latter is inverse to the dual of the first. Hence, we obtain natural direct sum decompositions

$$\mathfrak{a}_{\mathbf{P}_0} = \mathfrak{a}_{\mathbf{P}} \oplus \mathfrak{a}^{\mathbf{P}}_{\mathbf{P}_0} \quad \text{and} \quad \check{\mathfrak{a}}_{\mathbf{P}_0} = \check{\mathfrak{a}}_{\mathbf{P}} \oplus \check{\mathfrak{a}}^{\mathbf{P}}_{\mathbf{P}_0}.$$

and so, whenever $\mathbf{P} \subseteq \mathbf{P}'$, we may and will identify $\mathfrak{a}_{\mathbf{P}'}$ (respectively, $\check{\mathfrak{a}}_{\mathbf{P}'}$) with a natural subspace of $\mathfrak{a}_{\mathbf{P}}$ (respectively, $\check{\mathfrak{a}}_{\mathbf{P}}$). In particular, this allows us to define $\mathfrak{a}^{\mathbf{P}'}_{\mathbf{P}}$ as the intersection of $\mathfrak{a}_{\mathbf{P}}$ with $\mathfrak{a}^{\mathbf{P}'}_{\mathbf{P}_0}$ in $\mathfrak{a}_{\mathbf{P}_0}$ (and analogously, to define $\check{\mathfrak{a}}^{\mathbf{P}'}_{\mathbf{P}}$ as the intersection of $\check{\mathfrak{a}}_{\mathbf{P}}$ with $\check{\mathfrak{a}}^{\mathbf{P}'}_{\mathbf{P}_0}$ in $\check{\mathfrak{a}}_{\mathbf{P}_0}$). If we specify this to the case $\mathbf{P}' = \mathbf{G}$, then $\check{\mathfrak{a}}^{\mathbf{G}}_{\mathbf{P}}$ is nothing but the real vector space generated by all F-rational characters of $\mathbf{L}$ (or, by (4.23), equivalently $\mathbf{A_P}$) which vanish on our fixed maximal F-split central torus $\mathbf{A_G}$ in $\mathbf{G}$. Obviously, all these constructions carry through to the respective complexifications.

Recall that there is a well-known notion of positivity on $\check{\mathfrak{a}}_{\mathbf{P},\mathbb{C}}$: We first consider the action of $\mathbf{P}$ on $\mathbf{N}$ via the adjoint representation and let $\Delta(\mathbf{P}, \mathbf{A_P})$ denote the set of non-trivial weights of this action with respect to our chosen $\mathbf{A_P}$. We recall that the elements $\alpha \in \Delta(\mathbf{P}, \mathbf{A_P})$ are called the *roots* with respect to $(\mathbf{P}, \mathbf{A_P})$ and naturally identify with non-zero vectors in $\check{\mathfrak{a}}_{\mathbf{P}}$. Given a root $\alpha \in \Delta(\mathbf{P}, \mathbf{A_P})$, we denote by $\check{\alpha}$ its co-root (an element in $\mathfrak{a}_{\mathbf{P}}$).

Definition 4.24. An element $\nu \in \check{\mathfrak{a}}_{\mathbf{P},\mathbb{C}}$ is *positive*, in symbols $\nu \geq 0$, if

$$\mathrm{Re}(\langle \nu, \check{\alpha} \rangle) \geq 0 \qquad \text{for all } \alpha \in \Delta(\mathbf{P}, \mathbf{A_P}),$$

and *strictly positive*, in symbols $\nu > 0$, if the inequality is strict for all $\alpha \in \Delta(\mathbf{P}, \mathbf{A_P})$.

One easily checks that an element $\nu \in \check{\mathfrak{a}}_{\mathbf{P}}$ satisfies $\langle \nu, \check{\alpha} \rangle = 0$ for all $\alpha \in \Delta(\mathbf{P}, \mathbf{A_P})$, if and only if $\nu \in \check{\mathfrak{a}}_{\mathbf{G}}$. Moreover, it follows from the definition that $\Delta(\mathbf{G}, \mathbf{A_G})$ is empty (as $\mathbf{N_G}$ is trivial), whence every element of $\check{\mathfrak{a}}_{\mathbf{G},\mathbb{C}}$ is automatically strictly positive.

We also recall that a root $\alpha \in \Delta(\mathbf{P}, \mathbf{A_P})$ is called *simple*, if it cannot be written as the sum of two roots in $\Delta(\mathbf{P}, \mathbf{A_P})$. We will write $\Delta^0(\mathbf{P}, \mathbf{A_P})$ for the set of simple roots in $\Delta(\mathbf{P}, \mathbf{A_P})$ and abbreviate $\Delta^0 := \Delta^0(\mathbf{P}_0, \mathbf{A}_{\mathbf{P}_0})$ for the set of all simple F-roots of $\mathbf{G}$ (with respect to $\mathbf{A}_{\mathbf{P}_0}$). It is then well known, see [Bor91], Prop. 21.12, that the assignment

$$\theta \subseteq \Delta^0 \mapsto \mathbf{P}_\theta := C_{\mathbf{G}}\left(\left(\bigcap_{\alpha\in\theta} \ker \alpha\right)^{\circ}\right) \cdot \mathbf{N}_0, \tag{4.25}$$

($C_{\mathbf{G}}$ denoting the centralizer in $\mathbf{G}$) sets up a one-to-one correspondence between the subsets θ of Δ^0 and the set of standard parabolic F-subgroups $\mathbf{P}$ of $\mathbf{G}$. Obviously, the empty subset $\emptyset$ here corresponds to $\mathbf{P}_0 = \mathbf{P}_\emptyset$ and the full set of simple F-roots Δ^0 corresponds to $\mathbf{G} = \mathbf{P}_{\Delta^0}$.

In order to define parabolic induction, we shall need the following:

Definition 4.26. For a standard parabolic F-subgroup $\mathbf{P} = \mathbf{LN}$ of $\mathbf{G}$, the *Harish-Chandra height function* is the uniquely determined group homomorphism H_P from $L = \mathbf{L}(F)$ to $\mathfrak{a}_{\mathbf{P}}$, which satisfies

$$|\chi(\ell)|_F = e^{H_P(\ell)(\chi)} \qquad \text{for all } \chi \in \mathbf{X}_F(\mathbf{L}), \ell \in L.$$

Here, $|\cdot|_F$ is the normalized absolute value of F, cf. 2.1.

Remark 4.27. For non-archimedean local groups G, the reader will often find a different definition in the literature, which uses q – the order of the residue field of F – instead of Euler's constant e. However, this only leads to a minor renormalization, as our definition of H_P only differs from the one using q by a factor $\log(q)$, and we will hence choose to work with e in order to have a uniform base for all local fields F.

Thanks to the Harish-Chandra height function, for each $\nu \in \check{\mathfrak{a}}_{\mathbf{P},\mathbb{C}}$ we obtain a homomorphism

$$\begin{aligned} P &\to \mathbb{C}^\times, \\ p = \ell n &\mapsto e^{\langle \nu, H_P(\ell) \rangle}. \end{aligned} \tag{4.28}$$

Finally, we fix the following notation for a special element of $\check{\mathfrak{a}}_{\mathbf{P}}$, namely

$$\rho_P := \frac{1}{2} \sum_{\alpha \in \Delta(\mathbf{P}, \mathbf{A_P})} m(\alpha) \cdot \alpha,$$

where $m(\alpha)$ denotes the multiplicity of the root α in the adjoint action of $\mathbf{P}$ on $\mathbf{N}$, i.e., the dimension of its eigenspace.

We have now reached the moment, where we further specify our choice of a maximal compact subgroup K of G: To this end, we recall our fixed choice of a minimal parabolic subgroup $\mathbf{P}_0$ of $\mathbf{G}$ with maximal F-split central torus $\mathbf{A}_{\mathbf{P}_0}$. We will fix a maximal compact subgroup K of G *in good position* with respect to the pair $(\mathbf{P}_0, \mathbf{A}_{\mathbf{P}_0})$, i.e., a maximal compact subgroup K of G, which has the following two properties: For all standard parabolic F-subgroups $\mathbf{P}$ of $\mathbf{G}$

(1) $G = P \cdot K$ (and hence also, or equivalently, $G = K \cdot P$, by taking inverses).
(2) $K \cap P = (K \cap L)(K \cap N)$ and $K \cap L$ is a maximal compact subgroup of L, that satisfies the same two, previous properties with respect to the standard parabolic F-subgroups of $\mathbf{L}$ given by intersection.

In order to see, that there always exists such a maximal compact subgroup K of G, when F is non-archimedean, we may choose K to be a suitable *(good) special maximal compact subgroup*, a notion which was first defined by Bruhat-Tits in [Bru-Tit72], §4.4.1, in terms of the building associated to G and its apartment associated to A_{P_0}. We refer to [Ren10], Thm. V.5.1, or [Sil79], §0.6, which state the existence of such a maximal compact subgroup (even under more refined obstructions), but which both refer themselves to the just mentioned source [Bru-Tit72] for a proof. [The reader may find the complete argument for (1) (also) in [McD71][1]: Thm. 2.6.11.(2) in [McD71] together with [Bor91], Thm. 20.9 shows that there

[1]Once s/he has taken into account Deligne's suggestion for a correction of Axiom (V) on p. 27 in [McD71] as stated in [Cas12]:

Axiom (V). The commutator group $[U_\alpha, U_\beta]$ for $\alpha, \beta > 0$ is contained in the group generated by the U_γ with $\gamma > 0$ and not parallel to α or β.

exists a *maximal parahoric subgroup* K of G such that $G = K \cdot P_0$; Thm. 2.4.15 in [McD71] finally gives that any such maximal parahoric subgroup is maximal bounded (and hence maximal compact). As $P \supseteq P_0$, (1) follows. That one can even inforce (2) to hold, is a consequence of [Bru-Tit84], Prop. 4.2.17, which shows that for a special maximal compact subgroup K of G, the intersection $K \cap L$ is in fact a special maximal compact subgroup of L – a fact, from which the remaining assertions follow.]

If, in turn, F is archimedean, then – recalling [Bor91], Thm. 20.9 once more – the claim that $G = K \cdot P_0$ is the usual Iwasawa decomposition, see [Wal89], Lem. 2.2.8 or [Bor69], §11.19.(1), which is valid for any choice of K. Hence, (1) follows for any K. The same references show that for any choice of K, the intersection $K \cap N_0$ is the trivial group, whence so is $K \cap N$ for the unipotent radicals of all standard parabolic subgroups and consequently $K \cap P = K \cap L$. However, P being a Lie subgroup of G with finite component group, $K \cap P$ is a maximal compact subgroup of P and so $K \cap P = K \cap L$ is a maximal compact subgroup of L, which in summary shows that K also satisfies (2).

We may therefore make the following assumption, regardless of the nature of F:

We may and will henceforth assume that K is a maximal compact subgroup of G in good position with respect to our fixed choice of $\mathbf{P}_0$ and $\mathbf{A}_{\mathbf{P}_0}$.

As a simple consequence of this convention, for each $\nu \in \check{\mathfrak{a}}_{\mathbf{P},\mathbb{C}}$, the homomorphism in (4.28) gives rise to a well-defined homomorphism $G \to \mathbb{C}^\times$ by extending it trivially on K.

4.3.2 *Smooth induction*

We are now ready to define what we call *smooth parabolic induction.*

Definition 4.29 (Smooth parabolic induction). Let $\mathbf{P} = \mathbf{L} \cdot \mathbf{N}$ be a standard parabolic F-subgroup of $\mathbf{G}$, let (π, V) be an interesting representation of L, and let $\nu \in \check{\mathfrak{a}}_{\mathbf{P},\mathbb{C}}$. The space of all functions $f : G \to V^\infty$ satisfying the following two conditions:

(1) f is smooth, i.e., $f \in C^\infty(G, V^\infty)$ and

(2) for all $p = \ell n \in LN = P$ and all $g \in G$, it holds that

$$f(pg) = e^{\langle \nu + \rho_P, H_P(\ell) \rangle} \pi^\infty(\ell) f(g),$$

is denoted by $I_P^G(\pi^\infty, \nu)$. As a subspace of $C^\infty(G, V^\infty)$, we consider it equipped with the action of G given by right-translation. Moreover, if F is archimedean, then we endow it with the subspace topology coming from the Fréchet space $C^\infty(G, V^\infty)$ (see the discussion before Lem. 2.18). One easily checks from the definitions that $I_P^G(\pi^\infty, \nu)$ is closed inside $C^\infty(G, V^\infty)$, i.e., that $I_P^G(\pi^\infty, \nu)$ is itself a Fréchet space, cf. Lem. 1.11, and that the given action by right-translation is in fact continuous. Thus, $I_P^G(\pi^\infty, \nu)$ is always (regardless of the ground field) a representation of G. We will call the resulting G-representation on $I_P^G(\pi^\infty, \nu)$ the *smooth parabolic induction* of $\pi^\infty \otimes e^{\langle \nu, H_P(\cdot) \rangle}$, or, likewise, of π^∞ (with parameter ν).

Proposition 4.30. *$I_P^G(\pi^\infty, \nu)$ is a smooth and admissible representation of G and has a central character. Moreover, if the ground field is archimedean, then $I_P^G(\pi^\infty, \nu)$ is of moderate growth.*

Proof. For smoothness and admissiblity see [Bor-Wal00], III.2.1 and III.3.2 in the archimedean case (but we will resume this also below), and [Cas95], Thm. 2.4.1.(d) (recalling the compactness of $G/P \cong K/(K \cap P)$) in the non-archimedean case. For the existence of a central character one simply uses the fact that π^∞ has a central character, the very definition of smooth induction and that $Z_G \subseteq Z_L$. Moderate growth in the archimedean case follows from our Lem. 3.24 in combination with [Bor-Wal00], Cor. III.7.7. See also (4.34). □

Remark 4.31. Let $\nu \in \check{\mathfrak{a}}_{G,\mathbb{C}}$. Then it is easy to see that

$$I_P^G(\pi^\infty, \nu) \cong I_P^G(\pi^\infty, 0) \otimes e^{\langle \nu, H_G(\cdot) \rangle}.$$

Indeed, the natural map $I_P^G(\pi^\infty, \nu) \to I_P^G(\pi^\infty, 0) \otimes e^{\langle \nu, H_G(\cdot) \rangle}$, which assigns an $f \in I_P^G(\pi^\infty, \nu)$ to the tensor $f' \otimes 1$, where $f'(g) = e^{-\langle \nu, H_G(g) \rangle} f(g)$, defines an isomorphism of G-representations. (Exercise!)

4.3.3 L^2*-induction and questions of unitarity*

We start with the following observation in the case of a non-archimedean ground field F. It shows that for these F the concept of smooth parabolic induction goes well with unitary representations of G.

Proposition 4.32 (see [Cas95], Prop. 3.1.4). *Let F be non-archimedean. A representation induced from a unitary representation is again unitary. More precisely, suppose that $\pi \otimes e^{\langle \nu, H_P(\cdot)\rangle}$ is unitary with respect to some Hermitian form $\langle \cdot, \cdot \rangle_V$ on V. Then $I_P^G(\pi,\nu)$ is unitary with respect to the Hermitian form $(\cdot,\cdot)$ defined as*

$$(f_1, f_2) := \int_K \langle f_1(k), f_2(k)\rangle_V \, \mathrm{d}k\,.$$

Of course, Prop. 4.32 has to fail in the archimedean case by the sheer definition, since $I_P^G(\pi^\infty, \nu)$ will in general only be Fréchet and not Hilbert. A similar problem arises, even if F is non-archimedean and π is unitary: According to our definitions, it still makes no sense to speak about topological irreducibility of the unitary representation $I_P^G(\pi,\nu)$, simply because there is no topology (and therefore *a fortiori* no Hilbert space structure) on the latter, yet. We will now see how to solve these two issues by introducing the concept of L^2-induction.

To this end, let $\mathbf{P} = \mathbf{L} \cdot \mathbf{N}$ be a standard parabolic subgroup of $\mathbf{G}$, let (π, V) be a topologically irreducible unitary representation of L, equipped with a Hermitian form $\langle \cdot, \cdot \rangle_V$, and let ν be an element of $\check{\mathfrak{a}}_{\mathbf{P},\mathbb{C}}$. Consider the vector space $I = I(\pi,\nu)$ of all functions $f : G \to V$ satisfying the following two conditions:

(1) f is continuous, i.e., $f \in C(G, V)$, and
(2) for all $p = \ell n \in LN = P$ and all $g \in G$, it holds that
$$f(pg) = e^{\langle \nu + \rho_P, H_P(\ell)\rangle} \pi(\ell) f(g).$$

Obviously, I is stable under G by right-translation of functions.

Recall that $G = P \cdot K$. We may define a non-degenerate Hermitian from on I by setting

$$\langle f_1, f_2 \rangle_I := \int_K \langle f_1(k), f_2(k)\rangle_V \, \mathrm{d}k\,.$$

The Hilbert space closure of I with respect to this form shall be denoted by $\widehat{I_P^G(\pi,\nu)}$. We have the following.

Proposition 4.33. *Right-translation of G on I extends continuously to $\widehat{I_P^G(\pi,\nu)}$ and makes the latter into a pro-admissible representation of G on a Hilbert space with a central character. If $\pi \otimes e^{\langle \nu, H_P(\cdot)\rangle}$ is unitary, then so is $\widehat{I_P^G(\pi,\nu)}$. Moreover, there is an isomorphism of G-representations*

$$\widehat{I_P^G(\pi,\nu)}^{\infty} \cong I_P^G(\pi^\infty, \nu). \tag{4.34}$$

Proof. For the continuous extension of right-translation we refer to [Wal89], Lem. 1.5.3. It is now clear, that this way $\widehat{I_P^G(\pi,\nu)}$ is a representation of G on a Hilbert space, which has a central character (for this recall that $Z_G \subseteq Z_L$). Invoking, [Wal89], Lem. 1.5.3, once more, also shows preservation of unitarity.

The isomorphism of G-representations

$$\widehat{I_P^G(\pi,\nu)}^{\infty} \cong I_P^G(\pi^\infty,\nu)$$

is established in [Bor-Wal00], Cor. III.7.7 in the archimedean and on p. 148 of [Tad94] in the non-archimedean case. (Here, observe that the right hand side of (4.34) is well-defined according to our definitions: If F is archimedean, this holds by Cor. 3.34 and if F is non-archimedean by Lem. 2.18.(1) and Cor. 3.36.)

Finally, admissibility in the archimedean case (i.e., by definition, pro-admissibility) follows as in [Kna-Vog95], (11.42), while pro-admissibility is a consequence of (4.34) in the non-archimedean case: Indeed, let F be non-archimedean, let (ρ, W) be an irreducible, finite-dimensional representation of K and assume that there is a non-trivial $f \in \mathrm{Hom}_K(W, \widehat{I_P^G(\pi,\nu)})$ (otherwise there is obviously nothing to show). Then $f(W)$ is a finite-dimensional K-subrepresentation (in fact isomorphic to (ρ, W) by irreducibility of the latter and hence injectivity of f) of the restriction to K of $\widehat{I_P^G(\pi,\nu)}$. Moreover, by what we have just said about the continuous extension of the action of G on $\widehat{I_P^G(\pi,\nu)}$, the map $K \times f(W) \to f(W)$ is a continuous. We record the following intermediate Lemma.

Lemma 4.35. *Let F be non-archimedean and let (π, V) be a representation of G on a complete seminormed space V, with the property that the map $G \times V \to V$, $(g, v) \mapsto \pi(g)v$ is continuous. Then, the space of K-finite vectors $V_{(K)} := \{v \in V \,|\, \dim_{\mathbb{C}}\langle\pi(K)v\rangle < \infty\}$ and the space of smooth vectors V^∞ coincide and they are dense in V. In particular, every continuous finite-dimensional representation of K is smooth.*

Proof of the intermediate lemma. Equality of V^∞ and $V_{(K)}$ is a simple application of the usual "no-small-subgroups"-argument for Lie groups: Indeed, if $v \in V$ is smooth, then by definition $\pi(C)v = v$ for some open compact subgroup C of G. In particular, $C \cap K$ is an open compact subgroup of K under which v is invariant. The index of $C \cap K$ being finite in K

implies that $v \in V_{(K)}$. If on the contrary $v \in V_{(K)}$, then $V_v := \langle \pi(K)v \rangle$ is a finite-dimensional vector space, on which K acts continuously by restriction of π. Abbreviate $\dim_{\mathbb{C}} V_v =: n$ and identify $\mathrm{GL}(V_v)$ with $\mathrm{GL}_n(\mathbb{C})$. Then, π defines a continuous group homomorphism $\pi|_K : K \to \mathrm{GL}_n(\mathbb{C})$. As $\mathrm{GL}_n(\mathbb{C})$ is a Lie group, there exist neighborhoods of the identity id in $\mathrm{GL}_n(\mathbb{C})$, that do not contain a non-trivial subgroup of $\mathrm{GL}_n(\mathbb{C})$ (an assertion, which we leave as an exercise, cf. [Hel78], Ex. B.5 in Chp. II). The preimage $\pi|_K^{-1}(U)$ of such a neighborhood is hence, by continuity of the group-homomorphism $\pi|_K$, a neighborhood of the identity in K, so it contains an open compact subgroup C. As $\pi|_K(C)$ is a subgroup of $\mathrm{GL}_n(\mathbb{C})$, which is contained in U, necessarily $\pi|_K(C) = \{id\}$. Thus, C fixes every element of V_v, hence, in particular $v \in V^\infty$. Density of $V^\infty = V_{(K)}$ now follows as in [Bor72], Prop. 3.6. The remaining assertion is obvious by Thm. 1.8. □

Applying Lem. 4.35 to the continuous, finite-dimensional K-representation $f(W)$ implies that $f(W) = f(W)^\infty \subseteq \widehat{I_P^G(\pi,\nu)}^\infty$. Therefore, one has equality $\mathrm{Hom}_K(W, \widehat{I_P^G(\pi,\nu)}) = \mathrm{Hom}_K(W, \widehat{I_P^G(\pi,\nu)}^\infty)$ and so (4.34) together with the admissibility of $I_P^G(\pi^\infty,\nu)$, cf. Prop. 4.30, shows that $\widehat{I_P^G(\pi,\nu)}$ is pro-admissible. □

Definition 4.36 (L^2-parabolic induction). Let (π, V) be a topologically irreducible unitary representation of L and ν be an element of $\check{\mathfrak{a}}_{\mathbf{P},\mathbb{C}}$. We will call the continuous G-action (and hence representation) on the Hilbert space $\widehat{I_P^G(\pi,\nu)}$ introduced above the L^2-*induction* of $\pi \otimes e^{\langle \nu, H_P(\cdot)\rangle}$ or, likewise, of π (with parameter ν).

We obtain that $I_P^G(\pi^\infty,\nu)$ is "almost" the space of smooth vectors inside an interesting representation of G: Indeed, by Prop. 4.33, the only condition missing is irreducibility. It can be shown, however, that $I_P^G(\pi^\infty,\nu)$ is always generated by a finite set of elements as a G-representation, cf. [Bor-Wal00], III.3.2, when F is archimedean and [Ren10], Lem. VI.6.2, when F is non-archimedean.

4.3.4 $(\mathfrak{g}, K)$-*induction*

To complete the picture, let us also introduce one more concept of induction, namely that of "parabolic $(\mathfrak{g}, K)$-induction". To this end, let F be archimedean in this subsection and let (π, V) be again an interesting representation of L and $\nu \in \check{\mathfrak{a}}_{\mathbf{P},\mathbb{C}}$. Recall from the above that we have chosen a maximal compact subgroup K_P of L (and of P), namely $K_P := K \cap L = K \cap P$. By Lem. 3.6 and Thm. 3.16,

$$(\pi e^{\langle \nu+\rho_P, H_P(\cdot)\rangle})_{(K_P)} = (\pi e^{\langle \nu+\rho_P, H_P(\cdot)\rangle})_0$$

is a $(\mathfrak{l}, K_P)$-module, which, by the universal property of the universal enveloping algebra, defines a $\mathcal{U}(\mathfrak{p})$-module (with inherited K_P-action), denoted by $(\pi_{0,\nu}, V_{0,\nu})$. As a next step, let U_0 be the space of all linear maps $f : \mathcal{U}(\mathfrak{g}) \to V_{0,\nu}$, which satisfy the equation

$$f(YX) = \pi_{0,\nu}(Y)f(X) \text{ for all } Y \in \mathcal{U}(\mathfrak{p}), X \in \mathcal{U}(\mathfrak{g}).$$

Observe that $\mathcal{U}(\mathfrak{g})$ acts on U_0 by right-translation. Let us abbreviate this action by "$\star$". Then, we define U_1 to be the subspace of U_0 consisting of elements $f \in U_0$ satisfying the following conditions:

(1) $\dim_{\mathbb{C}}\langle X \star f : X \in \mathfrak{k}\rangle < \infty$;
(2) the subrepresentation $\mathcal{U}(\mathfrak{k}) \star f$ spanned by f is fully decomposable as a representation of $\mathfrak{k}$, i.e., there exist irreducible representations W_i of $\mathfrak{k}$ and a $\mathbb{C}$-linear, $\mathfrak{k}$-equivariant bijection

$$\mathcal{U}(\mathfrak{k}) \star f \xrightarrow{\sim} \bigoplus_i W_i.$$

Note that the action of $\mathfrak{k}$ on U_1 does not necessarily integrate to a representation of K. However, it integrates to an action (still denoted "$\star$") of the simply connected covering group of K, which we henceforth denote by $\widetilde{K}$. Note also that, if $\mathrm{pr} : \widetilde{K} \to K$ denotes the covering map, then there are inclusions $\ker(\mathrm{pr}) \subseteq \mathrm{pr}^{-1}(K_P) \subseteq \widetilde{K}$.

Now, consider now subspace of U_1 consisting of those $f \in U_1$, such that for all $k \in \mathrm{pr}^{-1}(K_P)$ and all $X \in \mathcal{U}(\mathfrak{g})$,

$$(k \star f)(X) = \pi_{0,\nu}(\mathrm{pr}(k))f(k^{-1}Xk).$$

This space is stable under $\mathfrak{g}$ and $\widetilde{K}$, but since $\ker(\mathrm{pr})$ acts trivially on it, the $\widetilde{K}$-action factors through the quotient $\widetilde{K}/\ker(\mathrm{pr}) \cong K$. Thus, we have finally obtained a $(\mathfrak{g}, K)$-module, which we shall denote by $I^{(\mathfrak{g},K)}_{(\mathfrak{p},K_P)}(\pi_{(K_P)}, \nu)$.

Definition 4.37 ($(\mathfrak{g}, K)$-parabolic induction). Let (π, V) be an interesting representation of L and let $\nu \in \check{\mathfrak{a}}_{P,\mathbb{C}}$. We will call the $(\mathfrak{g}, K)$-module $I^{(\mathfrak{g},K)}_{(\mathfrak{p},K_P)}(\pi_{(K_P)}, \nu)$ introduced above the $(\mathfrak{g}, K)$-*induction* of $(\pi e^{\langle \nu, H_P(\cdot)\rangle})_{(K_P)}$, or, likewise, of π_0, or even of π (with parameter ν).

Proposition 4.38. *Let (π, V) be an interesting representation of L. The $(\mathfrak{g}, K)$-module $I^{(\mathfrak{g},K)}_{(\mathfrak{p},K_P)}(\pi_{(K_P)}, \nu)$ is isomorphic to the $(\mathfrak{g}, K)$-module of K-finite vectors inside $I^G_P(\pi^\infty, \nu)$. In particular, $I^{(\mathfrak{g},K)}_{(\mathfrak{p},K_P)}(\pi_{(K_P)}, \nu)$ is admissible and finitely generated.*

Proof. The first assertion is [Bor-Wal00], Prop. III.2.4. For admissibility, combine Prop. 4.30 and Lem. 3.15, while $I^{(\mathfrak{g},K)}_{(\mathfrak{p},K_P)}(\pi_{(K_P)}, \nu)$ being finitely generated follows from $I^G_P(\pi^\infty, \nu)$ being a finitely generated G-representation, cf. the last paragraph of §4.3.3. □

Combining this result with Prop. 4.33 (for irreducible unitary representations), we obtain:

$$I^{(\mathfrak{g},K)}_{(\mathfrak{p},K_P)}(\pi_{(K_P)}, \nu) \cong \widehat{I^G_P(\pi, \nu)}_{(K)}. \tag{4.39}$$

(Note that, since $\widehat{I^G_P(\pi, \nu)}$ is admissible by Prop. 4.33, we can omit taking G-smooth vectors on the right-hand side by Thm. 3.16.) Moreover, combining Prop. 4.30 with Prop. 4.38 one gets the important corollary.

Corollary 4.40. *Let (π, V) be an interesting representation of L. Then $I^G_P(\pi^\infty, \nu)$ is a Casselman–Wallach representation.*

4.3.5 *Induction from unitary representation of* $\mathbf{GL}_n(D)$

Before we finally head on to our presentation of the proof of the Langlands classification in the next section, let us mention a particularly beautiful result on irreducibility of parabolic induction in the special case of the general linear group over a division algebra. More precisely, let F be a local field, let D be a finite-dimensional, central division algebra over F, cf. [Pla-Rap94], §1.4.1–§1.4.2, and consider the group of invertible $n \times n$-matrices with entries in D, $G = \mathrm{GL}_n(D)$. It is the group of F-points of a connected reductive algebraic group $\mathbf{G} = \mathbf{GL}_n/D$, see [Pla-Rap94], §2.3.1, whence an example of a local group as above. We recall that in the special case of $D = F$ (the so-called "split case") one simply retrieves the usual general linear group $\mathbf{GL}_n/F$.

The following result was first proved in the non-archimedean split case by Bernstein, [Ber84], (as pre-noticed by Jacquet, [Jac77]), while it follows in the archimedean split case from work of Vogan, [Vog86], and independently from a combination of Bernstein's approach and the work of Baruch, [Bar03]. The case of a general finite-dimensional, central division algebra over a non-archimedean field was finally established by Sécherre, [Séc09], building on results of Bushnell–Kutzko and Barbasch–Moy. See also [Lap-Mín15], Cor. 6.4, which provides a uniform, independent approach in the non-archimedean case. In turn, if the ground field F is archimedean, observe that a non-split finite-dimensional, central division algebra is necessarily isomorphic to $\mathbb{H}$, the Hamilton quaternions. The following result is then a consequence of the work of Vogan, [Vog86], and was re-established in more explicit form by Badulescu–Renard in [Bad-Ren10]. On a final note for the very accurate reader, we remark, that, in order to deduce the result in the form presented here from the just quoted references, we silently use Thm. 3.35 and Prop. 4.33 (in particular Eq. (4.34) therein) when F is non-archimedean.

Theorem 4.41. *Let F be a local field and suppose that $\mathbf{G} = \mathbf{GL}_n/D$ is the general linear group over D, where D is some finite-dimensional central division algebra over F. Let $\mathbf{P} = \mathbf{L} \cdot \mathbf{N}$ be a standard parabolic F-subgroup of $\mathbf{G}$ and let (π, V) be a topologically irreducible unitary representation of L. Then $\widehat{I_P^G(\pi,0)}$ is a topologically irreducible unitary representation of G.*

This theorem has the following.

Corollary 4.42. *Under the assumptions of Thm. 4.41, $I_P^G(\pi^\infty, 0)$ is irreducible. Moreover, if F is archimedean, then $I_{(\mathfrak{p},K_P)}^{(\mathfrak{g},K)}(\pi_{(K_P)}, 0)$ is an irreducible $(\mathfrak{g}, K)$-module.*

Proof. In order to deduce the first assertion from Thm. 4.41, we apply (4.34) and then use Prop. 2.26 in the archimedean, respectively, Thm. 3.35 in the non-archimedean case. Analogously, to see that $I_{(\mathfrak{p},K_P)}^{(\mathfrak{g},K)}(\pi_{(K_P)}, 0)$ is irreducible, we hark back to Prop. 4.38 (or (4.39)) and use Thm. 3.16. □

Chapter 5

Langlands Classification: Step 3

5.1 The result – tempered vs. square-integrable formulation

In this chapter, we finally give the main results pertaining to the Langlands classification and record a few consequences. We retain the notations and conventions from the last chapters, i.e., G denotes a local group and K denotes a fixed maximal compact subgroup of G in good position, cf. §4.3.1.

Now we turn to the statement and the proof of Langlands classification. If F is archimedean, the original result is due to Langlands (see Lemmas 3.13, 3.14, and 4.2. in [Lan89]) and Miličić (see [DMil77], Thm. 1, p. 75) with preliminary, important work by Harish-Chandra. It provides a one-to-one parameterization of irreducible admissible representations (π, V) of an archimedean local group G on a Banach space V up to infinitesimal equivalence. Using the theory of Casselman–Wallach completions, see Thm. 3.26, we shall present the corresponding reformulation for the representation on the space of smooth vectors in interesting representations up to equivalence of G-representations.

Exercise 5.1. In his original paper [Lan89], instead of the nowadays much more common notion of an irreducible admissible representation, Langlands has actually used the notion of an irreducible *quasi-simple* representation: That is an irreducible G-representation (π, V) on a Banach space V, which has an *infinitesimal character*, i.e., for which all $Z \in \mathcal{Z}(\mathfrak{g})$ act by multiplication by a scalar on V^∞. Show that an irreducible G-representation (π, V) on a Banach space V is admissible, if and only if it is quasi-simple. *Hint:* In order to show that quasi-simple implies admissible, use [WarII72], Prop. 9.1.3.1. For the converse, combine Thm. 3.16, [Vog81], Prop. 0.3.19.(a), the continuity of the action of $\mathfrak{g}$ on V^∞ (cf. Rem. 3.7) and the density of V_0 in V^∞ (for which you apply Lem. 3.8 to V^∞).

If F is non-archimedean, the key to the analogue of Langlands's classificaction in the archimedean case, is again provided by deep, but by himself unpublished work of Harish-Chandra, which has finally been revised and written up in modern language by Waldspurger, cf. [Wld03]. We also refer to [Kon03].

The one-to-one parameterization of the representations on the space of smooth vectors inside interesting representations provided by the Langlands classification makes use of the notion of *Langlands triples*: These are triples $(P, [\widehat{\tau}], \nu)$, where

- $\mathbf{P} = \mathbf{L} \cdot \mathbf{N}$ is a standard parabolic F-subgroup of $\mathbf{G}$, cf. §4.3.1,
- τ is a tempered representation of L (in particular, it is irreducible, admissible and unitary), and $[\widehat{\tau}]$ denotes its unitary class, cf. Def. 3.32,
- $\nu \in \check{\mathfrak{a}}_{\mathbf{P}}$ satisfies $\nu > 0$.

We may now state the following.

Theorem 5.2 (Langlands classification).

(1) Let $(P, [\widehat{\tau}], \nu)$ be a Langlands triple. Then $I_P^G(\tau^\infty, \nu)$ has a unique irreducible quotient, called the Langlands quotient *and denoted $J(I_P^G(\tau^\infty, \nu))$.*

(2) The assignment

$$(P, [\widehat{\tau}], \nu) \mapsto \left[J(I_P^G(\tau^\infty, \nu))\right]$$

defines a bijection between the family of all Langlands triples and the family of all equivalence classes $[\pi^\infty]$, where π runs through the interesting representations of G.

Proof. We will give a proof by references: Let first F be non-archimedean. Then an interesting representation is nothing but an irreducible admissible representation of G and hence the reader my find a complete proof of Thm. 5.2 in [Ren10], Thm. VII.4.2 (once s/he has mastered Exc. 4.22). See also [Kon03], Thm. 3.5.

We turn to the archimedean case. Let $(P, [\widehat{\tau}], \nu)$ be a triple as in Thm. 5.2.(1). Recalling our Exc. 4.16, it follows from the arguments presented in the proof of [Kna86], Thm. 8.53 (as applied to the group $G^{(1)}$) that any $\tau \in [\widehat{\tau}]$ satisfies the conditions of [Bor-Wal00], IV.3.6 (since $G^{(1)} = {}^0G$ in the notation of [Bor-Wal00], 0.1.2: See [Bor-Ser73], Prop. 1.2). Hence, [Bor-Wal00], Cor. IV.4.6 implies that there is a unique $(\mathfrak{g}, K)$-submodule U_0 of the $(\mathfrak{g}, K)$-module $I_P^G(\tau^\infty, \nu)_{(K)}$, such that $I_P^G(\tau^\infty, \nu)_{(K)}/U_0$ is irreducible. Thm. 5.2.(1) now follows essentially by invoking Harish-Chandra's theorem, Thm. 3.16. Indeed, applying it to the admissible representation $I_P^G(\tau^\infty, \nu)$, cf. Prop. 4.30, we consider the topological closure $U := \overline{U_0}$ of U_0 inside $I_P^G(\tau^\infty, \nu)$ and so obtain a G-subrepresentation, which is uniquely determined by U_0. Now recall that by Cor. 4.40, $I_P^G(\tau^\infty, \nu)$ is in fact a Casselman–Wallach representation. In particular, we may use (3.28), and obtain as $(\mathfrak{g}, K)$-modules,

$$\left(I_P^G(\tau^\infty, \nu)/U\right)_{(K)} \cong I_P^G(\tau^\infty, \nu)_{(K)}/U_{(K)},$$

which by Thm. 3.16 is furthermore isomorphic to $I_P^G(\tau^\infty, \nu)_{(K)}/U_0$. Therefore, the $(\mathfrak{g}, K)$-module $\left(I_P^G(\tau^\infty, \nu)/U\right)_{(K)}$ is irreducible and hence, again by Thm. 3.16, so is the ambient admissible G-representation $I_P^G(\tau^\infty, \nu)/U$. It follows that $I_P^G(\tau^\infty, \nu)$ admits an irreducible quotient.

We now show its uniqueness. To this end, suppose now U' is a G-subrepresentation of $I_P^G(\tau^\infty, \nu)$ such that $I_P^G(\tau^\infty, \nu)/U'$ is irreducible. As $I_P^G(\tau^\infty, \nu)/U'$ is admissible, cf. Prop. 4.30 and Prop. 2.34, Thm. 3.16 shows that its underlying $(\mathfrak{g}, K)$-module $\left(I_P^G(\tau^\infty, \nu)/U'\right)_{(K)}$ is irreducible as well. However, $I_P^G(\tau^\infty, \nu)$ is a Casselman–Wallach representation, cf. Cor. 4.40, so (3.28) shows that there is an isomorphism of $(\mathfrak{g}, K)$-modules

$$\left(I_P^G(\tau^\infty, \nu)/U'\right)_{(K)} \cong I_P^G(\tau^\infty, \nu)_{(K)}/U'_{(K)},$$

and so [Bor-Wal00], Cor. IV.4.6 implies that $U'_{(K)} = U_{(K)}$. Consequently, the topological completions of $U'_{(K)}$ and $U_{(K)}$ in $I_P^G(\tau^\infty, \nu)$ agree as well. But as U' and U are closed in $I_P^G(\tau^\infty, \nu)$, Lem. 3.8 shows that $U = U'$. This proves Thm. 5.2.(1).

We now turn to Thm. 5.2.(2). Let us first verify that the assignment $(P, [\widehat{\tau}], \nu) \mapsto \left[J(I_P^G(\tau^\infty, \nu))\right]$, as mentioned there, is well-defined. By what we have just observed above, to this end it is enough to show that each Langlands quotient $J(I_P^G(\tau^\infty, \nu))$ is equivalent to the representation on the space of smooth vectors of an interesting representation: Recall from Cor. 4.40 and Lem. 3.27 that $J(I_P^G(\tau^\infty, \nu))$ is an irreducible Casselman–Wallach representation. Hence, by Thm. 3.29, there exists an irreducible admissible representation $(\widehat{\pi}, \widehat{V})$ on a Hilbert space $\widehat{V}$, such that $J(I_P^G(\tau^\infty, \nu)) \cong \widehat{V}^\infty$. Now, since $I_P^G(\tau^\infty, \nu)$ has a central character, cf. Prop. 4.30, so does

$J(I_P^G(\tau^\infty,\nu))$. Therefore, by density of $J(I_P^G(\tau^\infty,\nu)) \cong \widehat{V}^\infty$ in $\widehat{V}$ (cf. [WarI72], Prop. 4.4.1.1)) and the continuity of the map $G \times \widehat{V} \to \widehat{V}$, also $(\widehat{\pi},\widehat{V})$ has a central character. Hence, $J(I_P^G(\tau^\infty,\nu))$ is equivalent to the representation on the space of smooth vectors of an interesting representation.

Let us now show that $(P,[\widehat{\tau}],\nu) \mapsto \left[J(I_P^G(\tau^\infty,\nu))\right]$ is in fact a bijection as claimed: If $\left[J(I_P^G(\tau^\infty,\nu))\right] = \left[J(I_{P'}^G(\tau'^\infty,\nu'))\right]$, then $J(I_P^G(\tau^\infty,\nu))_{(K)} \cong J(I_{P'}^G(\tau'^\infty,\nu'))_{(K)}$, so invoking (3.28) for the Casselman–Wallach representations $J(I_P^G(\tau^\infty,\nu))$ and $J(I_{P'}^G(\tau'^\infty,\nu'))$,

$$\begin{aligned} I_P^G(\tau^\infty,\nu)_{(K)}/U_{(K)} &\cong J(I_P^G(\tau^\infty,\nu))_{(K)} \\ &\cong J(I_{P'}^G(\tau'^\infty,\nu'))_{(K)} \\ &\cong I_{P'}^G(\tau'^\infty,\nu')_{(K)}/U'_{(K)}. \end{aligned}$$

Therefore, [Bor-Wal00], Lem. IV.4.9, implies that $P = P'$ $\nu = \nu'$ and $\tau^\infty_{(K)} \cong \tau'^\infty_{(K)}$ (recalling again Exc. 4.16 and [Kna86], Thm. 8.53, in order to see that the notion of "temperedness" as defined in [Bor-Wal00], IV.3.6, coincides with our notion of pro-temperedness for interesting representations admitting a unitary central character). As both τ and τ' are assumed to be unitary, $[\widehat{\tau}] = [\widehat{\tau'}]$ by Thm. 3.40 and so $(P,[\widehat{\tau}],\nu) = (P',[\widehat{\tau'}],\nu')$. This shows injectivity.

We now prove surjectivity: Let (π,V) be an interesting representation. By [Bor-Wal00], Thm. IV.4.11 and the arguments in the proof of [Kna86], Thm. 8.53, there exists a (in fact unique) triple $(P,[\tau']_{(\mathfrak{l},K_P)},\nu)$ – where P and ν are as in the definition of Langlands triples, but $[\tau']_{(\mathfrak{l},K_P)}$ denotes the infinitesimal equivalence class of an interesting L-representation τ', whose K_P-finite matrix coefficients are $2+\varepsilon$-integrable on $L^{(1)} = \ker H_P$ for all $\varepsilon > 0$, i.e., such that

$$\int_{L^{(1)}} |c_{v,v^\vee}(\ell)|^{2+\varepsilon}\, d\ell < \infty \qquad \text{for all } K_P\text{-finite } v \in \tau'^\infty \text{ and } v^\vee \in (\tau'^\infty)^\vee,$$

and all $\varepsilon > 0$ – such that the $(\mathfrak{g},K)$-module $\pi_{(K)}$ is isomorphic to the quotient $J(I_P^G(\tau'^\infty,\nu))_{(K)} \cong I_P^G(\tau'^\infty,\nu)_{(K)}/U_0$. By [Bor-Wal00], Cor. IV.3.7, $\tau'^\infty_{(K_P)}$ is isomorphic to a direct $(\mathfrak{l},K_P)$-summand X of an induced representation $I_Q^L(\delta^\infty,0)_{(K_P)}$, where $Q = L_Q N_Q$ is a standard parabolic subgroup of L and δ a square-integrable representation of L_Q. As $I_Q^L(\delta^\infty,0)_{(K_P)} \cong \widehat{I_Q^L(\delta,0)}_{(K_P)}$, see Prop. 4.38 and 4.39, and $\widehat{I_Q^L(\delta,0)}$ is a unitary, admissible representation, cf. Prop. 4.33, Thm. 3.16 implies that the topological completion $\tau := \overline{X}$ of X in $\widehat{I_Q^L(\delta, 0)}$ is an L-subrepresentation of $\widehat{I_Q^L(\delta, 0)}$, which is hence unitary and satisfies $\tau_{(K_P)} \cong X \cong \tau'^{\infty}_{(K_P)}$. Therefore, one sees that this infinitesimal equivalence class $[\tau']_{(\mathfrak{l}, K_P)}$ contains a unitary member and hence by Thm. 3.40 may in fact be equivalently replaced by the unitary equivalence class $[\widehat{\tau}]$ of the unitary representation τ. However, the K_P-finite matrix coefficients of τ are $2+\varepsilon$-integrable on $L^{(1)}$ for all $\varepsilon > 0$, cf. [Bor-Wal00], Cor. IV.3.7 and [Kna86], Thm. 8.53, so by our Exc. 4.16, τ is in fact tempered, whence Prop. 4.38 shows that $\pi_{(K)}$ is isomorphic to the quotient $J(I_P^G(\tau^\infty, \nu))_{(K)} \cong I_P^G(\tau^\infty, \nu)_{(K)}/U_0$, where now $(P, [\widehat{\tau}], \nu)$ is a proper Langlands triple. From the above we know already that $J(I_P^G(\tau^\infty, \nu))$ is a Casselman–Wallach representation. Therefore, Thm. 3.26 implies that necessarily $\pi^\infty \cong J(I_P^G(\tau^\infty, \nu))$ as G-representations. This proves the assertion in the archimedean case. □

The strength of this result is that it reduces the problem of understanding (smooth vectors in) interesting representations of G to the problem of understanding the *tempered* ones. Accordingly, we now turn our attention to tempered representations. We have the following important result, essentially due to Harish-Chandra (and which we partially already invoked in the proof of Thm. 5.2).

Theorem 5.3. *Let (τ, V) be an irreducible unitary admissible representation of G. Then (τ, V) is tempered if and only if there exist a standard parabolic subgroup $P = L \cdot N$ of G and a square-integrable representation (δ, W) of L such that τ^∞ is isomorphic to a direct summand of $I_P^G(\delta^\infty, 0)$. If so, the datum $(L, [\widehat{\delta}])$ is unique up to conjugacy, i.e., if $(L', [\widehat{\delta'}])$ is another pair satisfying the above condition, then there exists a $g \in G$ such that $gLg^{-1} = L'$ and $[\widehat{\delta}] = [\widehat{\delta'(g\_g^{-1})}]$.*

Proof. Again we argue by reducing the statement to references, distinguishing the archimedean and the non-archimedean case. If F is non-archimedean, then [Ren10] Thm. VII.2.6 (together with our Exc. 4.22) implies that (τ, V) is tempered if and only if it embeds as a subrepresentation of an induced representation $I_P^G(\delta, 0)$ as in the statement of the theorem, with $(L, [\widehat{\delta}])$ unique up to conjugacy. As $I_P^G(\delta, 0)$ is unitary by Prop. 4.32, any subrepresentation admits a G-stable complement, i.e., is a direct summand (cf. [Cas95], Prop. 2.1.5). Hence, we obtain the assertion in this case.

If F is archimedean, then by [Bor-Wal00], Prop. IV.3.7, an irreducible unitary representation (τ, V) is tempered if and only if $\tau_0 = \tau^\infty_{(K)}$ is isomorphic to a direct $(\mathfrak{g}, K)$-summand of $I_P^G(\delta^\infty, 0)_{(K)}$ with P and δ as in the statement of the theorem. (As in the proof of Thm. 5.2 above, here we also use Exc. 4.16 and the line of argumentation leading to [Kna86], Thm. 8.53, in order to see that the notion of "temperedness" as defined in [Bor-Wal00], IV.3.6, coincides with our notion of pro-temperedness for interesting representations admitting a unitary central character.) Let ϑ_0 be the image of τ_0 in $I_P^G(\delta^\infty, 0)_{(K)}$ and denote by c_0 a direct $(\mathfrak{g}, K)$-module complement of ϑ_0 in $I_P^G(\delta^\infty, 0)_{(K)}$. Moreover, let $\overline{\vartheta_0}$, $\overline{c_0}$ and $\overline{I_P^G(\delta^\infty, 0)_{(K)}}$ denote the respective topological completions in the Fréchet space $I_P^G(\delta^\infty, 0)$. As $I_P^G(\delta^\infty, 0)$ is admissible, cf. Prop. 4.30, $\overline{\vartheta_0}$ and $\overline{c_0}$ are G-subrepresentations of $I_P^G(\delta^\infty, 0) = \overline{I_P^G(\delta^\infty, 0)_{(K)}}$ by Thm. 3.16 and in fact admissible subrepresentations, cf. Lem. 2.30. Hence, invoking Thm. 3.16 once more, irreducibility of $\tau^\infty_{(K)} = \tau_0 \cong \vartheta_0 = (\overline{\vartheta_0})_{(K)}$ implies irreducibility of $\overline{\vartheta_0}$, whence directness of the algebraic sum $\vartheta_0 \oplus c_0$ implies $\overline{\vartheta_0} \cap \overline{c_0} = \{0\}$. Therefore, $\overline{\vartheta_0} \boxplus \overline{c_0}$ is a well-defined G-subrepresentation (see Exc. 2.22) of $I_P^G(\delta^\infty, 0)$. Since $I_P^G(\delta^\infty, 0)$ is a Casselman–Wallach representation (see Cor. 4.40), hence so are the subrepresentations $\overline{\vartheta_0}$, $\overline{c_0}$ and $\overline{\vartheta_0} \boxplus \overline{c_0}$, cf. Lem. 3.27. However,

$$(\overline{\vartheta_0} \boxplus \overline{c_0})_{(K)} = (\overline{\vartheta_0})_{(K)} \oplus (\overline{c_0})_{(K)} = \vartheta_0 \oplus c_0 \cong I_P^G(\delta^\infty, 0)_{(K)},$$

so

$$\overline{\vartheta_0} \boxplus \overline{c_0} \cong I_P^G(\delta^\infty, 0), \tag{5.4}$$

by Thm. 3.26. Similarly, τ^∞ is a Casselman–Wallach representation (Lem. 2.24, Lem. 2.30, Lem. 3.24, Thm. 3.16). As $(\overline{\vartheta_0})_{(K)} = \vartheta_0 \cong \tau^\infty_{(K)}$, Thm. 3.26 implies that $\overline{\vartheta_0} \cong \tau^\infty$ as G-representations. Therefore, τ^∞ is in fact isomorphic to a direct summand of $I_P^G(\delta^\infty, 0)$ by (5.4).

Conversely, if for an irreducible unitary representation (τ, V), τ^∞ is isomorphic to a direct summand of a representation of type $I_P^G(\delta^\infty, 0)$ as in the statement of the theorem, then obviously $\tau_0 = \tau^\infty_{(K)}$ is isomorphic to a direct $(\mathfrak{g}, K)$-summand of $I_P^G(\delta^\infty, 0)_{(K)}$. Therefore, (τ, V) is tempered by [Bor-Wal00], Prop. IV.3.7 as already observed above.

So it remains to prove that $(L, [\widehat{\delta}])$ is unique up to conjugacy. This is indeed a general fact which goes back to Langlands, cf. [Lan89]. We give a brief sketch of the argument, which we prefer to extract from [Spe-Vog80], pp. 233–234: Let (τ, V) be a tempered representation of G. If (L, δ) is a pair, such that τ^∞ is isomorphic to a direct summand of $I_P^G(\delta^\infty, 0)$, then in particular (τ_0, V_0) is equivalent with an irreducible quotient of $I_P^G(\delta^\infty, 0)_{(K)}$. It can now be shown that $L = C_G(A)$, i.e., L is the centralizer in G of a subgroup A, which is characterized by being a maximally non-compact factor of a (so-called) θ-stable Cartan subgroup $H = T^+ \cdot A$ of G. Moreover, H and (the Harish-Chandra parameter of) δ are both unique up to conjugation by some $g \in G$: All this follows from [Spe-Vog80], Thm. 2.9. Hence, by the topological property of being maximally non-compact, A and so $L = C_G(A)$ are both unique up to conjugation by this $g \in G$. This shows the claim. □

Remark 5.5. Obviously, the assumption in Thm. 5.3 that (τ, V) is an irreducible unitary *and* admissible representation is somewhat redundant: Indeed, if F is archimedean, then the word "admissible" can be omitted by Thm. 3.33, whereas assuming unitarity cannot (see Ex. 3.21). If, however, F is non-archimedean, then, at the opposite, assuming unitarity can be omitted, see [Ren10] Thm. VII.2.6 (in combination with our Exc. 4.22), while the assumption of admissibility in Thm. 5.3 is important. The reader is invited to compare this with Prop. 4.19.

Corollary 5.6. *Let* $\mathbf{G} = \mathbf{GL}_n/D$ *be as in the statement of Thm. 4.41. Then an irreducible unitary admissible representation* (τ, V) *of* G *is tempered if and only if* $\tau^\infty \cong I_P^G(\delta^\infty, 0)$*. If* F *is archimedean, then this is furthermore equivalent to* $\tau \cong \widehat{I_P^G(\delta, 0)}$*.*

Proof. The first assertion is a direct consequence of Cor. 4.42 and Thm. 5.3. The second assertion follows from the first, recalling Prop. 4.33 and Thm. 3.40. □

Thm. 5.3 shows that we can actually further refine our analysis of interesting representations by restricting to *square-integrable representations* (δ, W) of G (instead of tempered ones). Accordingly, we shall now consider triples $(P, [\widehat{\delta}], \lambda)$ similar to above, now called *square-integrable Langlands triples*, but with some modifications of the last two entries:

- $\mathbf{P} = \mathbf{L} \cdot \mathbf{N}$ is a standard parabolic F-subgroup of $\mathbf{G}$, cf. §4.3.1,
- δ is a square-integrable representation of L (in particular, it is irreducible, admissible and unitary), and $[\widehat{\delta}]$ denotes its unitary class,
- $\lambda \in \check{\mathfrak{a}}_{\mathbf{P}}$ satisfies $\lambda \geq 0$.

We shall also need to consider the following equivalence relation between triples: We write

$$(P, [\widehat{\delta}], \lambda) \sim (P', [\widehat{\delta'}], \lambda')$$

if there exists an element $g \in G$ such that:

- $gLg^{-1} = L'$,
- $[\widehat{\delta}] = [\widehat{\delta'(g\_g^{-1})}]$,
- $\lambda = g\lambda' g^{-1}$.

We agree to write $[(P, [\widehat{\delta}], \lambda)]$ for the equivalence class of the triple $(P, [\widehat{\delta}], \lambda)$ with respect to $\sim$. Then we obtain the following result

Theorem 5.7 (Square-integrable Langlands classification).

(1) Let $(P, [\widehat{\delta}], \lambda)$ be a square-integrable Langlands triple as above. Then $I_P^G(\delta^\infty, \lambda)$ has only finitely many irreducible quotients, denoted by $J_1(I_P^G(\delta^\infty, \lambda)), \ldots, J_n(I_P^G(\delta^\infty, \lambda))$, where $n = n([(P, [\widehat{\delta}], \lambda)])$.

(2) The family of equivalence classes $[\pi^\infty]$, where π runs through the interesting representations of G, can be written as a disjoint union

$$\coprod_{[(P, [\widehat{\delta}], \lambda)]} \{[J_1(I_P^G(\delta^\infty, \lambda))], \ldots, [J_n(I_P^G(\delta^\infty, \lambda))]\}$$

of finite "square-integrable Langlands packets", *indexed by the equivalence classes of triples $(P, [\widehat{\delta}], \lambda)$.*

Proof. This is essentially a consequence of Thm. 5.2 and Thm. 5.3: Let first F be non-archimedean and let $(P, [\widehat{\delta}], \lambda)$ be a square-integrable Langlands triple. We now prove (1). To this end, we first assume that $\lambda = 0$. Then $I_P^G(\delta, 0)$ decomposes as a finite direct sum of irreducible subrepresentations. Indeed, if is $I_P^G(\delta, 0)$ irreducible, we are done. So suppose that there is a proper subrepresentation $\tau \subset I_P^G(\delta, 0)$. Then, since $I_P^G(\delta, 0)$ is unitary and admissible (cf. Prop. 4.32 and Prop. 4.30), τ admits a direct G-invariant complement $\tau^\perp$ (see [Cas95], Prop. 2.1.5, for instance). If both τ and $\tau^\perp$ are irreducible, we are again done. So let us finally assume that (without loss of generality) $\tau^\perp$ is not irreducible. Then again by [Cas95], Prop. 2.1.5 any proper subrepresentation ϑ of $\tau^\perp$ has a direct G-invariant complement $\vartheta^\perp$ in $\tau^\perp$. However, as $I_P^G(\delta, 0)$ is finitely generated (as explained at the end of §4.3.3), this process of decomposing must stop after finitely many steps by [Cas95], Thm. 6.3.10, and therefore

$$I_P^G(\delta, 0) \cong \bigoplus_{i=1}^{n} \tau_i \tag{5.8}$$

decomposes as a finite direct sum of irreducible and in fact tempered representations by Thm. 5.3. Moreover, by the second assertion of Thm. 5.3 their number n only depends on the equivalence class $[(P, [\widehat{\delta}], 0)]$. This shows (1), if $\lambda = 0$. Applying Rem. 4.31 shows that (1) in fact holds more generally for all $\lambda \in \check{\mathfrak{a}}_{\mathbf{P}}$, which satisfy $\langle \lambda, \check{\alpha} \rangle = 0$ for all $\alpha \in \Delta(\mathbf{P}, \mathbf{A_P})$ (as those are the ones in $\check{\mathfrak{a}}_{\mathbf{G}}$). So, let finally be λ a general positive element of $\check{\mathfrak{a}}_{\mathbf{P}}^{\mathbf{G}}$. If λ is strictly positive, then (1) follows readily from Thm. 5.2 and Cor. 4.21. Hence, we are left with the case, when $\theta_\lambda := \{\alpha \in \Delta(\mathbf{P}, \mathbf{A_P}) | \langle \lambda, \check{\alpha} \rangle \neq 0\}$ is a *proper* subset of $\Delta(\mathbf{P}, \mathbf{A_P})$. Let $\mathbf{P}' = \mathbf{L}' \cdot \mathbf{N}'$ be the standard parabolic F-subgroup of $\mathbf{G}$, defined by the set of simple roots in θ_λ, cf. (4.25): Properness of θ_λ implies that $P \subset P' \subset G$. Then, combining [Ren10], Lem. VI.1.4 ("induction in stages") with (5.8), we obtain an isomorphism of G-representations of the form

$$I_P^G(\delta, \lambda) \cong \bigoplus_{i=1}^{n} I_{P'}^G(\tau_i, \nu), \tag{5.9}$$

with $\nu \in \check{\mathfrak{a}}_{\mathbf{P}'}$ strictly positive. Hence, (1) follows from Thm. 5.2 and the fact that P', ν and the unitary equivalence class of $I_{P\cap L'}^{L'}(\delta, 0)$ is uniquely determined by $[(P, [\widehat{\delta}], \lambda)]$.

The second assertion of Thm. 5.7 in the non-archimedean case is now a direct consequence of Thm. 5.2, Thm. 5.3 and (5.9).

The proof in the archimedean case is completely analogous, whence we focus on the topological intricacies, which occur along the way: Again, we first deal with the case, when $\lambda = 0$, but now we consider the Fréchet space $I_P^G(\delta^\infty, 0)$. We want to show that $I_P^G(\delta^\infty, 0)$ decomposes as the finite direct locally convex sum

$$I_P^G(\delta^\infty, 0) \cong \bigoplus_{i=1}^{n} \tau_i^\infty, \tag{5.10}$$

where the τ_i are irreducible tempered representations of G. To this end, we consider the (Hilbert space) representation $\widehat{I_P^G(\delta, 0)}$: If $\widehat{I_P^G(\delta, 0)}$ is irreducible, (5.10) holds by Prop. 4.33 and Thm. 5.3. So, assume that there is a proper subrepresentation $\tau \subset \widehat{I_P^G(\delta, 0)}$. Its orthogonal complement $\tau^\perp := \{v \in \widehat{I_P^G(\delta, 0)} | \langle v, w \rangle_I = 0 \ \forall w \in \tau\}$ is a closed, G-invariant[1] Hilbert

[1] This is easy to see: For arbitrary $v \in \tau$, $w \in \tau^\perp$ and $g \in G$ we have

$$\langle v, R(g)w \rangle_I = \langle R(g)R(g)^{-1}v, R(g)w \rangle_I = \langle R(g^{-1})v, w \rangle_I = 0,$$

as $R(g^{-1})v \in \tau$, τ being a G-subrepresentation of $\widehat{I_P^G(\delta, 0)}$. Hence, $R(g)w \in \tau^\perp$.

subspace of $\widehat{I_P^G(\delta,0)}$, such that $I_P^G(\delta^\infty,0) \cong \tau^\infty \oplus (\tau^\perp)^\infty$, cf. [Bou03], V, §1, Sect. 6, Thm. 2 in combination with Prop. 4.33 and Exc. 2.22. If τ and $\tau^\perp$ are irreducible, we are again done by Thm. 5.3. If one of these two unitary representations is not irreducible, then one may split off another proper subrepresentation by the same procedure – a process which has to stop after finitely many steps by (4.39), Prop. 4.38, Thm. 3.14 (which show that the underlying $(\mathfrak{g},K)$-module of $\widehat{I_P^G(\delta,0)}$ is of finite length) and Thm. 3.16. Therefore, (5.10) holds. Again, the number of irreducible summands n only depends on the equivalence class $[(P,[\widehat{\delta}],0)]$ by the second assertion of Thm. 5.3. In particular, this shows (1) in the archimedean case, if $\lambda=0$ and in fact, applying Rem. 4.31 and Lem. 1.15, even for all $\lambda \in \check{\mathfrak{a}}_{\mathbf{P}}$, which satisfy $\langle\lambda,\check{\alpha}\rangle = 0$ for all $\alpha \in \Delta(\mathbf{P},\mathbf{A_P})$. Moreover, invoking Thm. 5.2 and Cor. 4.21, assertion (1) also follows for all λ, which are strictly positive. In the remaining case, i.e., when $\theta_\lambda := \{\alpha \in \Delta(\mathbf{P},\mathbf{A_P}) | \langle\lambda,\check{\alpha}\rangle \neq 0\}$ is a proper subset of $\Delta(\mathbf{P},\mathbf{A_P})$, one again constructs a standard parabolic F-subgroup $\mathbf{P}' = \mathbf{L}'\cdot\mathbf{N}'$ of $\mathbf{G}$, satisfying $P \subset P' \subset G$, as for non-archimedean F. Now, combining [Vog81], Prop. 4.1.18 ("induction in stages") with [Bor-Wal00], Cor. III.7.7 and (5.10), we obtain an isomorphism of G-representations

$$I_P^G(\delta^\infty,\lambda) \cong \bigoplus_{i=1}^{n} I_{P'}^G(\tau_i^\infty,\nu), \tag{5.11}$$

with $\nu \in \check{\mathfrak{a}}_{\mathbf{P}'}$ strictly positive. Here, we observe that we could pull out the locally convex direct sum from the induction, because assigning each $f \in I_{P'}^G(\bigoplus_{i=1}^n \tau_i^\infty,\nu)$ the n-tuple of its component functions is obviously a continuous linear G-equivariant bijection onto the left hand side of (5.11), whence an isomorphism of G-representations by the Banach–Schauder theorem, cf. Thm. 1.7. Assertion (1) now finally follows in complete generality from (5.11), Thm. 5.2 and the fact that P', ν and the unitary equivalence class of $\widehat{I_{P\cap L'}^{L'}(\delta,0)}$ is uniquely determined by $[(P,[\widehat{\delta}],\lambda)]$.

The second assertion of Thm. 5.7 in the archimedean case is now again a direct consequence of Thm. 5.2, Thm. 5.3 and (5.11). □

Remark 5.12. Let $(P,[\widehat{\delta}],\lambda)$ be a square-integrable Langlands triple.

(1) If F is archimedean, then the equivalence classes $[J_i(I_P^G(\delta^\infty,\lambda))]$ of the irreducible quotients from Thm. 5.7.(1) are necessarily all distinct. Originally formulated for connected semisimple Lie groups G, this is a result of Knapp, cf. [Kna76], Thm. on p. 271, but may be deduced from [Spe-Vog80], Thm. 2.9 in the generality use here. See also [Kna82],

Main Theorem, p. 34, which applies to local archimedean groups G, such that Z_{G° is compact.
On the contrary, for F non-archimedean, this needs no longer be true. Indeed, it follows from [Key82], Thm. D_n, Thm. E_7, and Thm. E_8, that this type of "multiplicity one" for the irreducible quotients $J_i(I_P^G(\delta^\infty, \lambda))$ generally fails, if G is a simple group of type D_n, E_7 or E_8, the first counterexample being found for a group of type D_4 by Knapp–Zuckerman, see [Kna-Zuc80], Thm. 2.1.

(2) Let $\mathbf{G} = \mathbf{GL}_n/D$ be as in the statement of Thm. 4.41. Combining (the proof of) Thm. 5.7 with Cor. 5.6, we obtain that all square-integrable Langlands packets are singletons. This uniqueness result for $\mathbf{G} = \mathbf{GL}_n/D$ is the key to a special concept, which one calls *isobaric sum*. More on this in §6.2.2.

5.2 Some remarks on square-integrable representations

The results of this chapter tell us that, in order to understand interesting representations of G, it suffices to understand the square-integrable ones (and the attached composition series of their induced representations). We would therefore like to conclude this chapter with a few facts and remarks on square-integrable representations of G.

If the ground field F is archimedean, then square-integrable representations of a local group G are well-understood. Indeed, Harish-Chandra gave a complete classification of all discrete series representations of a connected semisimple Lie group (i.e., in our terminology, a connected archimedean local group with finite center) by means of so-called *Harish-Chandra parameters*, mirroring the well-known highest-weight theory for irreducible finite-dimensional representations of semisimple Lie algebras as well as the Cartan–Weyl theory for compact groups. We refer the reader to [Kna86], Thm. 12.21 and Thm. 9.20, *ibidem*, for details and a proof of this remarkable result. Another important result in this regard is the following.

Theorem 5.13 (See [Kna86], Thm. 12.20). *Let G be a connected semisimple Lie group. Then, in order for G to have square-integrable representations, it is necessary and sufficient that the (complex)* rank *of G (i.e., the complex dimension of a Cartan subalgebra of $\mathfrak{g}_{\mathbb{C}}$) be equal to the rank of its maximal compact subgroup K.*

Let now G be again an arbitrary local archimedean group, but let us still assume that the center Z_{G° of G° is compact. Then, Harish-Chandra's classification may be extended to these groups as well: This is implicitly carried out in the proof of Thm. 1.1 of [Kna-ZucI82] (which shows the irreducibility of the so-constructed representations and establishes the necessary criterion of equivalence in terms of generalized Harish-Chandra parameters) in combination with the proof of [Kna86], Prop. 12.32 (which shows exhaustion of the discrete series).

Finally, we invite the reader to think of how to deduce an extension of this latter result to the completely general case of an arbitrary archimedean local group G, i.e., when the center of G° is not necessarily compact: In fact, it can be shown that any archimedean local group G can be written as a direct product $G = G^{(1)} \times A$, where $G^{(1)} = \ker H_G$ and A is a so-called split-component (cf. [Bor-Ser73], Prop. 1.2) of G , i.e., in particular a connected Lie subgroup of Z_G. Hence, applying [Dei10], Thm. 7.5.29[2], to the decomposition $G = G^{(1)} \times A$, reduces the problem of classifying the discrete series representations of G to the analogous task for $G^{(1)}$ and A: The discrete series representation of the connected abelian Lie group A being simply the continuous group homomorphisms $\chi : A \to S^1 \subset \mathbb{C}^*$, it all boils down to understanding what are the discrete series representations of $G^{(1)}$. However, reconsidering the definitions of [Kna-ZucI82] and [Kna86], $G^{(1)}$ is one of the groups, to which the above mentioned classification of discrete series representations applies, thanks to [Var77], Prop. II.1.15.

From these observations, one can deduce that $\mathrm{GL}_n(\mathbb{R})$ has square-integrable representations if and only if either $n = 1$ or $n = 2$, while $\mathrm{GL}_n(\mathbb{C})$ has square-integrable representations precisely for $n = 1$.

In the non-archimedean case, the theory is much "richer" (i.e., more complicated). We will resume this discussion in the next chapter.

[2]In fact, referring to Deitmar's fairly general result is just done for convenience, but not necessary in the current situation: As the reader may easily verify him-/herself, once the decomposition $G = G^{(1)} \times A$ is established for $A \subsetneq Z_G$, the decomposition of a discrete series representation δ of G into a product $\delta \cong \delta^{(1)} \otimes \chi$ of irreducible unitary representations $\delta^{(1)}$ of $G^{(1)}$ and χ of A follows from the existence of a central character of δ.

Chapter 6

Special Representations: Part 1

6.1 Supercuspidal representations

In this and in the next chapter, we devote our attention to certain "special" classes of representations. We start off with so-called *supercuspidal* representations, and try to see how they could fit into the Langlands classification. *For the rest of this chapter, we assume the ground field F to be non-archimedean, cf. Rem. 6.2.*

Definition 6.1. Let (π, V) be an irreducible admissible representation of G. Then (π, V) is called *supercuspidal* if all its matrix coefficients are compactly supported modulo the center $Z = Z_G$ of G, i.e. if

$$(\operatorname{supp} c_{v,v^\vee})/Z \text{ is compact in } G/Z$$

for all $v \in V, v^\vee \in V^\vee$.

Remark 6.2. With a few modifications (namely, replacing "irreducible admissible" by "interesting", V by $V_{(K)}$ and $V^\vee$ by $(V_{(K)})^\vee$), the same notion would make sense also for an archimedean ground field F. However, it turns out that supercuspidal representations for archimedean local groups only exist if G/Z is compact itself.

Note that we obtain a "hierarchy" of representations that we can illustrate as follows:

$$\text{unitary supercuspidal} \Longrightarrow \text{square-integrable} \Longrightarrow \text{tempered} \Longrightarrow \text{unitary}.$$

Observe that by Prop. 4.19, a supercuspidal representation with a unitary central character is isomorphic to a unitary representation. Hence, we could equivalently have written "supercuspidal with unitary central character" as the first entry of the above line of implications.

In the previous chapters, we saw that we can describe all interesting representations in terms of the tempered ones, and that these in turn can be obtained as irreducible direct summands of an induced representation $I_P^G(\delta, 0)$ for a square-integrable representation δ. Reasoning along the same lines, one might wonder if it is possible to describe square-integrable representations in terms of supercuspidal ones. In some ways, this is made plausible by the observation that, if X is a locally compact Hausdorff space equipped with a "sufficiently nice" measure μ (e.g., Euclidean space together with the Lebesgue measure), then the space $C_c(X)$ of compactly supported functions on X is dense in $L^2(X, \mu)$. Indeed, turning back our attention the non-archimedean local representation theory, we have the following result:

Theorem 6.3 (Jacquet). *Let (δ, V) be a square-integrable representation of G. Then there exist a standard parabolic F-subgroup $\mathbf{P} = \mathbf{L} \cdot \mathbf{N}$ of $\mathbf{G}$ and a supercuspidal representation (σ, W) of L such that δ is equivalent to a subrepresentation of $I_{\mathbf{P}}^G(\sigma, 0)$. Moreover, if $\mathbf{P}' = \mathbf{L}' \cdot \mathbf{N}'$ and (σ', W') are another such choice for δ, then $[(P, [\widehat{\sigma}], 0)] = [(P', [\widehat{\sigma'}], 0)]$.*

Proof. Existence follows from [Ren10], Thm. VI.2.1 and Cor. VI.2.1. Uniqueness of equivalence classes $[(P, [\widehat{\sigma}], 0)]$ is established in [Mín-Séc14], Thm. 2.1. $\square$

Remark 6.4. (1) We owe the reader a little apology: In fact, in the formulation above, the last theorem can be considered as a bit misleading: Indeed, its assertion – although true, of course – holds not only for square-integrable representations, but far more generally even for *all* irreducible admissible representations. The reader may try to deduce this fact (not by looking up a direct proof in [Ren10], §VI.2.1, for instance, but) from Thm. 6.3 in its above formulation by reconsidering Thm. 5.7.

(2) We used a similar result in the archimedean case ("Casselman's subrepresentation theorem") to show that every Casselman–Wallach representation is isomorphic to the representation on the space of smooth vectors inside an interesting representation. The reader is invited to consult [Wal89], Prop. 4.2.3 in the light of Rem. 6.2, in order to see the analogy to Thm. 6.3.

6.2 A specification of the general theory: The Bernstein–Zelevinsky classification for GL_n

Until the end of this section, we restrict our attention to the general linear group, i.e., we set $\mathbf{G} = \mathbf{GL}_n/F$, in which case more explicit results can be obtained, known as the *Bernstein–Zelevinsky classification* (recall that F is still assumed non-archimedean). In order to be more in line with the general Langlands classification, we will formulate the respective results only in terms of quotients, rather than in terms of subrepresentations (as we did in Thm. 6.3).

We start with the simple observation that, if (σ, V) is a supercuspidal representation of $G = \mathrm{GL}_n(F)$ and $x \in \mathbb{R}$, then

$$\sigma(x) := |\det(\cdot)|_F^x \cdot \sigma$$

defines a supercuspidal representation on V. (More concretely, $\sigma(x)$ sends $g \in G$ to the linear automorphism $\sigma(x)(g) : v \mapsto |\det(g)|^x \cdot \sigma(g)v$ of V[1].) Now let r be a positive divisor of n, i.e., $n = r \cdot m$ for some $m \in \mathbb{Z}$, and consider the standard parabolic F-subgroup P of G consisting of the matrices of the form

$$\begin{pmatrix} \boxed{m} & & & & \\ & \boxed{m} & & \bigstar & \\ & & \boxed{m} & & \\ & 0 & & \ddots & \\ & & & & \boxed{m} \end{pmatrix},$$

where $\boxed{m}$ denotes a $\mathrm{GL}_m(F)$-block. Then, the Levi subgroup L of P consists of block-matrices of the form

$$\begin{pmatrix} \boxed{m} & & & & \\ & \boxed{m} & & 0 & \\ & & \boxed{m} & & \\ & 0 & & \ddots & \\ & & & & \boxed{m} \end{pmatrix}.$$

and is hence isomorphic to the r-fold product $L \cong \mathrm{GL}_m(F) \times \cdots \times \mathrm{GL}_m(F)$. For a supercuspidal representation (σ, V) of $\mathrm{GL}_m(F)$, consider the r-tuple of representations of $\mathrm{GL}_m(F)$, given by

$$(\sigma, \sigma(1), \ldots, \sigma(r-1)).$$

[1]Observe the different use of the brackets "$(\cdot)$" in $\sigma(x)$ and $\sigma(g)$. We trust that this will not cause any confusion.

This is denoted variously as $[\sigma, \sigma(r-1)]$, or $\Delta(\sigma, r)$ (or simply Δ, if no confusion can arise) and is called the *Bernstein–Zelevinsky segment* attached to σ and r. We leave it as an exercise to the reader to show that the representation $\sigma \otimes \sigma(1) \otimes \cdots \otimes \sigma(r-1)$ of L is supercuspidal (Hint: Use [Fla79], Thm. 1 and our Prop. 4.2.). In order to ease notation, we will denote the so-constructed representation again by Δ.

The following result, which is due to Bernstein and Zelevinsky (but to which one should prepend work of Harish-Chandra and Casselman (namely their "square-integrability criterion", cf. [Cas95], Thm. 6.5.1), and, in fact, append the diploma thesis [Aue04] of Auel), shows how Bernstein–Zelevinsky segments of irreducible supercuspidal representations can be used in order to classify square-integrable representations of $G = \mathrm{GL}_n(F)$ in a rather explicit manner:

Theorem 6.5 (Bernstein–Zelevinsky). *Let $\Delta = \Delta(\sigma, r)$ be a Bernstein–Zelevinsky segment. Then the representation $I_P^G(\Delta, 0)$ has a unique irreducible quotient, to be denoted $Q(\Delta)$. Furthermore, for an irreducible admissible representation (π, V) of $G = \mathrm{GL}_n(F)$ the following assertions are equivalent:*

(1) π is square-integrable.
(2) $\pi \cong Q(\Delta)$ for some segment $\Delta = [\sigma, \sigma(r-1)]$ such that $\sigma(\frac{r-1}{2})$ is unitary.

Proof. Existence and uniqueness of $Q(\Delta)$ is guaranteed by [Rod82], Prop. 9.(ii) (see also [Zel80], Prop. 2.10 and §3.1 therein). That (ii) implies (i) boils down to a rather simple consequence of Casselman's square-integrability criterion as applied to $Q(\Delta)$ as in the statement of the theorem: For the details see [Aue04], Thm. 6.12 and Ex. 6.13, therein, (or the original, but less explicit source [Zel80], Thm. 9.3, which refers to [Cas95], Thm. 6.5.1). Whereas that (i) implies (ii) is a theorem of Bernstein, whose assertion may be found as Prop. 11 in [Rod82], for instance, but whose proof Bernstein never published. We refer to [Aue04], Thm. 6.1, in particular §6.3 therein for the only reference we know, in which a complete argument was given. □

As a corollary of Thm. 6.5, Cor. 5.6 and Exc. 4.17, we obtain a classification of the tempered representations of G in terms of Bernstein–Zelevinsky segments.

Corollary 6.6. *For an irreducible admissible representation (π, V) of $G = \mathrm{GL}_n(F)$ the following are equivalent:*

(1) π is tempered.
(2) There exists a partition $n = \sum_{i=1}^k n_i$ of n and there exists for each i a Bernstein–Zelevinsky segment $\Delta_i = [\sigma_i, \sigma_i(r_i - 1)]$, such that $\sigma_i(\frac{r_i-1}{2})$ is unitary, and such that

$$\pi \cong I_P^G(Q(\Delta_1) \otimes \cdots \otimes Q(\Delta_k), 0),$$

where P is the standard parabolic subgroup of G with Levi subgroup isomorphic to $\mathrm{GL}_{n_1} \times \cdots \times \mathrm{GL}_{n_k}$.

Finally, (recalling Exc. 4.17 again) we can now reformulate the Langlands classification for $G = \mathrm{GL}_n(F)$ in terms of Bernstein–Zelevinsky segments. We shall do this in the two different ways, suggested by Thm. 5.2 and Thm. 5.7, i.e., according to which type of representations we choose as "building blocks". Thus, we obtain two different perspectives on the Langlands classification, now made concrete for the general linear group:

Corollary 6.7 (Tempered perspective). *Up to isomorphism, the irreducible admissible representations (π, V) of $G = \mathrm{GL}_n(F)$ are precisely the irreducible quotients of the induced representations $I_P^G(\tau_1 \otimes \cdots \otimes \tau_k, \nu)$, where:*

(1) P is a standard parabolic subgroup of G,
(2) $n_1 + \cdots + n_k = n$ is the partition of n corresponding to (the Levi subgroup of) P and for each i,

$$\tau_i \cong I_{P_i}^{\mathrm{GL}_{n_i}(F)}(Q(\Delta_{i_1}) \otimes \cdots \otimes Q(\Delta_{i_{k_i}}), 0)$$

is a tempered representation of $\mathrm{GL}_{n_i}(F)$, as described in Cor. 6.6 above, and
(3) $\nu \in \check{\mathfrak{a}}_{\mathbf{P}} \cong \mathbb{R}^k$ satisfies $\nu = (\nu_1, \ldots, \nu_k)$ with $\nu_1 > \nu_2 > \cdots > \nu_k$.

Here, the vector $\nu = (\nu_1, \ldots, \nu_k) \in \mathbb{R}^k$ is to be read in the coordinates given by the powers of the absolute value of the respective determinant $\left|\det_{\mathrm{GL}_{n_i}}(\cdot)\right|_F$, i.e.,

$$I_P^G(\tau_1 \otimes \cdots \otimes \tau_k, \nu) \cong I_P^G(\tau_1(\nu_1) \otimes \cdots \otimes \tau_k(\nu_k), 0),$$

where $\tau_i(\nu_i) := \left|\det_{\mathrm{GL}_{n_i}}(\cdot)\right|_F^{\nu_i} \cdot \tau_i$. This comment also applies to the next corollary, in which we present the Langlands classification in terms of square-integrable representations:

Corollary 6.8 (Square-integrable perspective). *Up to isomorphism, the irreducible admissible representations* (π, V) *of* $G = \mathrm{GL}_n(F)$ *are precisely the irreducible quotients of the induced representations* $I_P^G(\delta_1 \otimes \cdots \otimes \delta_k, \lambda)$*, where:*

(1) P *is a standard parabolic subgroup of* G*,*
(2) $n_1 + \cdots + n_k = n$ *is the partition of* n *corresponding to (the Levi subgroup of)* P *and for each* i*,* $\delta_i \cong Q(\Delta_i)$ *is a square-integrable representation of* $\mathrm{GL}_{n_i}(F)$*, as described in Thm. 6.5, and*
(3) $\lambda \in \check{\mathfrak{a}}_{\mathbf{P}} \cong \mathbb{R}^k$ *satisfies* $\lambda = (\lambda_1, \ldots, \lambda_k)$ *with* $\lambda_1 \geq \lambda_2 \geq \cdots \geq \lambda_k$*.*

This latter, somewhat more elegant "square-integrable perspective" will be important in §6.2.2.

6.2.1 *A step beyond* $\mathrm{GL}_n(F)$*: General D*

In passing, we remark that using the so-called *Jacquet–Langlands correspondence*, see [DKV84, Jac-Lan70], one may derive an analogously explicit description of the discrete spectrum (i.e., the equivalence classes of square-integrable representations) of $\mathbf{G} = \mathbf{GL}_n/D$, where D is now an arbitrary finite-dimensional central division algebra over our local non-archimedean field F, from Thm. 6.5 above and hence – again recalling the general results Cor. 5.6, Thm. 5.2 and Thm. 5.7 – one may finally obtain a description of the Langlands classification of the equivalence classes of all irreducible admissible representations of G in terms of Bernstein–Zelevinsky segments. In fact, such a translation of results from the split (i.e., $D = F$) to the general case, was the topic of Tadic's [Tad90]. However, even the reader, who seeks to find a description of the Langlands classification in terms of Bernstein–Zelevinsky segments *without* the detour to the Jacquet-Langlands correspondence, may be helped: Such a passionate person may distill the respective arguments from a combination of the techniques presented in [Mín-Séc13] and [Mín-Séc14], as translated to representations over the complex numbers. As this undertaking would definitively go beyond the scope of this book, we confine ourselves with this short remark.

6.2.2 *Local isobaric sums*

We conclude this chapter by giving the definition of *(local) isobaric sums* for the general linear group. To this end, let F be non-archimedean and let (π_1, V_1) and (π_2, V_2) be irreducible admissible representations of $\mathrm{GL}_{n_1}(F)$

and $\mathrm{GL}_{n_2}(F)$, respectively. By the above corollary, these two representations have corresponding classes of square-integrable Langlands triples $[(P_1, [\widehat{\delta_1}], \lambda_1)]$ and $[(P_2, [\widehat{\delta_2}], \lambda_2)]$, respectively, where $\lambda_1 = (a_1, \ldots, a_k)$ with $a_1 \geq \cdots \geq a_k$ and analogously $\lambda_2 = (b_1, \ldots, b_l)$ with $b_1 \geq \cdots \geq b_l$. Now, consider $\lambda := (c_1, ..., c_{k+l})$ with c_i running through the multiset of all the a_i's and b_j's, subject to the condition that $c_1 \geq c_2 \geq ... \geq c_{k+l}$. In other words, we order the merged entries of λ_1 and λ_2 decreasingly. As prescribed by this ordering of the $c_i's$, we may order the attached diagonal blocks of the Levis subgroups L_1 and L_2 and also the attached factors of the representations δ_1 and δ_2. In this way, we obtain a standard parabolic subgroup P of $G = \mathrm{GL}_n(F)$, $n := n_1 + n_2$, whose Levi subgroup L equals the direct product of the blocks of L_1 and L_2, permuted in the way prescribed above, as well as a square-integrable representation δ of L, obtained as the tensor product of the factors of δ_1 and δ_2, which are ordered accordingly (see Exc. 4.17 once more).

We observe that by construction, $\lambda \in \check{\mathfrak{a}}_{\mathbf{P}} \cong \mathbb{R}^{k+l}$, satisfies the condition of Cor. 6.8. Hence, applying said corollary, we obtain an irreducible admissible representations (π, V) of $G = \mathrm{GL}_n(F)$, uniquely determined (up to equivalence) as the unique irreducible quotient of $I_P^G(\delta, \lambda)$. We may now give the following.

Definition 6.9. The unique irreducible quotient (π, V) of the so-constructed representation $I_P^G(\delta, \lambda)$ is called the (local) *isobaric sum* of π_1 and π_2, denoted

$$\pi_1 \boxplus \pi_2.$$

Exercise 6.10. (1) Show that the concept of isobaric sums extends to an arbitrary finite number of summands $\pi_1 \boxplus \pi_2 \boxplus ... \boxplus \pi_k$ by providing the recipe of its construction.
(2) Show that the concept of isobaric sums works equally well for the spaces of smooth vectors π_1^∞, π_2^∞ of two interesting representations π_1 and π_2 of $\mathrm{GL}_{n_1}(F)$ and $\mathrm{GL}_{n_2}(F)$, respectively, over an *archimedean* local field F and provide the recipe of construction of what we shall denote $\pi_1^\infty \boxplus \pi_2^\infty$.
Hint: Firstly, observe that the description of the positive elements of $\check{\mathfrak{a}}_{\mathbf{P}}$ as the vectors $\lambda = (\lambda_1, \ldots, \lambda_n)$ with $\lambda_1 \geq \lambda_2 \geq \cdots \geq \lambda_n$ was purely formal and hence transfers verbatim from the non-archimedean case. Now combine Exc. 4.17 with (the proof of) Thm. 5.7 and Cor. 5.6. See also Rem. 5.12.

Chapter 7

Special Representations: Part 2

In this section, we introduce two further "special" classes of representations (namely *generic* and *unramified* ones), and again investigate how they fit into the Langlands classification. The ground field F is allowed to be an arbitrary local field, and $\mathbf{G}$ is again any connected reductive group over F, i.e., G is an arbitrary local group. We also fix an algebraic closure $\overline{F}$ of F.

7.1 Generic representations

From now on and throughout this section, we shall assume that $\mathbf{G}$ is *quasi-split*: This means that $\mathbf{G}$ has a Borel subgroup, i.e., a maximal connected solvable subgroup, $\mathbf{B}$ which is defined over F. In this case, $\mathbf{B}$ is automatically a minimal parabolic F-subgroup of $\mathbf{G}$, see [Bor91], Cor. 11.2, hence, by Thm. 20.9, *ibidem* we may assume that $\mathbf{B} = \mathbf{P}_0$. For its Levi decomposition we will adapt the more common notation $\mathbf{B} = \mathbf{T} \cdot \mathbf{U}$: Note that $\mathbf{T}$ is a *torus*, i.e., it is isomorphic to $\mathbf{GL_1} \times \cdots \times \mathbf{GL_1}$ over $\overline{F}$. In fact, $\mathbf{T}$ is a maximal torus in $\mathbf{G}$ (in terms of inclusion). See [Bor91], Thm. 10.6.(4) and Cor. 11.3.

Extending our notations from the archimedean case to all local fields, we let $\mathfrak{g}$ denote the Lie algebra (over F) of G and, similarly, we let $\mathfrak{t}$ be the Lie algebra of T. Moreover, we set $\mathfrak{g}_{\overline{F}} := \mathfrak{g} \otimes_F \overline{F}$ and $\mathfrak{t}_{\overline{F}} := \mathfrak{t} \otimes_F \overline{F}$. We further denote by $\Delta = \Delta(\mathfrak{g}_{\overline{F}}, \mathfrak{t}_{\overline{F}})$ the absolute ($\overline{F}$-)root system of $\mathfrak{g}$ with respect to $\mathfrak{t}$, i.e., the set of roots of $\mathfrak{g}_{\overline{F}}$ with respect to $\mathfrak{t}_{\overline{F}}$. It is well-known that $\mathbf{B}$ defines a choice of positivity on Δ: We briefly recall this in the following paragraph, mainly to fix notations and refer to [JMil17], §21.c for complete proofs.

Firstly, recall that the Lie algebra $\mathfrak{g}_{\overline{F}}$ decomposes as a direct sum $\mathfrak{g}_{\overline{F}} = \mathfrak{t}_{\overline{F}} \oplus \bigoplus_{\alpha\in\Delta} \mathfrak{g}_\alpha$, where $\mathfrak{g}_\alpha$ denotes the eigenspace of the linear functional α inside $\mathfrak{g}_{\overline{F}}$ and is hence isomorphic to $\overline{F}$ as a vector space. At the level of groups, there exists for each root $\alpha \in \Delta$ a unique connected $\overline{F}$-subgroup $U_\alpha \subseteq \mathbf{G}/\overline{F}$ such that U_α is normalized by $\mathbf{T}/\overline{F}$ and $\mathrm{Lie}(U_\alpha) = \mathfrak{g}_\alpha$. We call U_α the *root subgroup* attached to α. We then declare a root α to be *positive* (with respect to $\mathbf{B}$) if the corresponding root subgroup U_α is contained in $\mathbf{U}/\overline{F}$. The set of positive roots is denoted $\Delta^+ \subset \Delta$, and the corresponding set of simple $\overline{F}$-roots (which generate Δ^+) is denoted $\Delta^0 \subset \Delta^+$. We will need two more observations: Firstly, we note that $\mathbf{U}/\overline{F}$ is equal to the product $\prod_{\alpha\in\Delta^+} U_\alpha$. Secondly, observe that for each $\alpha \in \Delta$ there exists an $\overline{F}$-isomorphism $x_\alpha : \mathbf{G_a}/\overline{F} \to U_\alpha$ from the additive group $\mathbf{G_a}$ over $\overline{F}$ to U_α and one has

$$t \cdot x_\alpha(y) \cdot t^{-1} = x_\alpha(\alpha(t) \cdot y) \qquad \text{for all } t \in \mathbf{T}/\overline{F}, y \in \mathbf{G_a}/\overline{F}.$$

Definition 7.1. A homomorphism $\theta : \mathbf{U}/\overline{F} \to \mathbf{G_a}/\overline{F}$ of algebraic groups over $\overline{F}$ is called *generic*, or *non-degenerate*, if it induces a homomorphism $\mathbf{U}(\overline{F}) \to \overline{F}$ of the form

$$\theta\left(\prod_{\alpha\in\Delta^+} x_\alpha(y_\alpha)\right) = \sum_{\alpha\in\Delta^0} \lambda_\alpha y_\alpha \qquad \text{for all } y_\alpha \in \overline{F}$$

for suitable $\lambda_\alpha \in \overline{F}^*$.

Remark 7.2. At the cost of rescaling x_α, one may assume in the above definition that $\lambda_\alpha = 1$ for every simple root $\alpha \in \Delta^0$. We will assume to have normalized the morphisms x_α in this fashion henceforth.

Definition 7.3. A *generic character* of U is a non-trivial continuous group homomorphism $\psi : U \to \mathbb{C}^\times$ which can be written as a composition $\psi = \psi_F \circ \theta|_U$ where ψ_F is a unitary character of the field F and $\theta : \mathbf{U}(\overline{F}) \to \overline{F}$ is induced from a generic character as per the above definition. (Here we observe that since $\mathbf{G}$ is quasi-split, $\theta|_U$ takes values in F, hence the expression $\psi_F \circ \theta|_U$ is well-defined.)

Example 7.4. Consider $\mathbf{G} = \mathbf{GL}_n/F$ with the standard Borel subgroup $\mathbf{B}$ consisting of upper triangular matrices. Then, U consists of the unipotent matrices in B. By the well-known description of the (positive, resp. simple) roots for $\mathbf{GL}_n$ (together with our standing convention that $\lambda_\alpha = 1$ for all simple α), we see that $\psi : U \to \mathbb{C}^\times$ is generic if and only if it is of the form

$$\begin{pmatrix} 1 & u_{1,2} & & & \bigstar \\ & 1 & u_{2,3} & & \\ & & 1 & \ddots & \\ & 0 & & \ddots & u_{n-1,n} \\ & & & & 1 \end{pmatrix} \mapsto \psi_F(u_{1,2} + u_{2,3} + \cdots + u_{n-1,n})$$

for some non-trivial unitary character ψ_F of F.

For a generic character ψ, we may consider the following complex vector space:

$$\mathrm{Ind}_U^G[\psi] := \{W \in C^\infty(G, \mathbb{C}) : W(ug) = \psi(u)W(g) \text{ for all } u \in U, g \in G\}.$$

Clearly, G acts on $\mathrm{Ind}_U^G[\psi]$ by right-translation. It is easy to check that this turns $\mathrm{Ind}_U^G[\psi]$ into a representation of G: In the archimedean case, a small verification is necessary to establish that the subspace topology on $\mathrm{Ind}_U^G[\psi] \subseteq C^\infty(G, \mathbb{C})$ is Fréchet and that the action is continuous (Exercise!).

Definition 7.5. Let (π, V) be an interesting representation of G.

(1) The representation (π, V) is called *ψ-generic*, where ψ is a generic character of U, if there exists a nontrivial homomorphism of G-representations
$$\mathcal{W}_\psi : V^\infty \to \mathrm{Ind}_U^G[\psi].$$
(2) If (π, V) is ψ-generic, then (by irreducibility of π^∞) $\mathcal{W}_\psi$ is injective and its image $\mathcal{W}_\psi(\pi^\infty)$ is called the *ψ-Whittaker model* of (π^∞, V^∞).
(3) The representation (π, V) is *generic* if it is ψ-generic for some generic character ψ of U.

Proposition 7.6. *Let (π, V) be an interesting representation of G and let ψ be a generic character of U. Then (π, V) is ψ-generic if and only if there exists a nontrivial linear functional $\lambda : V^\infty \to \mathbb{C}$ such that*
$$\lambda(\pi(u) \cdot v) = \psi(u) \cdot \lambda(v) \text{ for all } v \in V^\infty, u \in U. \tag{7.7}$$
(Here, if F is archimedean, then λ is additionally assumed continuous.)

Sketch of a proof. Necessity is easy: Put $\lambda(v) := \mathcal{W}_\psi(v)(id)$ for $v \in V^\infty$. (To check continuity in the archimedean case, note that $\lambda = \mathrm{ev}_{id} \circ \mathcal{W}_\psi$, where evaluation at the identity $id \in G$
$$\mathrm{ev}_{id} : C^\infty(G, \mathbb{C}) \to \mathbb{C},$$
$$f \mapsto f(id)$$
is clearly a continuous linear map.)

As for sufficiency, suppose we are given a functional λ as in the statement of Prop. 7.6. For $v \in V^\infty$, consider the map

$$\mathcal{W}_\psi(v) : G \to \mathbb{C},$$
$$g \mapsto \lambda(\pi(g)v).$$

We want to show that $v \mapsto \mathcal{W}_\psi(v)$ defines a nontrivial G-equivariant linear map $V^\infty \to \mathrm{Ind}_U^G[\psi]$: We first check that the map $\mathcal{W}_\psi(v)$ is smooth for every $v \in V^\infty$. This is easy to see, since $\mathcal{W}_\psi(v)$ is obtained by composing λ with the smooth map $c_v : G \to V^\infty$, see Lem. 2.18 (and recalling in the archimedean case that a continuous, hence bounded, linear functional on a Fréchet space maps smooth curves to smooth curves, cf. [Kri-Mic97], Cor. 2.11, hence its composition with c_v will lead a smooth map by Cor. 3.14 *ibidem.*) Non-triviality, linearity and equivariance of $\mathcal{W}_\psi$ amount to a formal calculation, which we omit. Hence, in the non-archimedean case we are done. If F is archimedean, however, we still have to show that the map $\mathcal{W}_\psi$ is continuous. By Lem. 1.10, it suffices to show that it is bounded. In order to show boundedness, use:

- V^∞ has the *Heine-Borel property*, i.e., a subset of V^∞ is compact if and only if it is closed and bounded (see [Ber-Krö14], Cor. 5.6 and [Tre70], Cor. 1 on p. 520).
- The map $G \times V^\infty \to V^\infty$ is continuous.

□

The space of (continuous, if F is archimedean) linear functionals $\lambda : V^\infty \to \mathbb{C}$ satisfying (7.7) above is denoted $\Lambda(\pi^\infty, \psi)$ and its elements are often called *Whittaker functionals.* The content of the above proposition is that (π, V) is generic if and only if $\Lambda(\pi^\infty, \psi) \neq \{0\}$ for some generic character ψ.

The following results are meant to illustrate that generic representations are "nice" (hence, kind of motivating the name "generic"). The first one below shows that generic representations allow for a *canonical* realization: In other words, (ψ-)Whittaker models, if they exist, are unique. This fundamental and important result is due to Rodier in the non-archimedean case, while in the archimedean case it was first proved by Shalika for $G = \mathrm{GL}_n(F)$ and π unitary and by Casselman-Hecht-Miličić (and also by Jiang-Sun-Zhu) in the general case.

Theorem 7.8 (Cf. [Rod75], Thm. 2; [JSha74], Thm. 3.1, [CHM00], Thm. 9.2; [JSZ11], Thm. 11.4). *For an interesting representation (π, V) of G and a generic character ψ of U, the following are equivalent:*

(1) π *is* ψ*-generic.*
(2) $\dim_{\mathbb{C}} \Lambda(\pi^\infty, \psi) = 1$.

In other words, if π *is* ψ*-generic, the image in* $\mathrm{Ind}_U^G[\psi]$ *of any homomorphism of* G*-representations* $\mathcal{W}_\psi$ *as in Def. 7.5.(1) is the same.*

Remark 7.9. A way to think of this remarkable result is to compare it to the following, very basic statement: Every finite-dimensional complex vector space is isomorphic to $\mathbb{C}^n$ for some $n \geq 0$ and hence *admits a canonical realization*. Whittaker models – if they exist – allow for the analogous statement, but now for the space V^∞ of an interesting representation (π, V) of a local group.

The next result (previously known as the "standard module conjecture") shows that, in the framework of the Langlands classification, the generic representations are the ones fully induced from generic tempered representations. It is due to Shahidi, Casselman, Muić, H. H. Kim, Heiermann, W. Kim, and Opdam in the non-archimedean case and due to Vogan, Kostant, and Shahidi in the archimedean case.

Remark 7.10. For a precise statement of this deep result we shall need a certain technical assumption, which, however, can always be achieved, after a minor manipulation, cf. [Sha90], pp. 282–283, and which the reader may hence also safely skip, if s/he desires. In order to explain this assumption, let $\mathbf{P} = \mathbf{L} \cdot \mathbf{N}$ be the Levi decomposition of a standard parabolic F-subgroup $\mathbf{P}$ of $\mathbf{G}$, as in §4.3.1, and let ψ be a generic character of U as above. Firstly, one directly checks that $\psi_L := \psi|_{U \cap L}$ is a generic character for $U_L := U \cap L$, which is the group of F-points of the unipotent radical of the standard Borel subgroup of $\mathbf{L}$. Now, we recall from (4.25) that we may write $\mathbf{P} = \mathbf{P}_\theta$ for a unique subset of simple F-roots θ. Moreover, let $\widetilde{w}$ be the unique element of the Weyl group of $\mathbf{G}$ with respect to $\mathbf{A}_{\mathbf{P}_0}$, for which $\widetilde{w}(\theta) \subseteq \Delta^0$ and $-\widetilde{w}(\alpha) > 0$ for all $\alpha \in \Delta^0 \setminus \theta$, cf. [Sha10], p. 10. We choose and fix a representative $w \in G$ of $\widetilde{w}$. See [Sha10], last paragraph of 1.1. The technical assumption mentioned above now reads as

$$\psi_L(u_L) = \psi_L(w \cdot u_L \cdot w^{-1}) \tag{7.11}$$

for all $u_L \in U_L$.

Theorem 7.12. *Let (π, V) be an interesting representation of G and let $\pi^\infty \cong J(I_P^G(\tau^\infty, \nu))$ be written as its Langlands quotient. Let moreover ψ be a generic character of U and put $\psi_L := \psi|_{U\cap L}$. We may and will assume that ψ_L satisfies* (7.11)*. Then, the following are equivalent:*

(i) π is ψ-generic.
(ii) $\pi^\infty \cong I_P^G(\tau^\infty, \nu)$ and (τ, V_τ) is ψ_L-generic.

Proof. We will give a proof by references. If F is non-archimedean, this is Cor. 1.2 in [Hei-Opd13], which put a final and concluding milestone after a long series of deep, previous work by many authors, see above.

Let now F be archimedean. We sketch the argument by reducing the proof to results in the literature: In order to show that (ii) implies (i), suppose $\pi^\infty \cong I_P^G(\tau^\infty, \nu)$, as in (ii) of the statement of the theorem. Following Prop. 1.2 in [Bor-Ser73], we may write $L = L^{(1)} \times A$, where $L^{(1)} = \ker H_P$ and A is a connected Lie subgroup of the center Z_L. Therefore, $\tau \cong \tau^{(1)} \otimes \chi$, where $\tau^{(1)} = \tau|_{L^{(1)}}$ and $\chi = \omega_\tau|_A$, ω_τ denoting the central character of τ. Observe that connectivity of A implies that $\chi = e^{\langle \nu', H_P(\cdot)\rangle}$, for some $\nu' \in \check{\mathfrak{a}}_{P,\mathbb{C}}$. Now, recall that τ is assumed to be ψ_L-generic. Hence, by Prop. 7.6, there exists a non-trivial continuous linear functional $\lambda : V_\tau^\infty \to \mathbb{C}$, such that $\lambda(\tau(u_L)v) = \psi_L(u_L) \cdot \lambda(v)$ for all $u_L \in U_L$ and $v \in V_\tau^\infty$. However, one easily shows that $U_L \subset L^{(1)}$, whence by the equality of locally convex vector spaces $V_\tau^\infty = V_{\tau^{(1)}}^\infty$, this functional λ obviously is also a non-trivial continuous linear functional $V_{\tau^{(1)}}^\infty \to \mathbb{C}$, such that $\lambda(\tau^{(1)}(u_L)v) = \psi_L(u_L) \cdot \lambda(v)$ for all $u_L \in U_L$ and $v \in V_{\tau^{(1)}}^\infty$. Hence, putting $\sigma = \tau^{(1)}$ on p. 76 in [Sha10], Cor. 3.6.11 *ibidem* implies that there exists a non-trivial continuous linear functional

$$\lambda_\psi(\nu + \nu', \tau^{(1)}) : I_P^G(\tau^\infty, \nu) \to \mathbb{C}$$

satisfying $\lambda_\psi(\nu+\nu', \tau^{(1)})(I_P^G(\tau^\infty, \nu)(u)v) = \psi(u) \cdot \lambda_\psi(\nu+\nu', \tau^{(1)})(v)$ for all $u \in U$ and $v \in I_P^G(\tau^\infty, \nu)$. By assumption $\pi^\infty \cong I_P^G(\tau^\infty, \nu)$. So, there also exists a non-trivial continuous linear functional

$$\lambda_\pi : V^\infty \to \mathbb{C}$$

such that $\lambda_\pi(\pi(u)v) = \psi(u) \cdot \lambda_\pi(v)$ for all $v \in V^\infty$. Hence, π is ψ-generic by Prop. 7.6.

We now indicate how to see that (i) implies (ii). Suppose that π is ψ-generic. Since both π^∞ and $I_P^G(\tau^\infty, \nu)$ are Casselman–Wallach representations (by Cor. 4.40), it is enough to show that $\pi_{(K)}$ and $I_P^G(\tau^\infty, \nu)_{(K)}$

are isomorphic as $(\mathfrak{g}, K)$-modules, see Thm. 3.26. (After applying Weil's restriction of scalars from $\mathbb{C}$ to $\mathbb{R}$, cf. [Spr79], §3.3, if necessary) the reader may find the latter assertion as Thm. 2.1 in [Cas-Sha98], which the authors attribute to Vogan, [Vog78], and therefore omit a proof. However, in the very formulation of Thm. 2.1 in [Cas-Sha98], this result cannot be found in the quoted source [Vog78], so we at least sketch the idea of how to derive it from there: Firstly, one observes that the $(\mathfrak{g}, K)$-module $\pi_{(K)}$ is *large* in the sense of [Vog78], Def. 6.1, see Rem. 7.14 below. Hence, by the proof of "(a) $\Rightarrow$ (f)" in [Vog78], p. 93, there exists a standard parabolic F-subgroup $\mathbf{Q} = \mathbf{L_Q N_Q}$ in $\mathbf{G}$, a discrete series representation δ of L_Q and a $\lambda \in \check{\mathfrak{a}}_{\mathbf{Q}}$, $\lambda \geq 0$, such that $\pi_{(K)} \cong I_Q^G(\delta^\infty, \lambda)_{(K)}$, i.e., π is infinitesimally equivalent to the representation fully induced from its square-integrable Langlands datum. Eliminating those simple roots α, for which $\langle \lambda, \check{\alpha} \rangle = 0$ as in the proof of Thm. 5.7, one constructs a standard parabolic F-subgroup $\mathbf{P}' = \mathbf{L}' \cdot \mathbf{N}'$ of $\mathbf{G}$, satisfying $Q \subseteq P' \subseteq G$ and an isomorphism of G-representations

$$I_Q^G(\delta^\infty, \lambda) \cong I_{P'}^G(\tau'^\infty, \nu'), \tag{7.13}$$

with τ' a tempered representation of L' and $\nu' \in \check{\mathfrak{a}}_{\mathbf{P}'}$ strictly positive. See in particular (5.11). Observe that $n = 1$ in (5.11), as $\pi_{(K)} \cong I_Q^G(\delta^\infty, \lambda)_{(K)}$ is irreducible, cf. Thm. 3.16. Therefore, $(P', [\widehat{\tau'}], \nu')$ is a Langlands triple. As $I_{P'}^G(\tau'^\infty, \nu')$ is a Casselman–Wallach representation by Cor. 4.40, the isomorphism $\pi_{(K)} \cong I_{P'}^G(\tau'^\infty, \nu')_{(K)}$ extends to an isomorphism of irreducible G-representations $J(I_P^G(\tau^\infty, \nu)) \cong I_{P'}^G(\tau'^\infty, \nu')$ by Thm. 3.26, whence $P = P'$, $\tau \cong \tau'$ and $\nu = \nu'$ by the Langlands classification, cf. Thm. 5.2. In other words, $\pi^\infty \cong I_P^G(\tau^\infty, \nu)$. It hence only remains to argue why (τ, V_τ) is ψ_L-generic. This follows from our Prop. 7.6 in combination with [Sha85], Thm. 1.1.

□

Remark 7.14. For the sake of completeness, it would be desirable to provide a detailed proof of the statement that the underlying $(\mathfrak{g}, K)$-module of a generic representation is large. For complex semisimple Lie algebras $\mathfrak{g}$, the reader may want to consult Thm. 3.9 in [Kos78]. It seems to the author of this book, however, that a proof in the general case – although the statement has somehow been "in the air" ever since, see for instance the last paragraph on p. 98 in [Vog78] – has never been written down. Does the reader want to be the first to provide such a source?

7.2 Unramified representations

Until the end of this section, F is assumed to be non-archimedean and, unless otherwise stated, $\mathbf{G}$ is assumed to be quasi-split over F, cf. §7.1.

We fix a Borel subgroup $\mathbf{B}$ of $\mathbf{G}$, which is defined over F and a Levi decomposition $\mathbf{B} = \mathbf{T} \cdot \mathbf{U}$ as in the previous section. We may even choose the maximal torus $\mathbf{T}$ to be defined over F, cf. [Spr09], Thm. 13.3.6.(i). In passing we remind the reader that, if also the $\overline{F}$-isomorphism of algebraic groups $\mathbf{T} \xrightarrow{\sim} \mathbf{GL_1} \times \cdots \times \mathbf{GL_1}$ can be defined over F, i.e., be given by polynomials with coefficients in F, then $\mathbf{G}$ is called *F-split.* Moreover, we recall that a finite-degree extension F'/F of non-archimedean local fields is called *unramified* if $\mathfrak{p}_F \mathcal{O}_{F'} = \mathfrak{p}_{F'}$, or, equivalently, if $[\mathcal{O}_{F'}/\mathfrak{p}_{F'} : \mathcal{O}_F/\mathfrak{p}_F] = [F' : F]$. Here we have used the standard notations $\mathcal{O}_F$ for the valuation ring of F and $\mathfrak{p}_F$ for the unique maximal ideal of $\mathcal{O}_F$, and similarly for F'.

Definition 7.15. A quasi-split group $\mathbf{G}/F$ is *unramified*, if there exists an unramified extension F'/F for which $\mathbf{G}/F'$ is split over F'.

The following characterization of unramified local groups is widely known and its first mentioning seems to be in Tit's article [Tit79], there, however, in a slightly different language and without a complete proof. We provide a sketch of how to derive the result in its form presented here below. The reader, who is not firm in the theory of schemes may safely skip it and just take the result for granted.

Proposition 7.16. *A (connected reductive linear algebraic) group $\mathbf{G}/F$ is unramified if and only if $\mathbf{G}$ can be realized as a reductive, smooth $\mathcal{O}_F$-group scheme $\mathcal{G}$, i.e.,*

(i) There exists a smooth affine $\mathcal{O}_F$-group scheme $\mathcal{G}$, such that after extension of scalars from $\mathcal{O}_F$ to F, $\mathcal{G}/F \cong \mathbf{G}$ ("$\mathbf{G}$ has a smooth $\mathcal{O}_F$-model"), and

(ii) the special fibre of $\mathcal{G}$ over the residue field $k := \mathcal{O}_F/\mathfrak{p}_F$ is connected and reductive ("The $\mathcal{O}_F$-model is reductive").

Proof. Suppose first that there exists a smooth $\mathcal{O}_F$-group scheme $\mathcal{G}$ satsifying (i) and (ii) above. Then, since k is finite, the special fiber $\mathcal{G}_k$ of $\mathcal{G}$ over k admits a Borel subgroup $\mathcal{B}_k$ defined over k, cf. [Bor91], Prop. 16.6. Invoking [Dem-Gro70], Cor. XXII.5.8.3, the scheme of Borel subgroups of $\mathcal{G}_k$ is smooth, so $\mathcal{B}_k$ can be pulled back to an $\mathcal{O}_F$-Borel subgroup $\mathcal{B}$ of $\mathcal{G}$

and hence, applying extension of scalars, $\mathcal{B}/F$ defines a Borel subgroup of $\mathcal{G}/F$, which is defined over F. Applying the F-isomorphism of (i) to $\mathcal{B}/F$, we obtain a Borel subgroup $\mathbf{B}$ of $\mathbf{G}$, which is defined over F, hence $\mathbf{G}$ is quasisplit over F.

In order to see that $\mathbf{G}$ splits over a finite unramified extension F' of F, apply [Dem-Gro70], Cor. XXII.5.8.3 once more, but now to the scheme of maximal tori of $\mathcal{G}_k$, yielding a maximal $\mathcal{O}_F$-torus $\mathcal{T}$ of $\mathcal{G}$ by the same argument, which we have just used to obtain an $\mathcal{O}_F$-Borel subgroup $\mathcal{B}$ of $\mathcal{G}$. Completeness of $\mathcal{O}_F$ implies that after passing to a finite extension k' of k, over which $\mathcal{T}_k$ splits (cf. [Dem-Gro70], Cor. XXII.2.4), the $\mathcal{O}'$-torus $\mathcal{T}_{\mathcal{O}'}$ obtained from $\mathcal{T}$ after extension of scalars from $\mathcal{O}_F$ to the unramified extension $\mathcal{O}'$ corresponding to k'/k, splits over $\mathcal{O}'$, see [Con14], Thm. B.3.2.(2). Passing to the field of fractions F' of $\mathcal{O}'$, we finally obtain a finite unramified extension of F, such that $\mathcal{T}_{F'}$ is a maximal F'-split torus of $\mathcal{G}/F'$, i.e., recalling again the isomorphism provided in (i) above, $\mathbf{G}/F'$ is F'-split.

Now suppose that $\mathbf{G}/F$ is unramified and let F'/F be a finite unramified extension, such that $\mathbf{G}/F'$ is F'-split. Denote by $\mathcal{G}'_F$ the Chevalley group-scheme over F attached to $\mathbf{G}$, i.e., an F-split reductive affine algebraic F-group such that $\mathbf{G}$ and $\mathcal{G}'_F$ are F-forms of the same reductive $\overline{F}$-group $\mathbf{G}/\overline{F} \cong \mathcal{G}'_F/\overline{F}$, cf. [JMil17], Cor. 23.57. Any such F-form descends to a smooth affine $\mathcal{O}_F$-model $\mathcal{G}'$ in the sense of (i), by the following considerations: Firstly, we may assume, that $\mathcal{G}'_F$ is a subgroup of a general linear group GL_n/F, cf. [JMil17], Cor. 4.10, which provides a natural $\mathcal{O}_F$-structure $\mathcal{G}'$ of $\mathcal{G}'_F$ by taking the schematic closure of the image of $\mathcal{G}'_F$ in the $\mathcal{O}_F$-group $\mathrm{GL}_n/\mathcal{O}_F$. The $\mathcal{O}_F$-group scheme $\mathcal{G}'$ is flat with finite fibres. Applying [BLR90], Thm. 5 in §7.1, we may assume without loss of generality that $\mathcal{G}'$ is in fact smooth, hence a smooth $\mathcal{O}_F$-model in the sense of (i).

Having noted this, $\mathbf{G}$ corresponds to a unique class g in the pointed cohomology set $H^1(F'/F, \mathrm{Aut}_{\mathcal{G}'})$, see [Ser97], Cor. III, §1 (and recall that the extension F'/F is Galois by [Wei74], Cor. 2 & 3 to Thm. I.7). Here, $\mathrm{Aut}_{\mathcal{G}'}$ is a certain extension of $\mathrm{Out}_{\mathcal{G}'}$, i.e., of the constant $\mathcal{O}_F$-group scheme, which is given by our fixed choice of a basis of the geometric root datum of $\mathbf{G}$. In other words, $\mathrm{Aut}_{\mathcal{G}'}$ is given by a split exact sequence

$$\{0\} \to \mathcal{G}'/Z_{\mathcal{G}'} \to \mathrm{Aut}_{\mathcal{G}'} \to \mathrm{Out}_{\mathcal{G}'} \to \{0\}, \tag{7.17}$$

or, more precisely, after having chosen a pinning, cf. [JMil17], §23.d, on the level of $\mathcal{G}'$, $\mathrm{Aut}_{\mathcal{G}'}$ is a semi-direct product of the corresponding other terms in (7.17). Applying the functor $H^1(F'/F, \bullet)$ to this exact sequence, and recalling that the $\mathcal{O}_F$-group scheme $\mathrm{Out}_{\mathcal{G}'}$ is constant, we see that the image of g in $H^1(F'/F, \mathrm{Out}_{\mathcal{G}'})$ must be the image of a unique element $\overline{g}$ of $H^1(\mathcal{O}_{F'}/\mathcal{O}_F, \mathrm{Out}_{\mathcal{G}'})$. Applying our chosen section of (7.17) (which, as we recall, was defined over $\mathcal{O}_F$), we finally obtain an $\mathcal{O}_F$-form $\mathcal{G}$ of $\mathcal{G}'$: This is a smooth affine $\mathcal{O}_F$-group scheme, whose special fibre over the residue field k is connected and reductive. It hence remains to show that after extension of scalars from $\mathcal{O}_F$ to F, $\mathcal{G}/F \cong \mathbf{G}$. In order to see this, recall that by construction, the classes in $H^1(F, \mathrm{Aut}_{\mathcal{G}'})$ corresponding to $\mathcal{G}/F$ and $\mathbf{G}$ are mapped onto the same class in $H^1(F, \mathrm{Out}_{\mathcal{G}'})$. Consequently, $\mathcal{G}/F$ and $\mathbf{G}$ are inner forms of each other. However, since $\mathcal{G}/F$ and $\mathbf{G}$ are both quasi-split over F (the latter by assumption and the former by the argument from above, showing the other implication of the proposition), uniqueness of such forms, cf. [Con14], Prop. 7.2.12 or [JMil17], Thm. 23.51, implies that necessarily $\mathcal{G}/F \cong \mathbf{G}$ over F. Hence, $\mathcal{G}$ satisfies (i) and (ii) from above, i.e., it is a reductive $\mathcal{O}_F$-model of $\mathbf{G}$. □

If $\mathcal{G}$ is any $\mathcal{O}_F$-group scheme as in Prop. 7.16 above, then $\mathcal{G}(\mathcal{O}_F)$ is a maximal compact subgroup of $\mathcal{G}(F)$, and hence gives rise to a maximal compact subgroup of $G = \mathbf{G}(F)$, which we will denote again by $\mathcal{G}(\mathcal{O}_F)$: Indeed, it is easy to see that $\mathcal{G}(\mathcal{O}_F)$ is a compact subgroup of $\mathcal{G}(F)$. However, to verify that it is also *maximal* among such groups, is far from obvious in this generality and needs an argument: We refer to [MMac17], Thm. 1.1, for the most direct proof of this result, which we are aware of. (The classical arguments exploit profound knowledge of Bruhat-Tits theory, i.e., the strength of the results in [Bru-Tit72, Bru-Tit84] – which are formulated in the combinatorical language of buildings and so somehow different to our methodological approach here.)

Maximal compact subgroups of local groups over non-archimedean fields obtained in this way are called *hyperspecial maximal compact subgroups.*[1] Hence, Prop. 7.16 could be equivalently reformulated by saying that the linear algebraic F-group $\mathbf{G}$ underlying a local group G is unramified, if and only if G contains a hyperspecial maximal compact subgroup.

[1] A notion, which serves as an example for the Langlands Program's sense of humor at its most exuberant.

An important property of hyperspecial maximal compact subgroups is, that they are all conjugate to each other by an element of the F-points of the adjoint group of $\mathbf{G}$ (Exercise: Prove this fact by translating the corresponding result from [Tit79], §2.5). This is remarkable (and mimics the archimedean case) as general maximal compact subgroups of non-archimedean local groups are not necessarily conjugate to each other, cf. [Pla-Rap94], Prop. 3.14 for an example.

For an unramified group $\mathbf{G}$, we henceforth fix an $\mathcal{O}_F$-group scheme $\mathcal{G}$ as in Prop. 7.16, with the additional requirement that the (isomorphic image of the) corresponding hyperspecial maximal compact is in good position with respect to $\mathbf{P}_0 = \mathbf{B}$ and its fixed F-split central torus $\mathbf{A}_{\mathbf{P}_0}$ of $\mathbf{L}_0 = \mathbf{T}$, see the end of §4.3.1. This is always possible, which can be seen without using the full power of Bruhat–Tits theory as follows: Using rather basic/fundamental results from [Dem-Gro70], cf. Cor. XII.5.9.7, Cor. XXVI.5.2 and Prop. XXII.5.9.2, *ibidem*, one always finds $G = \mathcal{G}(\mathcal{O}_F) \cdot B = B \cdot \mathcal{G}(\mathcal{O}_F)$, as it is explained in [MMac17], Prop. 4.2.(3). Hence, (1) in our definition of being in good position holds. Moreover, $\mathcal{G}(\mathcal{O}_F) \cap T$ is then a maximal compact subgroup of T, from which one deduces that also (2) holds.

When $\mathbf{G}$ *is unramified, we shall assume from now on that* $K = \mathcal{G}(\mathcal{O}_F)$.[2]

Definition 7.18. Let $\mathbf{G}$ be unramified. An irreducible representation (π, V) of G is *unramified*, if the space of K-invariant vectors in V is nonzero, i.e. if $V^K = \{v \in V : \pi(k)v = v \text{ for all } k \in K\}$ contains a non-zero vector.

Remark 7.19. As the choice of $K = \mathcal{G}(\mathcal{O}_F)$ depends on the very choice of the $\mathcal{O}_F$-group scheme $\mathcal{G}$, so does the notion of (π, V) being unramified depend on K, i.e., it would be more precise to highlight the role of K in Def. 7.18 and say that an irreducible representation (π, V) of G satisfying its conditions is K-unramified. However, we decided to follow the common use of the notion of being unramified as it can be found in the literature, which suppresses the dependence on K.

Proposition 7.20. *Let* $\mathbf{G}$ *be unramified and let* (π, V) *be an unramified representation of* G. *Then:*

(i) (π, V) *is admissible (and hence, interesting).*
(ii) $\dim_{\mathbb{C}}(V^K) = 1$.

[2]Of course, this convention depends on the choice if the $\mathcal{O}_F$-group scheme $\mathcal{G}$ as well as on the isomorphism $\mathcal{G}/F \cong \mathbf{G}$.

Proof. Assertion (i) follows form the very general fact, that an irreducible representation having a non-zero smooth vector must be smooth (because otherwise V^∞ would be a proper subrepresentation) and hence admissible by Jacquet's theorem, cf. Thm. 3.1.

For (ii), observe that by the admissibility, implied by (i), $V^K = V(\mathbf{1}_K)$ must be (non-zero and) finite-dimensional. Moreover, by a general principle, see [Bus-Hen06], Prop. 4.3, for instance, V^K becomes an irreducible simple module of the convolution-algebra of smooth, compactly supported K-biinvariant $\mathbb{C}$-valued functions $C_c^\infty(K\backslash G/K)$ acting on V^K by convolution:

$$\pi(f)v := \int_G f(g) \cdot \pi(g)v \; dg \qquad \forall v \in V^K, f \in C_c^\infty(K\backslash G/K).$$

However, as K is hyperspecial, $C_c^\infty(K\backslash G/K)$ is commutative: This is the most difficult part of the proof, as it relies on the so-called Satake isomorphism, which describes the algebra $C_c^\infty(K\backslash G/K)$ explicitly: See [Car79], Thm. 4.1 for a precise statement and detailed sketch of a proof[3] establishing the Satake isomorphism in the generality used here, and Cor. 4.1 *ibidem* for the thereby implied commutativity of $C_c^\infty(K\backslash G/K)$. Therefore, V^K being now identified as a finite-dimensional simple module of a commutative $\mathbb{C}$-algebra, V^K must be one-dimensional by Schur's lemma, cf. [EGHLSVY11], Cor. 2.3.12, for instance. □

Exercise 7.21. Show Prop. 7.20.(ii), but replace the references to Schur's lemma and Jacquet's theorem, by using Hilbert's Nullstellensatz and a more detailed feature of the Satake isomorphism: The algebra $C_c^\infty(K\backslash G/K)$ is not only commutative but also finitely generated over $\mathbb{C}$ and isomorphic to a subalgebra of a polynomial algebra $\mathbb{C}[X_1^\pm, ..., X_m^\pm]$ in finitely many variables, cf. [Car79], Thm. 4.1 and Cor. 4.1, again. (*Hint*: Follow our line or argumentation from above to identify V^K as a (non-zero) simple $C_c^\infty(K\backslash G/K)$-module. This only needs irreducibility and smoothness of (π, V), not admissibility. The algebra $C_c^\infty(K\backslash G/K)$ being commutative implies that V^K must be isomorphic to a quotient of $C_c^\infty(K\backslash G/K)$ by a

[3]Cartier's survey article from 1979 is a meanwhile classical and very convenient source. For a complete proof (including a detailed argument for the somewhat technically laborious "triangularity relation", mentioned without proof in italic letters in [Car79], p. 148 below (27)) we refers to [Hai-Ros10], Thm. 1.0.1 in combination with Cor. 11.1.2 and Prop. 11.1.4 (and Lem. 10.2.1), *ibidem*, where the Satake isomorphism is established for all special maximal compact subgroups.

maximal ideal. Now use that $C_c^\infty(K\backslash G/K)$ is a finitely generated subalgebra of some $\mathbb{C}[X_1^\pm, ..., X_m^\pm]$ and apply Hilbert's Nullstellensatz.)

We conclude this section (and indeed our whole investigations of representations of local groups) with a result of how unramified representations fit into the framework of the square-integrable Langlands classification. To this end, observe that if $\mathbf{G}$ is unramified, the maximal torus $\mathbf{T}$ is unramified as well: Indeed, the F-torus $\mathbf{T}$ is obviously its own Borel subgroup, which is hence defined over F and so $\mathbf{T}$ is quasisplit over F. Moreover, it is in the same way obvious, that $\mathbf{T}$ is its own maximal torus, which, as $\mathbf{G}$ is assumed to be unramified, splits over a finite unramified extension F' of F, hence the connected reductive linear algebraic F-group $\mathbf{T}$ is indeed unramified.

Therefore, it makes sense to apply Def. 7.18 to irreducible representations of T, i.e., to talk about unramified characters $\chi : T \to \mathbb{C}^\times$. Let us denote $K_T := K \cap T$, which is a hyperspecial maximal compact subgroup of T. It is now a trivial observation that a character is K_T-unramified, if and only if $\chi|_{K_T} \equiv 1$, i.e., if χ is trivial on the hyperspecial subgroup $K_T = K \cap T$ of T.

The relevant result, which describes unramified local representations in terms of their square-integrable Langlands datum is due to Borel, Casselman, Matsumoto, Satake *et al.*, and reads as follows:

Theorem 7.22. *Let* $\mathbf{G}/F$ *be an unramified group. Let* (π, V) *be an irreducible admissible representation of* G *and let* $[(P, [\widehat{\delta}], \lambda)]$ *be its square-integrable Langlands datum, cf. Thm. 5.7. Then, the following are equivalent:*

(i) (π, V) *is* K*-unramified.*
(ii) $P = B$, $[\widehat{\delta}] = [\widehat{\chi}]$, *for a* K_T*-unramified unitary character* χ *of* T *and* π *is isomorphic to the unique* K*-unramified member* $J^K(I_P^G(\chi, \lambda))$ *of the square-integrable Langlands packet attached to* $I_P^G(\chi, \lambda)$.

Proof. Recalling that the functor of taking K-invariant vectors is exact, any induced representation $I_P^G(\chi, \lambda)$ as in (ii) has a unique irreducible quotient with a K-invariant vector. Therefore, (ii) is a well-stated assertion. As obviously, "(ii) $\Rightarrow$ (i)" is trivial, we only argue why (i) implies (ii).

To this end, observe that the natural inclusion of the (fixed) maximal F-split torus $\mathbf{A}_{\mathbf{P}_0}$ of the Levi-subgroup $\mathbf{T}$ of $\mathbf{B}$ (cf. §4.3.1) induces an isomorphism $A_{P_0}/(K \cap A_{P_0}) \xrightarrow{\sim} T/(K \cap T)$, whence every K_T-unramified character can be viewed as a $K \cap A_{P_0}$-invariant character of A_{P_0}. As $A_{P_0} \cong F^* \times ... \times F^*$, any such character may furthermore be viewed as the product of characters of the form $| \cdot |^{s_i}$, $s_i \in \mathbb{C}$, one for each copy of F^*, see Exc. 4.17 and [Wei74], Chp. VII, §3, Prop. 8. Hence, every K_T-unramified character χ of T admits a factorization $\chi = \chi^\circ \cdot e^{\langle \lambda, H_B(\cdot) \rangle}$, with χ° a unitary unramified character of T and $\lambda \in \check{\mathfrak{a}}_0$.

We now use [Mín11], Prop. 2.6 (which uses [Car79], Cor. 4.2: This is [Mín11], Prop. 2.5.) and obtain, that every isomorphism class of (K-)unramified representations (π, V) of G gives rise to a certain (K_T-)unramified character χ_π of T, which is unique up to conjugation by the Weyl group. In other words, using our above observation, writing $\chi_\pi = \chi_\pi^\circ \cdot e^{\langle \lambda_\pi, H_B(\cdot) \rangle}$, with χ_π° a unitary (K_T-)unramified character of T and $\lambda_\pi \in \check{\mathfrak{a}}_0$, each isomorphism class of (K-)unramified representations (π, V) of G gives rise to a unitary (K_T-)unramified character χ_π° of T, which is unique up to association, and a $\lambda_\pi \in \check{\mathfrak{a}}_0$, which is positive. Hence, $[(B, \widehat{[\chi_\pi^\circ]}, \lambda_\pi)]$ is a well-defined square-integrable Langlands triple, which is now uniquely determined by the isomorphism class of (π, V). By what we observed in the beginning of this proof, the attached square-integrable Langlands packet contains a unique (K-)unramified member $J^K(I_B^G(\chi_\pi^\circ, \lambda_\pi))$. Indeed, $\pi \cong J^K(I_B^G(\chi_\pi^\circ, \lambda_\pi))$, which is a direct consequence of [Mín11], Prop. 2.6 and (2.2), *ibidem*. This completes the proof. □

Part 2
Global Groups

We are now going to pass to the representation theory of global groups, that is, to the representation theory of the groups of adelic points of a connected reductive group over an algebraic number field. In a certain sense, this will mean to consider all local groups (and their representations) "at the same time", but – as already predicted and explained by Aristotle (cf. [Aristoteles, Metaphysics], Z 17, 1041b, 12–19) – the resulting whole will be greater than the sum of its parts. Our efforts throughout Part II will largely be focused on the attempt to set up a theory of "smooth-automorphic representations" of global groups and to compare it to the usual approach to automorphic forms.

Chapter 8

Basic Notions and Concepts from Functional Analysis ("Global")

8.1 Inductive limits

We shall now introduce and recall the notion of a *(strict) inductive limit* of seminormed spaces and its special case of an *LF-space*, a concept which shall be used quite extensively in the sequel. In our presentation below, we will freely use, what we recalled in §1.

8.1.1 *Inductive limits and strict inductive limits*

According to the author's opinion, one of the very best references on inductive limits, strict inductive limits and LF-spaces are still Grothendieck's "Brazilian lectures" from 1954, see [AGro73], in which the notion of an inductive limit of seminormed spaces is just synonymous with the final seminormed structure relative to the same data. Hence, in other, more precise words, a vector space[1] V is the *inductive limit* of an (arbitrarily large) family $(u_i, V_i)_{i\in I}$ of seminormed spaces V_i and linear maps $u_i : V_i \to V$, if and only if V carries the final structure with respect to the linear maps u_i, i.e., the finest locally convex topology with respect to which each linear map u_i is continuous, or, equivalently, the topology generated by the family of seminorms

$$\mathcal{P} := \{p \text{ is a seminorm on } V \mid \forall i \in I \; p\circ u_i \text{ is a continuous seminorm of } V_i\}. \tag{8.1}$$

In this generality, however, it is not true – although each seminormed space V_i is assumed to be Hausdorff by our standing conventions, cf. Def. 1.3 – that $\mathcal{P}$ is point-separating, i.e., that the resulting locally convex topology on V is Hausdorff (Exercise: Give an example for this phenomenon, based

[1] Recall that we assume all vector spaces to be defined over the complex numbers.

on the basic observation made after Lem. 1.11, or see Exc. 2 in [AGro73], Chp. 4, Part 1, §4). Hence, again according to our standing conventions, it is untrue that this construction makes V necessarily into a seminormed space. This is not too surprising, as the topology on V and the topologies on the V_i's do not have any more in common than the arbitrarily chosen linear maps u_i (whose images $u_i(V_i)$ could span a subspace, which is just much smaller than V; nor must these images have anything to do with each other.)

In order to overcome this problem, we shall from now on work with the following more refined notion of a *strict inductive limit*: Suppose that we are given a vector space V of the form $V = \bigcup_{n\in\mathbb{N}} V_n$ for an increasing sequence $(V_n)_{n\in\mathbb{N}}$ of seminormed spaces, i.e., each V_n is a seminormed space, which is a closed, topological vector subspace of V_{n+1} (not necessarily a proper one[2]). Put $u_n : V_n \to V$ to be the natural inclusion $\iota_n : V_n \hookrightarrow V$, $\iota_n(v) := v$. Then, viewing V as the inductive limit of the collection $(V_n, \iota_n)_{n\in\mathbb{N}}$, the resulting family of seminorms $\mathcal{P}$, as defined in (8.1), is indeed point-separating, hence V becomes a seminormed space, when given the inductive limit topology with respect to the just described collection $(V_n, \iota_n)_{n\in\mathbb{N}}$, see [AGro73], Chp. 4, Part 1, Prop. 3.

A seminormed space V obtained in this way shall henceforth be called the *strict inductive limit with defining sequence* $(V_n, \iota_n)_{n\in\mathbb{N}}$, denoted in symbols $V = \varinjlim V_n$. Likewise, the seminormed spaces V_n shall also be called the *steps* of the limit V. We obtain the following summarizing lemma.

Lemma 8.2. *Let V be the inductive limit of the family $(V_i, u_i)_{i\in I}$. A linear map $u : V \to W$ to a seminormed space W is continuous if and only if all composite maps $u \circ u_i$, $i \in I$, are continuous. If V is Hausdorff (e.g., if V is a strict inductive limit) and each seminormed space V_i is bornological (respectively, barreled), then so is V. If V is even the strict inductive limit of the family $(V_n, \iota_n)_{n\in\mathbb{N}}$, then each step V_n is closed in V, V induces on each V_n its original locally convex topology and the bounded subsets of V are those, which are contained and bounded in some step V_n. Moreover, the seminormed space V is complete if and only if each V_n is complete.*

Proof. The easy parts are left to the reader, cf. Lem. 1.10. For the remaining assertions, see [AGro73], Chp. 4, Part 1, Cor. 1 and Prop. 3. □

[2] I.e., although the limit is called strict, the inclusions $V_n \subseteq V_{n+1}$ are not required to be strict, but rather V_n (with its locally convex topology) to be a *closed* subspace of V_{n+1} (with its locally convex topology).

The following corollary is now a direct consequence of Lem. 8.2 and Thm. 1.8:

Corollary 8.3. *Let $V = \varinjlim V_n$ be a strict inductive limit. If all V_n are finite-dimensional, then V carries the finest locally convex topology.*

As another consequence of Lem. 8.2 we also record the following.

Corollary 8.4. *Let $V = \varinjlim V_n$ and $W = \varinjlim W_n$ be strict inductive limits, for which there is a strictly increasing sequence $(k_n)_{n\in\mathbb{N}}$ of natural numbers, such that $V_n \subseteq W_n \subseteq V_{k_n}$ as seminormed spaces for all $n \in \mathbb{N}$. Then, V is isomorphic to W as a seminormed space.*

Proof. Obviously, $V_n \subseteq W_n \subseteq V_{k_n}$ implies $V_n \subseteq W_n \subseteq V_{k_n} \subseteq W_{k_n}$, whence $V = W$ as vector spaces. So, by Lem. 8.2 it is enough to show that the identity map induces continuous maps $V_n \hookrightarrow W$ and $W_n \hookrightarrow V$ for all $n \in \mathbb{N}$. However, this is clear since the first map $V_n \hookrightarrow W$ is just the composition of the continuous inclusions $V_n \hookrightarrow W_n \hookrightarrow W$ and the latter map $W_n \hookrightarrow V$ is the composition of the continuous inclusions $W_n \hookrightarrow V_{k_n} \hookrightarrow V$ (all of which are given by the identity map). Hence, V is isomorphic to W as a seminormed space. □

8.1.2 *LF-spaces*

We define an *LF-space* to be a seminormed space V of the form $V = \varinjlim V_n$ for a sequence $(V_n)_{n\in\mathbb{N}}$ of Fréchet spaces (the abbreviation "LF" coming from "limit Fréchet"). As it follows from Lem. 8.2 and Lem. 1.10, an LF-space is complete, bornological, and barreled.

Remark 8.5. An LF-space is not Fréchet, unless in the trivial case, when the defining sequence $(V_n)_{n\in\mathbb{N}}$ becomes stationary, i.e., $V_n = V$ for some $n \in \mathbb{N}$: Indeed, strict inductive limits, which are not finally stationary, cannot have the Baire-property[3] and hence cannot be Fréchet, see, e.g., [Jar81], Sect. 5.1, Thm. 1 (together with our Rem. 1.6).

As an easy but illustrative example, consider the sequence $(V_n)_{n\in\mathbb{N}}$, where $V_n = \bigoplus_{i=1}^n \mathbb{C}$. The limit space is then the space $V = \bigoplus_{n\in\mathbb{N}} \mathbb{C}$, topologized as in §1.2.2, for which we just saw in Cor. 8.3 that it carries its finest locally convex topology, hence admits no countable subbasis of seminorms.

[3]This is because the closed steps V_n have empty interior, as otherwise they would be absorbing and hence equal to the whole space.

This example is paradigmatic: Indeed, following the short argument presented in [Alc85], Lem. 2, the reader may easily convince her-/himself that any LF-space, which is not by the above explained trivial reason Fréchet, must contain a copy of the space $\bigoplus_{n\in\mathbb{N}}\mathbb{C}$ as a direct summand.

Warning 8.6. It is important to notice that *a closed subspace $W \subseteq V$ of an LF-space V is not necessarily itself an LF-space.* In particular, it is not true in general that, if $W \subset V$ is closed and $(V_n)_{n\in\mathbb{N}}$ is a defining sequence for V, then $W \cong \varinjlim(W \cap V_n)$. See [AGro54], p. 89, for a counterexample (involving Köthe spaces) and the main result of [Kas-Rot70] for another counterexample (involving the theory of distributions).

Warning 8.7. Let $V = \varinjlim V_n$ and $W = \varinjlim W_n$ be LF-spaces such that for each n there is a continuous injection $V_n \hookrightarrow W_n$ mapping V_n onto a *closed* subspace of W_n. This induces a continuous inclusion $V \hookrightarrow W$ by definition of the inductive limit topology. However, *the image of this map is not necessarily closed in W.* A counterexample, relevant to our consideration of global representation theory, will be provided in Ex. 10.47.

8.2 Tensor products involving an infinite number of seminormed spaces

8.2.1 *The inductive tensor product*

In order to motivate this section, we start with the following.

Warning 8.8 ("LF-spaces and projective tensor products do not go well together"). Let V be a Fréchet space and $W = \varinjlim W_n$ be an LF-space. Even if each W_n is finite-dimensional (and even if V has the – here undefined, cf. [Tre70], Def. 50.1 – property of being nuclear), then, in general,

$$V \,\overline{\otimes}_{\mathsf{pr}}\, (\varinjlim W_n) \ncong \varinjlim(V \,\overline{\otimes}_{\mathsf{pr}}\, W_n).$$

Consider, as a counterexample, $V = \prod_{n\in\mathbb{N}}\mathbb{C}$ (with the product topology) and $W = \bigoplus_{n\in\mathbb{N}}\mathbb{C}$ with defining sequence $W_n = \bigoplus_{i=1}^{n}\mathbb{C}$: We leave it to the reader to show that V is indeed a Fréchet space (and also to look up [Tre70], Prop. 50.1, to convince her-/himself that V is indeed nuclear, if the reader is interested in this property as well). It follows now from a combination of [Tre70], Thm. 50.1, [Köt79], §44, 5, p. 283, and Lem. 1.15 that the resulting seminormed spaces $V \,\overline{\otimes}_{\mathsf{pr}}\, (\varinjlim W_n)$ and $\varinjlim(V \,\overline{\otimes}_{\mathsf{pr}}\, W_n)$ are not isomorphic.

Remark 8.9. The situation considered in Warn. 8.8 above is in fact typical, when working with smooth-automorphic representations and hence a quite crucial point for global aims. We refer to Thm. 12.16, where we will (have to) reconsider this issue.

Thus, the completed projective tensor product is not the right concept of a topologized tensor product, when dealing with inductive limits. For this reason, we will have to introduce yet another notion of tensor product – the *inductive tensor product* alluded to in the title of this section – which behaves well with respect to inductive limits. We refer to [AGro66], 1, §3, n° 1, Prop. 13 for the following definition being well-posed:

Definition 8.10. Let V and W be seminormed spaces. The *inductive tensor product* $V \otimes_{\mathsf{in}} W$ of V and W consists of the (algebraic) tensor product $V \otimes W$ equipped with the finest locally convex topology which makes the natural bilinear map $V \times W \to V \otimes W$, $(v, w) \mapsto v \otimes w$ separately continuous (i.e., continuous in each variable v and w). Its completion is denoted $V \overline{\otimes}_{\mathsf{in}} W$ and called the *completed inductive tensor product.*

We remark that the inductive tensor product (and hence its completion) are Hausdorff, as obviously for all seminormed spaces V and W the identity map $V \otimes_{\mathsf{in}} W \to V \otimes_{\mathsf{pr}} W$ is continuous (and $V \otimes_{\mathsf{pr}} W$ is Hausdorff, cf. §1.2.3). Hence $V \otimes_{\mathsf{in}} W$ (and $V \overline{\otimes}_{\mathsf{in}} W$) are seminormed spaces according to our conventions.

The next result, which is an application of the Banach-Steinhaus theorem, shows that, for Fréchet spaces, the notions of the (completed) inductive and the (completed) projective tensor product in fact coincide:

Proposition 8.11. *If V and W are Fréchet, then the identity map $V \otimes_{\mathsf{in}} W \to V \otimes_{\mathsf{pr}} W$ is an isomorphism of seminormed spaces. Therefore, their completions are isomorphic Fréchet spaces.*

Proof. By [Bou03], III, §5, Sect. 2, Cor. 1, the natural bilinear map $V \times W \to V \otimes_{\mathsf{in}} W$, $(v, w) \mapsto v \otimes w$ is not only separately but jointly continuous. Hence, the identity map $V \otimes_{\mathsf{in}} W \to V \otimes_{\mathsf{pr}} W$ is an isomorphism of seminormed spaces, cf. 1.2.3. Consequently, $V \overline{\otimes}_{\mathsf{in}} W$ and $V \overline{\otimes}_{\mathsf{pr}} W$ are isomorphic Fréchet spaces, see Lem. 1.15. □

Now let V_i, W_i, $i = 1, 2$ be seminormed spaces and let $f_i : V_i \to W_i$ be continuous linear maps. Then the natural map $V_1 \otimes V_2 \to W_1 \otimes W_2$, given by the linear extension of the assignment $v_1 \otimes v_2 \mapsto f_1(v_1) \otimes f_2(v_2)$ is continuous with respect to the inductive tensor product topologies and hence extends to a continuous linear map of the respective completions, see [AGro66], 1, §3, n° 1, p. 75. We will denote these maps by $f_1 \otimes_{\mathsf{in}} f_2$ and $f_1 \overline{\otimes}_{\mathsf{in}} f_2$, respectively.

The main advantage in considering inductive tensor products instead of projective ones is that they are compatible with inductive limits (which also justifies the terminology in retrospect). More precisely, we have the following theorem.

Theorem 8.12. *Let V and W be seminormed spaces, which are inductive limits of the families $(V_i, u_i)_{i\in I}$ and $(W_j, v_j)_{j\in J}$, respectively. Then $V \otimes_{\mathsf{in}} W$ is isomorphic to the seminormed space $V \otimes W$ given the inductive limit topology with respect to the family $(V_i \otimes_{\mathsf{in}} W_j, u_i \otimes_{\mathsf{in}} v_j)_{(i,j)\in I\times J}$. Moreover, on the subspace of $V \overline{\otimes}_{\mathsf{in}} W$, spanned by the images of the spaces $V_i \overline{\otimes}_{\mathsf{in}} W_j$, the subspace topology from $V \overline{\otimes}_{\mathsf{in}} W$ is the same as the seminormed topology given by the inductive limit of the family $(V_i \overline{\otimes}_{\mathsf{in}} W_j, u_i \overline{\otimes}_{\mathsf{in}} v_j)_{(i,j)\in I\times J}$.*

Proof. We refer to [AGro66], 1, §3, n° 1, Prop. 14.I, observing that by the choice of topologies on V and W, the *a priori* only linear maps u_i and v_j are in fact continuous, hence $u_i \otimes_{\mathsf{in}} v_j$ and $u_i \overline{\otimes}_{\mathsf{in}} v_j$ make sense. □

In view of Prop. 8.11 and Thm. 8.12 and the negative result mentioned in Warn. 8.8, the reader may want to ask, why we have not just introduced the completed inductive tensor product from the very beginning and forgot about the projective tensor product completely. Indeed, if all seminormed spaces in sight are Fréchet spaces, then we could have done so by means of Prop. 8.11. However, for more general seminormed spaces the completed inductive tensor product also has some severe disadvantages as compared to the projective one: For instance, in general, $\overline{\otimes}_{\mathsf{in}}$ is *not associative*. Indeed, one may even have

$$V \overline{\otimes}_{\mathsf{in}} (\mathbb{C} \overline{\otimes}_{\mathsf{in}} W) \not\cong (V \overline{\otimes}_{\mathsf{in}} \mathbb{C}) \overline{\otimes}_{\mathsf{in}} W$$

This is because separately continuous maps do not in general extend to the completions.

This finally led us to present both tensor products in all their advantages and disadvantages, making the projective tensor product more natural for local considerations (where sometimes even more general seminormed spaces than Fréchet spaces are to be considered) and, as we will see below, the inductive tensor product more natural for global representation theory.

8.2.2 *The restricted Hilbert space tensor product*

Global representation theory will require one more concept of how to topologize tensor products that involve an infinite number of seminormed spaces. This further concept is referred to as the *restricted Hilbert space tensor product*, which we shall now introduce.

Let $((H_n, \langle\cdot,\cdot\rangle_n))_{n\in\mathbb{N}}$ be a sequence of Hilbert spaces, such that for all but finitely many n, we are given a fixed choice of a unit vector $h_n^\circ \in H_n$. That is, we suppose to have fixed a finite subset $N_0 \subseteq \mathbb{N}$ and for each $n \in \mathbb{N}\setminus N_0$, an element $h_n^\circ \in H_n$ with $\|h_n^\circ\|_n = 1$. We observe that the set

$$\mathcal{N} := \{M \subseteq \mathbb{N} : N_0 \subseteq M \text{ and } M \text{ is finite}\}$$

is a directed set with respect to inclusion. If $M, M' \in \mathcal{N}$, such that $M \subseteq M'$, we define a map

$$\lambda_{M,M'} : \widehat{\bigotimes}_{n\in M} H_n \to \widehat{\bigotimes}_{n\in M'} H_n$$

by continuous linear extension of the assignment

$$\bigotimes_{n\in M} h_n \mapsto \bigotimes_{n\in M} h_n \otimes \bigotimes_{n\in M'\setminus M} h_n^\circ,$$

which is obviously an isometry of Hilbert spaces by the choice of the vectors h_n°. The set $\mathcal{N}$ together with the isometries $\lambda_{M,M'}$ hence forms a direct system of Hilbert spaces, and we consider its algebraic direct limit $\varinjlim_{M\in\mathcal{N}}\left(\widehat{\bigotimes}_{n\in M} H_n\right)$, i.e., the formal disjoint union $\coprod_{M\in\mathcal{N}} \widehat{\bigotimes}_{n\in M} H_n$ modulo the equivalence relation that two elements $h_M \in \widehat{\bigotimes}_{n\in M} H_n$ and $h'_{M'} \in \widehat{\bigotimes}_{n\in M'} H_n$ are to be identified, if and only if there is an $M'' \supseteq M \cup M'$ such that $\lambda_{M,M''}(h_M) = \lambda_{M',M''}(h'_{M'})$. It is clear that $\varinjlim_{M\in\mathcal{N}}\left(\widehat{\bigotimes}_{n\in M} H_n\right)$ carries in a natural way a non-degenerate Hermitian form, which is given on pure tensors by the (obviously well-defined) pairing

$$\langle \bigotimes_{n\in\mathbb{N}} h_n, \bigotimes_{n\in\mathbb{N}} h'_n \rangle := \prod_{n\in\mathbb{N}} \langle h_n, h'_n\rangle_n. \tag{8.13}$$

The Hilbert space closure of the algebraic direct limit $\varinjlim_{M\in\mathcal{N}}\left(\widehat{\bigotimes}_{n\in M} H_n\right)$ with respect to this form is denoted

$$\widehat{\bigotimes}'_{n\in\mathbb{N}} H_n$$

and called the *restricted Hilbert space tensor product of the* H_n*'s* (with respect to $(h_n^\circ)_{n\in\mathbb{N}\setminus N_0}$).

Remark 8.14. Although suppressed in notation, $\widehat{\bigotimes}'_{n\in\mathbb{N}} H_n$ actually depends on the choice of the vectors h_n°, $n \in \mathbb{N} \setminus N_0$. We leave it to the reader to convince her-/himself that different choices for a finite number of indices lead to equivalent, i.e., isomorphic, Hilbert spaces, whereas choices differing at an infinite number of indices (in general) lead to inequivalent Hilbert spaces. However, in all of our applications this issue will turn out to be irrelevant, see Rem. 10.39.

Chapter 9

First Adelic Steps

9.1 Global fields and global groups

From now on, F will denote a *global field* of characteristic 0, i.e., an algebraic number field. There are many excellent introductions to the theory of algebraic number fields, among which we especially refer the reader to [Swi01, Neu99, Wei74] (of which we shall assume a very basic knowledge in what follows).

We let $\mathcal{O}_F$ be the ring of integers of F and we let S be the set of (non-trivial) places of F. Hence, we may write $S = S_\infty \cup S_f$, where S_∞ and S_f denote the sets of archimedean and non-archimedean places of F, respectively. We also write "$\mathsf{v} \mid \infty$" in order to say that $\mathsf{v} \in S_\infty$ and analogously, "$\mathsf{v} < \infty$" in order to express that $\mathsf{v} \in S_f$. By definition, each $\mathsf{v} \in S$, corresponds to an equivalence class of (non-trivial, multiplicative) valuations of F, represented by the *normalized absolute value*, denoted $|\cdot|_\mathsf{v}$. More explicitly, this correspondence is given as follows:

- If $\mathsf{v} \in S_\infty$ is a real place, i.e., if v corresponds to an embedding $\imath_\mathsf{v} : F \hookrightarrow \mathbb{R}$, then let
$$|\cdot|_\mathsf{v} := |\imath_\mathsf{v}(\cdot)|,$$
where the absolute value on the right-hand side is the usual absolute value on $\mathbb{R}$.
- If $\mathsf{v} \in S_\infty$ is a complex place, i.e., if v corresponds to a pair $(\imath_\mathsf{v}, \overline{\imath_\mathsf{v}})$ of complex-conjugate embeddings $F \hookrightarrow \mathbb{C}$, then let
$$|\cdot|_\mathsf{v} := |\imath_\mathsf{v}(\cdot)|^2,$$
where the absolute value on the right-hand side is the usual absolute value on $\mathbb{C}$. Observe that $|\cdot|_\mathsf{v}$ is *not* a norm in this case as it does not satisfy the triangle inequality.

- If $\mathsf{v} \in S_f$ is a non-archimedean place, corresponding to a prime ideal $\mathfrak{p} \lhd \mathcal{O}_F$, then

$$|\cdot|_{\mathsf{v}} := [\mathcal{O}_F : \mathfrak{p}]^{-\nu_{\mathfrak{p}}(\cdot)},$$

  where $\nu_{\mathfrak{p}}(x)$ is the exact power of $\mathfrak{p}$ which divides the ideal $(x) := x\mathcal{O}_F$.

From now on, F_{v} shall stand for the topological completion of F with respect to $|\cdot|_{\mathsf{v}}$, whose continuous extension to F_{v} shall be denoted by the same symbol.

Remark 9.1. There is a uniform approach to the normalized absolute value on the locally compact topological fields F_{v}: Let $\mathrm{d}x_{\mathsf{v}}$ by any choice of a Haar measure on F_{v} and let $a_{\mathsf{v}} \in F_{\mathsf{v}}^{\times}$. Then, the map $x_{\mathsf{v}} \mapsto a_{\mathsf{v}}x_{\mathsf{v}}$ is an automorphism of the topological field F_{v}, so $\mathrm{d}(a_{\mathsf{v}}x_{\mathsf{v}})$ is a Haar measure on F_{v}, and hence there exists a unique positive real number $y_{\mathsf{v}} = y_{\mathsf{v}}(a_{\mathsf{v}})$, independent of the choice of $\mathrm{d}x_{\mathsf{v}}$, such that $\mathrm{d}(a_{\mathsf{v}}x_{\mathsf{v}}) = y_{\mathsf{v}}\,\mathrm{d}x_{\mathsf{v}}$. It turns out that $y_{\mathsf{v}} = |a_{\mathsf{v}}|_{\mathsf{v}}$.

We refer for instance to [Swi01], Sect. 2, §6 for all the above assertions.

The following observation is as obvious as crucial: *The field F_{v} is a local field as defined in §2, which is archimedean (respectively, non-archimedean) if and only if $\mathsf{v} \mid \infty$ (respectively, $\mathsf{v} < \infty$).*

If v is non-archimedean, we shall denote by $\mathcal{O}_{\mathsf{v}}$ the topological completion (i.e., closure) of $\mathcal{O}_F$ in F_{v}.

The *adèles* of F are the elements of

$$\mathbb{A} := \prod_{\mathsf{v} \in S}{}' F_{\mathsf{v}},$$

where $\prod'$ denotes the restricted product with respect to the $\mathcal{O}_{\mathsf{v}}$'s, i.e., $\mathbb{A}$ is the set of all $(a_{\mathsf{v}})_{\mathsf{v} \in S} \in \prod_{\mathsf{v} \in S} F_{\mathsf{v}}$ for which $a_{\mathsf{v}} \in \mathcal{O}_{\mathsf{v}}$ for *almost all*, – that means, for all but finitely many – $\mathsf{v} \in S$. We equip $\mathbb{A}$ with the topology given by the base of open sets of the form

$$\prod_{\mathsf{v} \in T} U_{\mathsf{v}} \times \prod_{\mathsf{v} \notin T} \mathcal{O}_{\mathsf{v}}$$

as T runs through the finite subsets of S containing S_∞ and $U_{\mathsf{v}} \subseteq F_{\mathsf{v}}$ runs through the open sets in F_{v}. One easily convinces oneself that this topological space becomes a locally compact, Hausdorff topological ring with the ring operations induced by the inclusion $\prod' F_v \hookrightarrow \prod F_v$. Having said this, it is now clear that its topology does not coincide with the subspace topology coming from the full direct product $\prod F_v$, as the latter is not locally compact.

One shows that the diagonal map

$$F \hookrightarrow \mathbb{A}$$
$$x \mapsto (x)_{v \in S}$$

is well-defined and is an embedding of F as a discrete subgroup of $\mathbb{A}$ (Exercise!). We understand that from now on F is identified with its image in $\mathbb{A}$. The quotient $F\backslash\mathbb{A}$ is compact and in particular has finite invariant measure. See, for instance, [Swi01], Sect. 2, §9 for the above.

For $a = (a_v)_{v \in S} \in \mathbb{A}$, we define

$$\|a\|_{\mathbb{A}} := \prod_v |a_v|_v \in \mathbb{R}_{\geq 0}$$

to be called the *adèlic norm.*[1] One easily sees that $\|x\|_{\mathbb{A}} = 1$ for all $x \in F^*$. Indeed, let $dx := \prod_v dx_v$ for some Haar measures dx_v on F_v, then dx is a Haar measure on $\mathbb{A}$. Now, for any $a \in \mathbb{A}^\times$, the map $\mathbb{A} \to \mathbb{A}$, $x \mapsto ax$ is an automorphism of the topological ring $\mathbb{A}$, so $d(ax)$ is again a Haar measure on $\mathbb{A}$ and $d(ax) = \|a\|_{\mathbb{A}} \cdot dx$ by Rem. 9.1. If in particular $a \in F^\times$, then this map induces a homeomorphism $F\backslash\mathbb{A} \xrightarrow{\sim} F\backslash\mathbb{A}$ of the compact space $F\backslash\mathbb{A}$ which preserves the quotient measure, and so $\|a\|_{\mathbb{A}} = 1$ for all $a \in F^\times$.

The "smaller brother" of $\mathbb{A}$ is the topological subring of the *"finite" adèles*,

$$\mathbb{A}_f := \prod_{v<\infty}' F_v,$$

which we may think of being embedded into $\mathbb{A}$ by adding zeros at all $v \in S_\infty$. Obviously, both $\mathbb{A}$ and $\mathbb{A}_f$ are commutative F-algebras.

For the rest of this book, we let $\mathbf{G}/F$ be a Zariski-connected reductive linear algebraic group over F, cf. §2.1, with F taking the role of a global field now.

[1] However, $\|\cdot\|_{\mathbb{A}}$ is not a norm in the sense of functional analysis, as $\|a\|_{\mathbb{A}} = 0$ does not imply $a = 0$.

Hence, the group $\mathbf{G}(\mathbb{A})$ is defined. As $\mathbf{G}$ is linear, i.e., is considered as a Zariski-closed subgroup of some general linear group $\mathbf{GL}_n/F$, $\mathbf{G}(\mathbb{A})$ is in fact a group of $n \times n$-matrices, whose entries are adèles. From this point of view, $\mathbf{G}(\mathbb{A})$ inherits a natural topology, by considering the embedding $\mathbf{G}(\mathbb{A}) \hookrightarrow \mathbb{A}^{n^2+1}$, $g \mapsto (g, \det(g)^{-1})$, which maps $\mathbf{G}(\mathbb{A})$ onto a closed set. Endowing $\mathbf{G}(\mathbb{A})$ with the pulled-back subspace topology makes it into a locally compact, Hausdorff topological group, which contains $\mathbf{G}(F)$ via the obvious diagonal embedding as a discrete subgroup. However, in practice one needs an alternative description of the topological group $\mathbf{G}(\mathbb{A})$, which is in many regards better suited in order to deal with questions of representation theory, and which we now recall.

Firstly, for every $\mathsf{v} \in S$, the extension of scalars $\mathbf{G}/F_\mathsf{v}$ from F to the completion F_v is a Zariski-connected (see [JMil17], Prop. 1.34) reductive group over the local field F_v, i.e., $G_\mathsf{v} := \mathbf{G}(F_\mathsf{v})$ is a local group in the sense of §2.1. It now follows from [Spr79], Lem. 4.9 and Prop. 7.16 that we have

Proposition 9.2. *$\mathbf{G}/F_\mathsf{v}$ is unramified for almost all v.*

Consequently, there is a minimal finite subset $T_0 \subset S$, such that for all $\mathsf{v} \notin T_0$, there exists a smooth reductive $\mathcal{O}_\mathsf{v}$-model $\mathcal{G}_\mathsf{v}$ of $\mathbf{G}/F_\mathsf{v}$ in the sense of Prop. 7.16. We choose and fix such a model henceforth, keeping in mind that there are different choices for them.

Definition 9.3. We set

$$G(\mathbb{A}) := \prod_{\mathsf{v}\in S}{}' G_\mathsf{v},$$

where the restricted product is taken with respect to the (fixed choice of the) hyperspecial maximal compact subgroups $\mathcal{G}_\mathsf{v}(\mathcal{O}_\mathsf{v})$ of G_v.

Then, $G(\mathbb{A})$ becomes a locally compact, Hausdorff topological group with group operation given by the inclusion $\prod' G_\mathsf{v} \hookrightarrow \prod G_\mathsf{v}$ and a base of topology given by the products

$$\prod_{\mathsf{v}\in T} U_\mathsf{v} \times \prod_{\mathsf{v}\notin T} \mathcal{G}_\mathsf{v}(\mathcal{O}_\mathsf{v}) \tag{9.4}$$

as T runs over the finite subsets of S containing T_0 and U_v runs over the open sets in G_v. One gets:

Lemma 9.5. *There is an isomorphism $\mathbf{G}(\mathbb{A}) \xrightarrow{\sim} G(\mathbb{A})$ of topological groups.*

Remark 9.6. This isomorphism amounts to a certain change of point of view: Whereas the elements of $G(\mathbb{A})$ are $|S|$-tuples of matrices, each matrix having entries in a particular completion F_{v}, by contrast, the group $\mathbf{G}(\mathbb{A})$ of adèlic points of $\mathbf{G}$ is a group of matrices whose entries are $|S|$-tuples.

Analogously, we may set

$$G(\mathbb{A}_f) := \prod_{\mathsf{v}<\infty}' G_{\mathsf{v}},$$

$$G_\infty := \prod_{\mathsf{v}|\infty} G_{\mathsf{v}}$$

and embed them into $G(\mathbb{A})$ by putting $id_{\mathsf{v}} \in G_{\mathsf{v}}$ at the missing places. Doing so, $G(\mathbb{A}_f)$ and G_∞ become locally compact, Hausdorff topological groups with the subspace topology: It turns out that $G(\mathbb{A}_f)$ is totally disconnected, whereas G_∞ is a real Lie group with finitely many connected components, which satisfies what we had mentioned in Rem. 2.1. Indeed, G_∞ is isomorphic to a local archimedean group as it is isomorphic to the group of $\mathbb{R}$-points of the linear algebraic $\mathbb{R}$-group $\mathrm{Res}_{F/\mathbb{Q}}(\mathbf{G})/\mathbb{R}$, obtained from extension of scalars from $\mathbb{Q}$ to $\mathbb{R}$ from Weyl's restriction of scalars $\mathrm{Res}_{F/\mathbb{Q}}(\mathbf{G})$ from F to $\mathbb{Q}$ of $\mathbf{G}$: $G_\infty \cong \mathrm{Res}_{F/\mathbb{Q}}(\mathbf{G})(\mathbb{R})$. (Exercise: Proof these claims on $G(\mathbb{A}_f)$ and G_∞.) Of course, we could have equivalently started from the natural locally compact, Hausdorff topology on the local groups G_{v} and topologized G_∞ and $G(\mathbb{A}_f)$ according to their product structures relative to the local groups G_{v} appearing, and we would have obtained the same topology as above. By construction of $G(\mathbb{A})$, there is equality $G(\mathbb{A}) = G_\infty \times G(\mathbb{A}_f)$.

Obviously, there are two more types of natural embeddings, the first being

$$G_{\mathsf{v}} \hookrightarrow G(\mathbb{A}),$$

which is given by forming an $|S|$-tuple by adding the local identity matrix at all places different from v and the pushed-forward diagonal embedding

$$\mathbf{G}(F) \hookrightarrow \mathbf{G}(\mathbb{A}) \xrightarrow{\sim} G(\mathbb{A}), \tag{9.7}$$

which are both embeddings of topological groups. We will write $G(F)$ to denote the image of $\mathbf{G}(F)$ in $G(\mathbb{A})$. A property built into (9.7) is that the v-component of the image of each $g \in \mathbf{G}(F)$ in $G(\mathbb{A})$ lies in $\mathcal{G}_{\mathsf{v}}(\mathcal{O}_{\mathsf{v}})$ for all $\mathsf{v} \notin T_0$. In general, $G(F)$ is a discrete subgroup of $G(\mathbb{A})$, which follows directly from F being a discrete subgroup of $\mathbb{A}$ along the diagonal embedding, but it does not have finite covolume in general (although $F\backslash\mathbb{A}$ was even compact). We will return to the latter point in the next section.

We may also define the adelic group norm $\|\cdot\| = \|\cdot\|_{G(\mathbb{A})} : G(\mathbb{A}) \to \mathbb{R}_{>0}$,

$$\|g\| := \prod_{\mathsf{v}\in S} \max_{1\leq i,j\leq n} \{|(g_\mathsf{v})_{ij}|_\mathsf{v}, |(g_\mathsf{v}^{-1})_{ij}|_\mathsf{v}\}.$$

We note that $\|g\| = \|g_\infty\| \cdot \|g_f\|$ for all $g = (g_\infty, g_f) \in G(\mathbb{A})$ and that there exist $c_0, C_0 \in \mathbb{R}_{>0}$ such that

$$\|g\| \geq c_0 \qquad \text{and} \qquad \|gh\| \leq C_0\|g\|\|h\| \tag{9.8}$$

for all $g, h \in G(\mathbb{A})$ Also, obviously, $\|g_\infty\| = \|g_\infty\|_\infty$, where the latter group norm is the one defined in Def. 3.22 for the local archimedean group G_∞.

Exercise 9.9. Prove (9.8). *Hint*: First show the second inequality and then use that $\|g\| = \|g^{-1}\|$.

9.2 Parabolic subgroups and attached data

We fix a minimal parabolic F-subgroup $\mathbf{P}_0$ of $\mathbf{G}$ and a Levi decomposition $\mathbf{P}_0 = \mathbf{L}_0 \cdot \mathbf{N}_0$. Furthermore, let $\mathbf{A}_{\mathbf{P}_0} \subseteq \mathbf{L}_0$ denote a maximal F-split torus in the center of $\mathbf{L}_0$. The *standard* parabolic F-subgroups $\mathbf{P}$ of $\mathbf{G}$ are those, which contain $\mathbf{P}_0$. We may and will also supposed that we are given a Levi decomposition $\mathbf{P} = \mathbf{L} \cdot \mathbf{N}$, where $\mathbf{L} \supseteq \mathbf{L}_0$, $\mathbf{N} \subseteq \mathbf{N}_0$ and $\mathbf{A}_\mathbf{P} \subseteq \mathbf{A}_{\mathbf{P}_0}$. Of course, for every $\mathsf{v} \in S$, $\mathbf{P}/F_\mathsf{v}$ is then a parabolic F_v-subgroup of the connected reductive linear algebraic group $\mathbf{G}/F_\mathsf{v}$ and we may hence assume that a minimal parabolic F_v-subgroup of $\mathbf{G}/F_\mathsf{v}$ has been fixed making $\mathbf{P}/F_\mathsf{v}$ into a standard parabolic F_v-subgroup.

If $\mathbf{H} \in \{\mathbf{P}, \mathbf{L}, \mathbf{N}\}$, we shall denote by $H(\mathbb{A})$ the image of $\mathbf{H}(\mathbb{A})$ in $G(\mathbb{A})$, as provided by Lem. 9.5, and use the notations H_v, $H(\mathbb{A})$, $H(\mathbb{A}_f)$ and H_∞ accordingly. Of course, by our conventions for maximal compact subgroups of local groups, fixed at the end of §4.3.1 and in the course of §7.2, the locally compact group $L(\mathbb{A})$ (defined by reference to $G(\mathbb{A})$) is identical to what we would have obtained, if we had applied the procedures of §9.1 directly to the connected reductive linear algebraic group $\mathbf{L}/F$.

We now go about fixing maximal compact subgroups of $G(\mathbb{A})$, $G(\mathbb{A}_f)$ and G_∞. First, we fix some (any) maximal compact subgroup K_∞ of G_∞: By definition of $G_\infty = \prod G_\mathsf{v}$, it must hold that $K_\infty = \prod K_\mathsf{v}$ for some maximal compact subgroup $K_\mathsf{v} \subseteq G_\mathsf{v}$, for each $\mathsf{v} \in S_\infty$.

Analogously, we fix once and for all a maximal compact subgroup $K_{\mathbb{A}_f} \subset G(\mathbb{A}_f)$. Again, $K_{\mathbb{A}_f} = \prod_{\mathsf{v}<\infty} K_\mathsf{v}$ for maximal compact subgroups $K_\mathsf{v} \subset G_\mathsf{v}$ (which are, as we recall, automatically open as well, cf. §2.1, or directly [Pey87], Cor. 1 to Thm. 2). For $\mathsf{v} \notin T_0$ (i.e., when $\mathbf{G}/F_\mathsf{v}$ is unramified) we abbreviate $K_\mathsf{v} = \mathcal{G}_\mathsf{v}(\mathcal{O}_\mathsf{v})$, i.e., K_v equals our fixed choice of a hyperspecial maximal compact subgroup of G_v, used in order to define $G(\mathbb{A})$, whereas if $\mathsf{v} \in T_0 \cap S_f$, we take K_v to be a special maximal compact subgroup of G_v, as at the end of §4.3.1, which is in good position with respect to the fixed minimal parabolic F_v-subgroup of $\mathbf{G}/F_\mathsf{v}$, which makes $\mathbf{P}_0/F_\mathsf{v}$ standard parabolic.

Finally we set $K_\mathbb{A} = K_\infty \times K_{\mathbb{A}_f}$. By our choices made, $K_\mathbb{A}$ is a maximal compact subgroup of $G(\mathbb{A})$, which is in good position with respect to $\mathbf{P}_0$ and $\mathbf{A}_{\mathbf{P}_0}$, i.e., for all standard parabolic F-subgroups $\mathbf{P}$ of $\mathbf{G}$, it holds that

(1) $G(\mathbb{A}) = P(\mathbb{A}) \cdot K_\mathbb{A} = K_\mathbb{A} \cdot P(\mathbb{A})$.

(2) $P(\mathbb{A}) \cap K_\mathbb{A} = (L(\mathbb{A}) \cap K_\mathbb{A})(N(\mathbb{A}) \cap K_\mathbb{A})$ and $(L(\mathbb{A}) \cap K_\mathbb{A})$ is a maximal compact subgroup of $L(\mathbb{A})$, which satisfies the same two, previous conditions with respect to the standard parabolic F-subgroups of $\mathbf{L}$, obtained by intersection.

The careful reader will have observed the subtlety that in §4.3.1 we had fixed a minimal parabolic subgroup first and only then chose a hyperspecial maximal compact subgroup, which was in good position with respect to it, whereas now, we are actually already given a hyperspecial maximal compact subgroup $K_\mathsf{v} = \mathcal{G}_\mathsf{v}(\mathcal{O}_\mathsf{v})$ and *then* we took the freedom to arbitrarily fix a minimal parabolic F-subgroup $\mathbf{P}_0$ of $\mathbf{G}$. This is consistently possible, as any two minimal parabolic F-subgroups of $\mathbf{G}$ are conjugate by an element $g \in \mathbf{G}(F)$, cf. [Bor91], Thm. 20.9, which maps into $\mathcal{G}_\mathsf{v}(\mathcal{O}_\mathsf{v})$ for all $\mathsf{v} \notin T_0$ by means of Lem. 9.5, no matter which choice of reductive $\mathcal{O}_\mathsf{v}$-model $\mathcal{G}_\mathsf{v}$ we had fixed before.

Having fixed $K_\mathbb{A}$, we also fix once and for all a cofinal sequence $(K_n)_{n\in\mathbb{N}}$ of compact open subgroups of $K_n = \prod_{\mathsf{v}<\infty} K_{n,\mathsf{v}}$ of $K_{\mathbb{A}_f}$, satisfying $K_n \supset K_{n+1}$ and $\bigcap_n K_n = \{id_f\}$, which forms a neighbourhood base of $id_f \in G(\mathbb{A}_f)$. Moreover, having fixed $K_\mathbb{A}$, gives us a way to fix a Haar measure on $G(\mathbb{A})$, $G(\mathbb{A}_f)$ and G_∞ in a consistent way: For each $\mathsf{v} \in S$, let

dg_v be the unique Haar measure on G_v satisfying

$$\int_{K_v} dg_v = 1.$$

As Haar measure on G_∞ (respectively, $G(\mathbb{A}_f)$) we simply take $dg_\infty := \prod_{v|\infty} dg_v$ (respectively, $dg_f := \prod_{v<\infty} dg_v$) and hence putting $dg := dg_\infty \cdot dg_f$ yields a Haar measure on $G(\mathbb{A})$. Similarly, we will denote by $dk_\infty := dg_\infty|_{K_\infty}$ (respectively, $dk_f := dg_f|_{K_f}$) the respective restricted Haar measure and just note here that the analogous notation will be used without further mention also for other subgroups of $G(\mathbb{A})$.

The final task for this section is to construct an analogue of the Harish-Chandra height function for global fields. To this end, let $\mathbf{P} = \mathbf{L} \cdot \mathbf{N}$ be a standard parabolic F-subgroup of $\mathbf{G}$ and let $\mathfrak{a}_\mathbf{P} := \mathrm{Hom}(\mathbf{X}_F(\mathbf{L}), \mathbb{R})$, the group-homomorphisms from the group of F-rational characters $\mathbf{X}_F(\mathbf{L})$ of $\mathbf{L}$ into the additive group of real numbers. Then

$$H_P : L(\mathbb{A}) \to \mathfrak{a}_\mathbf{P}$$

is defined to be the unique group-homomorphism satisfying

$$\|\chi(\ell)\|_\mathbb{A} = e^{H_P(\ell)(\chi)}$$

for all $\chi \in \mathbf{X}_F(\mathbf{L})$ and $\ell \in L(\mathbb{A})$. Here, we composed the isomorphism of topological groups $L(\mathbb{A}) \xrightarrow{\sim} \mathbf{L}(\mathbb{A})$, provided by Lem. 9.5, with the canonical functorial extension of the F-rational character $\chi : \mathbf{L} \to \mathbf{GL}_1$ to a character $\chi : \mathbf{L}(\mathbb{A}) \to \mathbb{A}^\times$. As $K_\mathbb{A}$ is in good position, we can extend H_P to a function on $G(\mathbb{A})$: In suggestive notation, if $g \in G(\mathbb{A})$, $g = \ell \cdot n \cdot k$, set $H_P(g) := H_P(\ell)$.

The height function allows one to "repair" the aforementioned problem of $G(F)$ not having finite covolume in $G(\mathbb{A})$ in general. Indeed, specialize the above definition to $\mathbf{P} = \mathbf{G}$ (and hence $\mathbf{L} = \mathbf{G}$), in order to obtain a character $H_G : G(\mathbb{A}) \to \mathfrak{a}_\mathbf{G}$, and set

$$G(\mathbb{A})^{(1)} := \ker H_G.$$

Example 9.10. In the special case $\mathbf{G} = \mathbf{GL}_n/F$ one has $\mathbf{X}_F(\mathbf{G}) = \{\det^k \mid k \in \mathbb{Z}\}$ (no matter what was our global ground-field F) and hence $G(\mathbb{A})^{(1)} = \{g \in \mathrm{GL}_n(\mathbb{A}) : \|\det g\|_\mathbb{A} = 1\}$.

We have the following

Proposition 9.11. *(1) There exists a connected Lie subgroup $A_G^{\mathbb{R}}$, isomorphic to $\mathbb{R}_{>0}^{\dim_F \mathbf{A_G}}$, of the center of G_∞ such that multiplication within $G(\mathbb{A})$ induces an isomorphism of topological groups*

$$G(\mathbb{A}) \cong A_G^{\mathbb{R}} \times G(\mathbb{A})^{(1)}.$$

(2) The group $G(F)$ is a closed, discrete subgroup of the unimodular group $G(\mathbb{A})^{(1)}$ and the homeomorphic spaces

$$G(F)\backslash G(\mathbb{A})^{(1)} \cong A_G^{\mathbb{R}} G(F)\backslash G(\mathbb{A})$$

have finite volume induced by dg.

Proof. Leaving (1) as an exercise for the reader, we note that $G(F) \subset G(\mathbb{A})^{(1)}$ as $\|F\|_{\mathbb{A}} = 1$. It follows from [Oes84], §I.5.5 that $G(F)$ is closed in $G(\mathbb{A})^{(1)}$ and from §I.3.5 *ibidem*, that it is in fact discrete. Finiteness of its covolume in the unimodular group $G(\mathbb{A})^{(1)}$ is a meanwhile classical theorem of Borel, see [Bor63], Thm. 5.8. □

Warning 9.12. Despite what is sometimes stated in the literature, $A_G^{\mathbb{R}}$ is *not* always isomorphic to $\mathrm{Res}_{F/\mathbb{Q}}(\mathbf{A_G})(\mathbb{R})^\circ$. For a simple counterexample, consider $\mathbf{G} = \mathbf{GL}_1/F$, where F is a real quadratic extension of $\mathbb{Q}$. Then $\mathbf{G} = \mathbf{A_G}$, hence $A_G^{\mathbb{R}} \cong \mathbb{R}_{>0}^{\dim_F \mathbf{A_G}} = \mathbb{R}_{>0}$. However,

$$\mathrm{Res}_{F/\mathbb{Q}}(\mathbf{A_G})(\mathbb{R})^\circ = \mathrm{Res}_{F/\mathbb{Q}}(\mathbf{G})(\mathbb{R})^\circ \cong (\mathrm{GL}_1(\mathbb{R}) \times \mathrm{GL}_1(\mathbb{R}))^\circ = (\mathbb{R}_{>0})^2.$$

In the subsequent chapters, we will use the notation $[G] := A_G^{\mathbb{R}} G(F)\backslash G(\mathbb{A})$.

Chapter 10

Representations of Global Groups – The Very Basics

10.1 Global representations

We start off with the definition of a representation of $G(\mathbb{A})$.

Definition 10.1. (1) A pair (π, V) is called a *representation* of $G(\mathbb{A})$, if V is a complete seminormed space and $\pi : G(\mathbb{A}) \to \mathrm{Aut}_{\mathbb{C}}(V)$ is a homomorphism of groups, such that the map

$$\begin{aligned} G(\mathbb{A}) \times V &\to V \\ (g, v) &\mapsto \pi(g)v \end{aligned}$$

is continuous.

(2) A representation (π, V) of $G(\mathbb{A})$ is *irreducible*, if $V \neq \{0\}$ and V does not admit any proper, closed $G(\mathbb{A})$-invariant subspace. It is called *finitely generated*, if there are finitely many $v_1, ..., v_n \in V$ such that $V = \mathrm{Cl}_V(\langle \pi(G(\mathbb{A}))v_1, ..., \pi(G(\mathbb{A}))v_n \rangle)$, where we remind the reader that "Cl_V" denotes the topological closure in V and $\langle X \rangle$ the $\mathbb{C}$-linear span of X.

(3) Two representations of $G(\mathbb{A})$ are *isomorphic*, or *equivalent*, if there exists a bicontinuous, $\mathbb{C}$-linear, $G(\mathbb{A})$-equivariant bijection between them.

Of course, given a representation (π, V) of $G(\mathbb{A})$, we may restrict the action of $G(\mathbb{A})$ to its (commuting) subgroups G_∞ and $G(\mathbb{A}_f)$. Let us denote the corresponding group homomorphism $\pi|_{G_\infty}$ and $\pi|_{G(\mathbb{A}_f)}$, respectively. If V is Fréchet, then $\pi|_{G_\infty}$ is a representation of the local archimedean group $G_\infty \cong \mathrm{Res}_{F/\mathbb{Q}}(\mathbf{G})(\mathbb{R})$ according to our definitions, cf. Def. 2.2, and hence a remark is in order, why we did not make our life easy and just supposed that a $G(\mathbb{A})$-representation is to be defined on a Fréchet-space.

It will turn out that our global needs demand a certain compromise between G_∞ and $G(\mathbb{A}_f)$, whose product constitutes $G(\mathbb{A})$, in the definitions of our global notions: Indeed, a representation (π, V) of $G(\mathbb{A})$ also giving rise to an action of $G(\mathbb{A}_f)$, forces us to allow G_∞ to act on more general complete seminormed spaces V, than Fréchet spaces (such as LF-spaces, see Rem. 10.22), i.e., here we have to be more generous "in favour of $G(\mathbb{A}_f)$", whereas – vice versa – a representation (π, V) of $G(\mathbb{A})$ also giving rise to an action of G_∞, forces us to introduce a topological obstruction on the pair $(\pi|_{G(\mathbb{A}_f)}, V)$ (namely that V is a complete seminormed space and that $G(\mathbb{A}_f) \times V \to V$, $(g_f, v) \mapsto \pi|_{G(\mathbb{A}_f)}(g_f)v$ is continuous), i.e., here we have to be more restrictive "in favour of G_∞".

This need for a reconciliation between the two factors G_∞ and $G(\mathbb{A}_f)$ in $G(\mathbb{A})$ comes at a certain price, though: Whereas a closed $G(\mathbb{A})$-stable subspace U of a $G(\mathbb{A})$-representation V – henceforth to be called a $G(\mathbb{A})$-*subrepresentation of* V – is quite obviously a $G(\mathbb{A})$-representation by restriction (since closed subspaces of complete seminormed spaces are complete), it is not necessarily the case any more that the quotient V/U together with the obvious linear action of $G(\mathbb{A})$, given by $G(\mathbb{A}) \times V/U \to V/U$, $(g, v + U) \mapsto \pi(g)v + U$ is a representation. Indeed, the reader shall be warned that it may fail that this map is jointly continuous as well as that the quotient V/U is complete (and invited the provide her-/himself a counterexample using, for instance, [Kha82], p. 108). However, if for some reason V/U is complete and barreled, then V/U becomes a $G(\mathbb{A})$-representation by the above linear action, as it follows from [BouII04], VIII, §2, Prop. 1.

Nevertheless, there is an important class of representations, which are in perfect analogy with our local observations, cf. §3.4:

Definition 10.2. A representation (π, V) of $G(\mathbb{A})$ is called *unitary*, if V is a Hilbert space, whose Hermitian form $\langle \cdot, \cdot \rangle$ is invariant under the action of $G(\mathbb{A})$, i.e., for all $g \in G(\mathbb{A})$ and $v, w \in V$, we have $\langle \pi(g)v, \pi(g)w \rangle = \langle v, w \rangle$.

In passing we note that – analogous to the case of local groups, cf. 2.4 – the concept of a (likewise: irreducible or finitely generated) representation and its attached notions of isomorphy and of unitarity translate verbatim from $G(\mathbb{A})$ to its closed subgroup $K_\mathbb{A}$, simply by replacing these groups with one another in the respective definitions.

10.2 Smooth representations

Definition 10.3. Let V be a seminormed space. A function $\varphi : G(\mathbb{A}) \to V$ will be called *smooth* if

(1) for each $g_f \in G(\mathbb{A}_f)$, the function $\varphi(\_\_ \cdot g_f)$ is smooth as a function on the smooth manifold G_∞,
(2) φ is uniformly locally constant, i.e., there exists an open compact subgroup $K_f \subseteq G(\mathbb{A}_f)$ such that

$$\varphi(gk_f) = \varphi(g) \qquad \text{for all } g \in G(\mathbb{A}), k_f \in K_f.$$

Obviously, recalling our cofinal sequence of open compact subgroups $(K_n)_{n\in\mathbb{N}}$, fixed in course of §9.2, the second condition in the definition of smoothness is equivalent to assuming that φ is right K_n-invariant for some $n \in \mathbb{N}$. We observe that each K_n having finite index in $K_{\mathbb{A}_f}$, a smooth function φ is necessarily *right* $K_{\mathbb{A}_f}$*-finite*, i.e.,

$$\dim_{\mathbb{C}}\langle \varphi(\_\_ \cdot k_f) : k_f \in K_{\mathbb{A}_f}\rangle < \infty.$$

The vector-space of smooth functions $\varphi : G(\mathbb{A}) \to V$ will be denoted $C^\infty(G(\mathbb{A}), V)$. If $V = \mathbb{C}$, then we use the simpler notation $C^\infty(G(\mathbb{A})) := C^\infty(G(\mathbb{A}), \mathbb{C})$. We record the following easy lemma, whose point, however does not seem to be completely clear in the literature:

Lemma 10.4. *Every* $\varphi \in C^\infty(G(\mathbb{A}))$ *is continuous.*

Proof. We need to show that given a $g = (g_\infty, g_f) \in G_\infty \times G(\mathbb{A}_f) = G(\mathbb{A})$ and an $\varepsilon > 0$, there exists an open neighbourhood U of $id \in G(\mathbb{A})$, such that $|\varphi(gu) - \varphi(g)| < \varepsilon$ for all $u \in U$. By assumption, $\varphi|_{G_\infty \times g_f} := \varphi(\_\_ \cdot g_f)$ is a smooth, hence continuous function $G_\infty \to \mathbb{C}$. In particular, it is continuous at our given g_∞. Hence, there exists an open neighbourhood U_∞ of id_∞ in G_∞, such that $|\varphi|_{G_\infty \times g_f}(g_\infty u_\infty) - \varphi|_{G_\infty \times g_f}(g_\infty)| < \varepsilon$ for all $u_\infty \in U_\infty$. Inserting into the definitions, we obtain $|\varphi(gu_\infty) - \varphi(g)| < \varepsilon$ for all $u_\infty \in U_\infty$. Now, let U_f be a compact open subgroup of $G(\mathbb{A}_f)$ such that $\varphi(hu_f) = \varphi(h)$ for all $h \in G(\mathbb{A}), u_f \in U_f$. But then $U := U_\infty \times U_f$ is an open neighbourhood of id in $G(\mathbb{A})$, such that $|\varphi(gu) - \varphi(g)| = |\varphi(gu_\infty) - \varphi(g)| < \varepsilon$ for all $u = (u_\infty, u_f) \in U$. This shows the claim. □

Definition 10.5. Let (π, V) be a representation of $G(\mathbb{A})$. A vector $v \in V$ is called *smooth* if its orbit map

$$c_v : G(\mathbb{A}) \to V,$$
$$g \mapsto \pi(g)v$$

is smooth. The set of all smooth vectors in V is a linear subspace of V and will be denoted $V^{\infty_{\mathbb{A}}}$.

Lemma 10.6. *The subspace $V^{\infty_{\mathbb{A}}}$ is stable under the action of $G(\mathbb{A})$, i.e., $\pi(g)v \in V^{\infty_{\mathbb{A}}}$ for all $g \in G(\mathbb{A})$, $v \in V^{\infty_{\mathbb{A}}}$.*

Proof. Let $g = (g_\infty, g_f) \in G(\mathbb{A})$, $v \in V^{\infty_{\mathbb{A}}}$ be arbitrary. For $h_f \in G(\mathbb{A}_f)$, consider the function $c_{\pi(g)v}(\text{—} \cdot h_f) : G_\infty \to V$: Commutativity of G_∞ and $G(\mathbb{A}_f)$ in $G(\mathbb{A})$ implies that

$$c_{\pi(g)v}(\text{—} \cdot h_f) = c_v(\text{—} \cdot h_f g_f) \circ R_{g_\infty},$$

where R_{g_∞} denotes right-translation $G_\infty \xrightarrow{\sim} G_\infty$, $h_\infty \mapsto h_\infty g_\infty$, whence $c_{\pi(g)v}(\text{—} \cdot h_f)$ is smooth as it is the composition of two smooth functions. Now, let K_f be an open compact subgroup of $G(\mathbb{A}_f)$ under which c_v is right-invariant. Then, $gK_fg^{-1} = g_fK_fg_f^{-1}$ is an open compact subgroup of $G(\mathbb{A}_f)$, which leaves $c_{\pi(g)v}$ invariant on the right. Hence, $c_{\pi(g)v}$ is smooth, which shows the claim. □

Having identified the space of globally smooth vectors $V^{\infty_{\mathbb{A}}}$ in a $G(\mathbb{A})$-representation (π, V) as a $G(\mathbb{A})$-stable subspace, we would like to make it into a $G(\mathbb{A})$-representation, i.e., we have to specify a complete, locally convex topology on $V^{\infty_{\mathbb{A}}}$, which makes the natural map $G(\mathbb{A}) \times V^{\infty_{\mathbb{A}}} \to V^{\infty_{\mathbb{A}}}$, $(g, v) \mapsto \pi(g)v$ continuous. As one should already expect from the local archimedean case, simply taking the subspace topology from V does not provide a solution to this problem, as the latter makes $V^{\infty_{\mathbb{A}}}$ into a dense (and hence usually incomplete) subspace of V, see [Mui-Žun20], Cor. 2(3).

In order to overcome this problem, we shall denote by $c_v|_{G_\infty}$ (respectively, $c_v|_{G(\mathbb{A}_f)}$) the restriction of c_v to G_∞ (respectively, $G(\mathbb{A}_f)$) and by $V^{\infty_{\mathbb{R}}}$ (respectively, V^{∞_f}) the subspace of all vectors in V, for which $c_v|_{G_\infty}$ (respectively, $c_v|_{G(\mathbb{A}_f)}$) is smooth. For the totally disconnected locally compact group $G(\mathbb{A}_f)$, smoothness of $c_v|_{G(\mathbb{A}_f)}$ just amounts to the usual notion, already invoked in the case of non-archimedean local groups, namely being locally constant.

We remark that if $v \in V^{\infty_{\mathbb{R}}}$, then c_v automatically satisfies condition (1) in Def. 10.3, i.e., $c_v(\text{—} \cdot g_f) : G_\infty \to V$ is smooth for all $g_f \in G(\mathbb{A}_f)$; whereas, if $v \in V^{\infty_f}$, then c_v automatically satisfies condition (2) in Def. 10.3.

Exercise 10.7. Show these two claims. (*Hint*: The second assertion being trivial, recall for your proof of the first claim that for each given $g_f \in G(\mathbb{A}_f)$, the continuous linear operator $\pi(g_f) : V \xrightarrow{\sim} V$ is bounded and therefore maps smooth curves to smooth curves, see [Kri-Mic97], Cor. 2.11. Hence, so does the composition $\pi(g_f) \circ c_v|_{G_\infty}$, which equals $c_v(\_\_ \cdot g_f)$ as G_∞ commutes with $G(\mathbb{A}_f)$ in $G(\mathbb{A})$. However, by [Kri-Mic97], Cor. 3.14, this means that $c_v(\_\_ \cdot g_f)$ is smooth.)

As a consequence, we obtain $V^{\infty_{\mathbb{A}}} = V^{\infty_{\mathbb{R}}} \cap V^{\infty_f}$, or, in other words and recalling our fixed cofinal sequence of open compact subgroups $(K_n)_{n\in\mathbb{N}}$ from §9.2,

$$V^{\infty_{\mathbb{A}}} = \bigcup_{n\in\mathbb{N}} (V^{\infty_{\mathbb{R}}})^{K_n},$$

where, as usual, an upper index "K_n" denotes the subspace of K_n-fixed vectors.

Of course, $V^{\infty_{\mathbb{A}}}$ being dense in V, also the "sandwiched" space $V^{\infty_{\mathbb{R}}}$ would be dense in V, when given the subspace topology. However, generalizing our observations from §2.2, we observe that it is still true that the assignment $V^{\infty_{\mathbb{R}}} \to C^\infty(G_\infty, V)$, $v \mapsto c_v|_{G_\infty}$, maps $V^{\infty_{\mathbb{R}}}$ onto a closed subspace of the complete seminormed space $C^\infty(G_\infty, V)$, see [Warl72], p. 253, so pulling back the subspace-topology, inherited from $C^\infty(G_\infty, V)$, makes $V^{\infty_{\mathbb{R}}}$ into a complete seminormed space. Again, an alternative description of the so-obtained locally convex topology on $V^{\infty_{\mathbb{R}}}$ is provided by considering the seminorms $p_{\alpha,D}(v) := p_\alpha(\pi(D)v)$, where p_α runs through the continuous seminorms on V and D through the elements of the universal enveloping algebra $\mathcal{U}(\mathfrak{g}_\infty)$. See [Cas89I], Lem. 1.2 for a proof.

A short moment of thought shows that same procedure may be applied to the subspace V^{K_n}: Indeed, by the continuity of the linear operators $\pi(k)$, $k \in K_n$, the space V^{K_n} being the intersection of the kernels of the continuous operators $\pi(k) - id$, $k \in K_n$, is a closed and hence complete subspace of V. Obviously, it is also G_∞-stable, as the actions of G_∞ and $G(\mathbb{A}_f)$ commute. It therefore makes sense to consider the complete seminormed space $\left(V^{K_n}\right)^{\infty_{\mathbb{R}}}$, which again allows an alternative description of its continuous seminorms as the $p_{\alpha,D}$, where p_α now runs through the continuous seminorms on V^{K_n}. However, as the latter are nothing else than the restrictions of the continuous seminorms on V, we immediately get

$$(V^{\infty_{\mathbb{R}}})^{K_n} = \left(V^{K_n}\right)^{\infty_{\mathbb{R}}} \tag{10.8}$$

as seminormed spaces. This implies that $(V^{\infty_{\mathbb{R}}})^{K_n}$ must be closed in $V^{\infty_{\mathbb{R}}}$ (and hence, obviously also in $(V^{\infty_{\mathbb{R}}})^{K_{n+1}}$), and so is a complete seminormed space. We may hence equip $V^{\infty_{\mathbb{A}}}$ with the strict inductive limit topology

$$V^{\infty_{\mathbb{A}}} = \varinjlim (V^{\infty_{\mathbb{R}}})^{K_n},$$

cf. §8.1.1. Following our Lem. 8.2, in this way $V^{\infty_{\mathbb{A}}}$ becomes a complete seminormed space.

Exercise 10.9. This one is for the purists among the readers, who think that – as V^{K_n} is not a representation of $G(\mathbb{A})$ – the symbol $\left(V^{K_n}\right)^{\infty_{\mathbb{R}}}$ strictly speaking is an abuse of notation, whence should rather be avoided: Give a direct proof of the fact that $(V^{\infty_{\mathbb{R}}})^{K_n}$ is closed in $V^{\infty_{\mathbb{R}}}$, which obviates introducing the space $\left(V^{K_n}\right)^{\infty_{\mathbb{R}}}$. (*Hint*: In fact you will see that you will have to use the same ingredients as we did above – at least if you follow the same ideas which came to the author's mind – whence this exercise basically amounts to reassembling our above argument. You will need: The description of the locally convex topology on $V^{\infty_{\mathbb{R}}}$ by the seminorms $p_{\alpha,D}$; the fact that the action of the $\pi(k)$ and the $\pi(D)$ commute; completeness of V; continuity of the operators $\pi(k)$; and the fact that $V^{\infty_{\mathbb{R}}}$ is Hausdorff.)

We may hence restrict our given action of $G(\mathbb{A})$ to the stable, complete seminormed space $V^{\infty_{\mathbb{A}}}$. Let $(\pi^{\infty_{\mathbb{A}}}, V^{\infty_{\mathbb{A}}})$ denote this restricted action.

Proposition 10.10. *The pair* $(\pi^{\infty_{\mathbb{A}}}, V^{\infty_{\mathbb{A}}})$ *is a representation of* $G(\mathbb{A})$.

Proof. After all of our preparatory work, we are left to show that $G(\mathbb{A}) \times V^{\infty_{\mathbb{A}}} \to V^{\infty_{\mathbb{A}}}$, $(g, v) \mapsto \pi^{\infty_{\mathbb{A}}}(g)v$ is continuous. We proceed in various steps, following closely the proof of Prop. 2(1) in [Mui-Žun20]. In our first two steps, we will show that this map is separately continuous.

Step 1: Let $g = (g_\infty, g_f) \in G(\mathbb{A})$ be arbitrary. We have $\pi^{\infty_{\mathbb{A}}}(g) = \pi^{\infty_{\mathbb{A}}}(g_f) \circ \pi^{\infty_{\mathbb{A}}}(g_\infty)$, whence, in order to see that $\pi^{\infty_{\mathbb{A}}}(g) : V^{\infty_{\mathbb{A}}} \to V^{\infty_{\mathbb{A}}}$ is continuous, it suffices to prove continuity of $\pi^{\infty_{\mathbb{A}}}(g_f)$ and $\pi^{\infty_{\mathbb{A}}}(g_\infty)$. For the continuity of the linear map $\pi^{\infty_{\mathbb{A}}}(g_\infty)$, recall from [Warl72], p. 253, that $G_\infty \times V^{\infty_{\mathbb{R}}} \to V^{\infty_{\mathbb{R}}}$, $(g_\infty, v) \mapsto \pi^{\infty_{\mathbb{A}}}(g_\infty)v$ is continuous. Hence, so is $\pi^{\infty_{\mathbb{A}}}(g_\infty) : (V^{\infty_{\mathbb{R}}})^{K_n} \to (V^{\infty_{\mathbb{R}}})^{K_n}$ for each $n \in \mathbb{N}$. Invoking Lem. 8.2, also $\pi^{\infty_{\mathbb{A}}}(g_\infty) : V^{\infty_{\mathbb{A}}} \to V^{\infty_{\mathbb{A}}}$ is continuous for every $g_\infty \in G_\infty$.

For the continuity of the linear map $\pi^{\infty_{\mathbb{A}}}(g_f)$, we again use Lem. 8.2 in order to reduce the problem to showing that $\pi^{\infty_{\mathbb{A}}}(g_f) : (V^{\infty_{\mathbb{R}}})^{K_n} \to V^{\infty_{\mathbb{A}}}$ is continuous for all $n \in \mathbb{N}$. However, the image $\pi^{\infty_{\mathbb{A}}}(g_f)\Big((V^{\infty_{\mathbb{R}}})^{K_n}\Big)$ is obviously contained in $(V^{\infty_{\mathbb{R}}})^{K_m}$, for every $m \in \mathbb{N}$, such that $K_m \subseteq g_f K_n g_f^{-1}$

(such m's exist because the cofinal sequence $(K_n)_{n\in\mathbb{N}}$ is a basis of neighborhoods of the identity in $G(\mathbb{A}_f)$), and each such space $(V^{\infty_\mathbb{R}})^{K_m}$ inherits from $V^{\infty_\mathbb{A}}$ its original topology as a subspace of $V^{\infty_\mathbb{R}}$, cf. Lem. 8.2 once more. Therefore, to show continuity of $\pi^{\infty_\mathbb{A}}(g_f) : (V^{\infty_\mathbb{R}})^{K_n} \to V^{\infty_\mathbb{A}}$ it suffices to show that $\pi^{\infty_\mathbb{A}}(g_f) : (V^{\infty_\mathbb{R}})^{K_n} \to V^{\infty_\mathbb{R}}$ is continuous. To this end, we need to show that for each continuous seminorm p on $V^{\infty_\mathbb{R}}$, there exists a continuous seminorm q on $(V^{\infty_\mathbb{R}})^{K_n}$, such that

$$p(\pi^{\infty_\mathbb{A}}(g_f)v) \leq q(v) \tag{10.11}$$

for all $v \in (V^{\infty_\mathbb{R}})^{K_n}$. However, knowing that the continuous seminorms on $V^{\infty_\mathbb{R}}$ (and hence also on $(V^{\infty_\mathbb{R}})^{K_n}$) are the $p_{\alpha,D}$ and knowing that $\pi^{\infty_\mathbb{A}}(g_f)$ commutes with $\pi(D)$ for each $D \in \mathcal{U}(\mathfrak{g}_\infty)$, (10.11) holds by the continuity of $\pi(g_f) : V \to V$. Therefore, $\pi^{\infty_\mathbb{A}}(g_f) : V^{\infty_\mathbb{A}} \to V^{\infty_\mathbb{A}}$ is continuous for every $g_f \in G(\mathbb{A}_f)$.

In summary, we have shown that $\pi^{\infty_\mathbb{A}}(g) : V^{\infty_\mathbb{A}} \to V^{\infty_\mathbb{A}}$ is continuous for all $g \in G(\mathbb{A})$.

Step 2: Now, let $v \in V^{\infty_\mathbb{A}}$ be arbitrary. We want to show that $c_v : G(\mathbb{A}) \to V^{\infty_\mathbb{A}}$, $g \mapsto \pi^{\infty_\mathbb{A}}(g)v$ is continuous. To this end, fix $n \in \mathbb{N}$ such that $v \in (V^{\infty_\mathbb{R}})^{K_n}$. As $(V^{\infty_\mathbb{R}})^{K_n}$ inherits from $V^{\infty_\mathbb{A}}$ its original topology as a closed subspace of $V^{\infty_\mathbb{R}}$, see Lem. 8.2, it suffices to prove that the restricted orbit map $c_v : G_\infty \times K_n \to V^{\infty_\mathbb{R}}$ is continuous at $id \in G(\mathbb{A})$. By our description of the continuous seminorms on $V^{\infty_\mathbb{R}}$ as the seminorms $p_{\alpha,D}$, this hence amounts to proving that for every continuous seminorm p_α on V and $D \in \mathcal{U}(\mathfrak{g}_\infty)$,

$$p_\alpha(\pi(D)(\pi^{\infty_\mathbb{A}}(g)v - v)) \to 0 \quad \text{as} \quad g \to id \in G(\mathbb{A}). \tag{10.12}$$

Being given a p_α and a D as above, note that, since the vector space, spanned by $\mathrm{Ad}(G_\infty)D$ is finite-dimensional, there exists an $m \in \mathbb{Z}_{>0}$, linearly independent $D_1, \ldots, D_m \in \mathcal{U}(\mathfrak{g}_\infty)$, and functions $c_1, \ldots, c_m \in C^\infty(G_\infty)$ such that

$$\mathrm{Ad}(g_\infty^{-1})D = \sum_{i=1}^{m} c_i(g_\infty)D_i, \tag{10.13}$$

holds for all $g_\infty \in G_\infty$. In particular, at the identity $id_\infty \in G_\infty$, we obtain

$$\sum_{i=1}^{m} c_i(id_\infty)D_i = \mathrm{Ad}(id_\infty)D = D. \tag{10.14}$$

Hence, for all $g=(g_\infty,g_f)\in G_\infty\times K_n$, we have

$$\begin{aligned}p_\alpha(\pi(D)(\pi^{\infty_{\mathbb{A}}}(g)v-v)) &= p_\alpha(\pi(D)(\pi^{\infty_{\mathbb{A}}}(g_\infty)v-v))\\ &= p_\alpha\big(\pi^{\infty_{\mathbb{A}}}(g_\infty)\pi\big(\mathrm{Ad}(g_\infty^{-1})D\big)v-\pi(D)v\big)\\ &= p_\alpha\left(\sum_{i=1}^m c_i(g_\infty)\pi^{\infty_{\mathbb{A}}}(g_\infty)\pi(D_i)v-c_i(id_\infty)\pi(D_i)v\right),\end{aligned}$$

where the last line follows from combining the linearity of the operators $\pi^{\infty_{\mathbb{A}}}(g_\infty)$ with (10.13) and (10.14). However, as each summand $c_i(g_\infty)\pi^{\infty_{\mathbb{A}}}(g_\infty)\pi(D_i)v-c_i(id_\infty)\pi(D_i)v$ tends to $0\in V$ as $g\to id\in G(\mathbb{A})$, this shows (10.12). Therefore, for every $v\in V^{\infty_{\mathbb{A}}}$, $c_v: G(\mathbb{A})\to V^{\infty_{\mathbb{A}}}$, $g\mapsto\pi^{\infty_{\mathbb{A}}}(g)v$ is continuous.

Consequently, $G(\mathbb{A})\times V^{\infty_{\mathbb{A}}}\to V^{\infty_{\mathbb{A}}}$, $(g,v)\mapsto\pi^{\infty_{\mathbb{A}}}(g)v$ is separately continuous.

Interlude: Therefore, whenever V was a Fréchet space, we are done by [BouII04], VIII, §2, Prop. 1. Indeed, if V is Fréchet, then – following the explanations of §2.3 – so is $V^{\infty_{\mathbb{R}}}$, and hence so are all the (closed) subspaces $(V^{\infty_{\mathbb{R}}})^{K_n}$. Consequently, they are all barreled by our Lem. 1.10. It then follows from our Lem. 8.2, that $V^{\infty_{\mathbb{A}}}$ is barreled, whence one may apply [BouII04], VIII, §2, Prop. 1 to get joint continuity of $G(\mathbb{A})\times V^{\infty_{\mathbb{A}}}\to V^{\infty_{\mathbb{A}}}$, $(g,v)\mapsto\pi^{\infty_{\mathbb{A}}}(g)v$ from its just verified separate continuity.

Step 3: In order to obtain the general case, we still need to prove that $G(\mathbb{A})\times V^{\infty_{\mathbb{A}}}\to V^{\infty_{\mathbb{A}}}$, $(g,v)\mapsto\pi^{\infty_{\mathbb{A}}}(g)v$ is continuous at $(id,0)$. We observe that without loss of generality, we may suppose that our fixed cofinal sequence $(K_n)_{n\in\mathbb{N}}$ of open compact subgroups in $G(\mathbb{A}_f)$ satisfies

$$K_{n+1}\subseteq\bigcap_{k\in K_{\mathbb{A}_f}/K_n} kK_nk^{-1},\qquad n\in\mathbb{N}.$$

Doing so, this implies that for every $g\in G_\infty\times K_{\mathbb{A}_f}$,

$$\pi(g)(V^{\infty_{\mathbb{R}}})^{K_n}=(V^{\infty_{\mathbb{R}}})^{g_fK_ng_f^{-1}}\subseteq(V^{\infty_{\mathbb{R}}})^{K_{n+1}}. \tag{10.15}$$

By the definition of the strict inductive limit topology on $V^{\infty_{\mathbb{A}}}$, a basis of open neighborhoods $\mathscr{U}$ of 0 in $V^{\infty_{\mathbb{A}}}$ is given by the family of open sets

$$U_{(\sigma_n)_{n\in\mathbb{N}},(\varepsilon_n)_{n\in\mathbb{N}}}:=\mathrm{AConv}\left(\bigcup_{n\in\mathbb{N}}B_{\sigma_n}(\varepsilon_n)\right),$$

where $(\sigma_n)_{n\in\mathbb{N}}$ is a sequence of continuous seminorms on $V^{\infty_\mathbb{R}}$, $(\varepsilon_n)_{n\in\mathbb{N}}$ is a sequence of strictly positive real numbers,

$$B_{\sigma_n}(\varepsilon_n) := \left\{v \in (V^{\infty_\mathbb{R}})^{K_n} : \sigma_n(v) < \varepsilon_n\right\}, \qquad n \in \mathbb{N}. \tag{10.16}$$

and "AConv" denotes taking the absolute convex hull. See [AGro73], Chp. 4, Part 1, Prop. 1. An even smaller basis of neighborhoods $\mathscr{U}_0 \subseteq \mathscr{U}$ of 0 in $V^{\infty_\mathbb{A}}$ may be obtained by requiring that the seminorms σ_n are of the form

$$\sigma_n(v) = \sum_{i=1}^{r_n} p_{n,i}(\pi(D_{n,i})v), \qquad v \in V^{\infty_\mathbb{R}},$$

where $r_n \in \mathbb{Z}_{>0}$, $p_{n,1}, \dots, p_{n,r_n}$ are continuous seminorms on V, and $D_{n,1}, \dots, D_{n,r_n} \in \mathcal{U}(\mathfrak{g}_\infty)$.

Now, let us fix a compact neighbourhood C of id_∞ in G_∞ and a $U = U_{(\sigma_n),(\varepsilon_n)} \in \mathscr{U}_0$. To prove continuity of $G(\mathbb{A}) \times V^{\infty_\mathbb{A}} \to V^{\infty_\mathbb{A}}$, $(g, v) \mapsto \pi^{\infty_\mathbb{A}}(g)v$ at $(id, 0)$, it suffices to find a set $U' = U_{(\sigma'_n),(\varepsilon'_n)} \in \mathscr{U}$ such that for all $g \in C \times K_{\mathbb{A}_f}$, the following implication holds:

$$v \in U' \quad \Rightarrow \quad \pi^{\infty_\mathbb{A}}(g)v \in U,$$

i.e.,

$$v \in \mathrm{AConv}\left(\bigcup_{n\in\mathbb{N}} B_{\sigma'_n}(\varepsilon'_n)\right) \quad \Rightarrow \quad \pi^{\infty_\mathbb{A}}(g)v \in \mathrm{AConv}\left(\bigcup_{n\in\mathbb{N}} B_{\sigma_n}(\varepsilon_n)\right).$$

To prove this claim, it is enough to find, given an open ball $B_{\sigma_{n+1}}(\varepsilon_{n+1})$ as above, a continuous seminorm σ'_n on $(V^{\infty_\mathbb{R}})^{K_n}$ and $\varepsilon'_n \in \mathbb{R}_{>0}$ such that for all $g \in C \times K_{\mathbb{A}_f}$,

$$v \in B_{\sigma'_n}(\varepsilon'_n) \quad \Rightarrow \quad \pi^{\infty_\mathbb{A}}(g)v \in B_{\sigma_{n+1}}(\varepsilon_{n+1}),$$

i.e., by (10.16) and (10.15), such that for all $g \in C \times K_{\mathbb{A}_f}$ and $v \in (V^{\infty_\mathbb{R}})^{K_n}$,

$$\sigma'_n(v) < \varepsilon'_n \quad \Rightarrow \quad \sigma_{n+1}(\pi^{\infty_\mathbb{A}}(g)v) < \varepsilon_{n+1}. \tag{10.17}$$

By our description of the continuous seminorms on $V^{\infty_\mathbb{R}}$ we may assume that σ_{n+1} is of the form

$$\sigma_{n+1}(v) = p(\pi^{\infty_\mathbb{A}}(D)v), \qquad v \in V^{\infty_\mathbb{R}},$$

for some continuous seminorm p on V and $D \in \mathcal{U}(\mathfrak{g}_\infty)$. We fix $m \in \mathbb{Z}_{>0}$, $c_1, \dots, c_m \in C^\infty(G_\infty)$ and $D_1, \dots, D_m \in \mathcal{U}(\mathfrak{g}_\infty)$ as in (10.13) and, invoking the continuity of $G(\mathbb{A}) \times V \to V$, $(g, v) \mapsto \pi(g)v$, we may also fix a continuous seminorm p' on V such that

$$p(\pi(g)v) \le p'(v), \qquad g \in C \times K_{\mathbb{A}_f},\ v \in V. \tag{10.18}$$

Now, for all $g \in C \times K_{\mathbb{A}_f}$ and $v \in (V^{\infty_\mathbb{R}})^{K_n}$, we have

$$\begin{aligned}
\sigma_{n+1}(\pi^{\infty_\mathbb{A}}(g)v) &= p(\pi(D)\pi^{\infty_\mathbb{A}}(g)v) \\
&= p\big(\pi^{\infty_\mathbb{A}}(g)\pi\big(\mathrm{Ad}(g_\infty^{-1})D\big)v\big) \\
&\overset{(10.13)}{=} p\left(\pi^{\infty_\mathbb{A}}(g)\pi\left(\sum_{i=1}^{m} c_i(g_\infty)D_i\right)v\right) \\
&= p\left(\sum_{i=1}^{m} c_i(g_\infty)\pi^{\infty_\mathbb{A}}(g)\pi(D_i)v\right) \\
&\le \sum_{i=1}^{m}\left(\max_{w\in C}|c_i(w)|\right)p(\pi^{\infty_\mathbb{A}}(g)\pi(D_i)v) \\
&\overset{(10.18)}{\le} \sum_{i=1}^{m}\left(\max_{w\in C}|c_i(w)|\right)p'(\pi(D_i)v) =: \sigma'_n(v).
\end{aligned}$$

Hence, the just-defined continuous seminorm σ'_n and $\varepsilon'_n := \varepsilon_{n+1}$ have the property (10.17). Therefore, $G(\mathbb{A}) \times V^{\infty_\mathbb{A}} \to V^{\infty_\mathbb{A}}$, $(g, v) \mapsto \pi^{\infty_\mathbb{A}}(g)v$ is continuous at $(id, 0)$. This completes the proof of Step 3 and hence of the whole lemma. □

Knowing that $(\pi^{\infty_\mathbb{A}}, V^{\infty_\mathbb{A}})$ is a representation of $G(\mathbb{A})$, the expression $((\pi^{\infty_\mathbb{A}})^{\infty_\mathbb{A}}, (V^{\infty_\mathbb{A}})^{\infty_\mathbb{A}})$ makes sense and is a representation of $G(\mathbb{A})$. In complete analogy to the case of local groups, cf. Lem. 2.18, we obtain

Lemma 10.19. *Let (π, V) be a representation of $G(\mathbb{A})$. Then,*

(1) $(\pi^{\infty_\mathbb{A}})^{\infty_\mathbb{A}} \cong \pi^{\infty_\mathbb{A}}$.
(2) The natural linear inclusions $V^{\infty_\mathbb{A}} \hookrightarrow V^{\infty_\mathbb{R}} \hookrightarrow V$ defined by the identity map are continuous.

Proof. The first assertion may be proved by repeated use of (10.8) together with Lem. 8.2 (which guarantees that $(V^{\infty_\mathbb{A}})^{K_n} = (V^{\infty_\mathbb{R}})^{K_n}$ topologically), and the fact that $((V^{K_n})^{\infty_\mathbb{R}})^{\infty_\mathbb{R}} = (V^{K_n})^{\infty_\mathbb{R}}$ is a meaningful equation of seminormed paces (as explained in [WarI72], first paragraph on p. 254).

Indeed,

$$\begin{aligned}
(V^{\infty_{\mathbb{A}}})^{\infty_{\mathbb{A}}} &= \varinjlim((V^{\infty_{\mathbb{A}}})^{\infty_{\mathbb{R}}})^{K_n} \\
&\overset{(10.8)}{=} \varinjlim\left((V^{\infty_{\mathbb{A}}})^{K_n}\right)^{\infty_{\mathbb{R}}} \\
&\overset{\text{Lem. 8.2}}{=} \varinjlim\left((V^{\infty_{\mathbb{R}}})^{K_n}\right)^{\infty_{\mathbb{R}}} \\
&\overset{(10.8)}{=} \varinjlim\left((V^{K_n})^{\infty_{\mathbb{R}}}\right)^{\infty_{\mathbb{R}}} \\
&\overset{\text{[WarI72], p. 254}}{=} \varinjlim(V^{K_n})^{\infty_{\mathbb{R}}} \\
&\overset{(10.8)}{=} \varinjlim(V^{\infty_{\mathbb{R}}})^{K_n} \\
&= V^{\infty_{\mathbb{A}}}.
\end{aligned}$$

In order to obtain the second assertion, recall that by our explicit description of the locally convex topology on $V^{\infty_{\mathbb{R}}}$ using the seminorms $p_{\alpha,D}$ (and the fact that $1 \in \mathbb{R} \subset \mathcal{U}(\mathfrak{g}_\infty)$), the identity map $V^{\infty_{\mathbb{R}}} \hookrightarrow V$ is continuous. Moreover, so are all the identity maps $(V^{\infty_{\mathbb{R}}})^{K_n} \hookrightarrow V^{\infty_{\mathbb{R}}}$, as their domains were given the subspace topology from $V^{\infty_{\mathbb{R}}}$. It follows now from Lem. 8.2 that the identity map $V^{\infty_{\mathbb{A}}} = \varinjlim(V^{\infty_{\mathbb{R}}})^{K_n} \hookrightarrow V^{\infty_{\mathbb{R}}}$ is continuous. □

Definition 10.20. A representation (π, V) of $G(\mathbb{A})$ is called *smooth*, if the identity map $V^{\infty_{\mathbb{A}}} \hookrightarrow V$ is an isomorphism of representations.

Of course, this just means that $V = V^{\infty_{\mathbb{A}}}$ as seminormed spaces, or, as indicated by Lem. 10.19.(2) above, equivalently, that we have two identities of seminormed spaces

$$V = V^{\infty_{\mathbb{R}}} = \varinjlim V^{K_n}. \tag{10.21}$$

Moreover, $(\pi^{\infty_{\mathbb{A}}}, V^{\infty_{\mathbb{A}}})$ is smooth for every representation (π, V) of $G(\mathbb{A})$, as we have just shown in Lem. 10.19.

Remark 10.22. By the very construction, $V^{\infty_{\mathbb{A}}}$ is an LF-space, if V was a Fréchet space. Indeed, if V is Fréchet, then – following the explanations of §2.3 – so is $V^{\infty_{\mathbb{R}}}$ and hence so are the closed subspaces $(V^{\infty_{\mathbb{R}}})^{K_n}$, cf. Lem. 1.11. Consequently, $V^{\infty_{\mathbb{A}}} = \varinjlim(V^{\infty_{\mathbb{R}}})^{K_n}$ is an LF-space by definition. It is this necessary complication of topologies, which forces us to go beyond the realm of Fréchet spaces, when talking about representations of global groups $G(\mathbb{A})$: Passing from a Fréchet space V to its "smoothification", i.e., the representation $(\pi^{\infty_{\mathbb{A}}}, V^{\infty_{\mathbb{A}}})$ on the space of globally smooth vectors $V^{\infty_{\mathbb{A}}}$ needs the wider class of LF-spaces.

The following proposition shows that smooth $G(\mathbb{A})$-representations behave (quite) well under taking subrepresentations and quotients. It shall be observed that this is by no means clear *a priori* as – for instance – closed subspaces of LF-spaces are not automatically LF-spaces, cf. Warn. 8.6.

Proposition 10.23. *Let (π, V) be a smooth $G(\mathbb{A})$-representation and let $U \subseteq V$ be a subrepresentation. Then, U is smooth and, if V/U satisfies the conditions of Def. 10.1 to be a $G(\mathbb{A})$-representation, then it is also smooth.*

Proof. We start off with the following general observation: For a representation (π, V) of $G(\mathbb{A})$ and a K_n in our fixed cofinal sequence of open compact subgroups of $G(\mathbb{A}_f)$, define $E^{K_n} : V \to V$ to be the standard normalized continuous projection onto V^{K_n}:

$$E^{K_n}(v) := \frac{1}{\mathrm{vol}_{dk_f}(K_n)} \int_{K_n} \pi(k_f)v \; dk_f\,.$$

We set $V_n := \ker\big(E^{K_{n-1}}|_{V^{K_n}}\big)$ and obtain by induction that

$$V^{K_n} \cong \bigoplus_{i \leq n} V_i, \qquad \forall n \in \mathbb{N}. \tag{10.24}$$

Indeed, if n_0 is the smallest element of $\mathbb{N}$, then $V_{n_0} = V^{K_{n_0}}$ by definition and the claim holds trivially. Now, suppose that the claim was shown for n. Then, since $E^{K_n}|_{V^{K_{n+1}}}$ is a continuous projection onto V^{K_n}, we have

$$V^{K_{n+1}} \cong V^{K_n} \oplus \ker\big(E^{K_n}|_{V^{K_{n+1}}}\big) \cong \left(\bigoplus_{i \leq n} V_i\right) \oplus V_{n+1} = \bigoplus_{i \leq n+1} V_i,$$

hence, (10.24) holds. Consequently, letting (π, V) be a smooth representation as in the statement of the proposition and using our characterization (10.21), we get

$$V = \varinjlim V^{K_n} \cong \varinjlim \left(\bigoplus_{i \leq n} V_i\right) \cong \bigoplus_{n \in \mathbb{N}} V_n. \tag{10.25}$$

It is worth noting that here the yet abstract isomorphism $V \cong \bigoplus_{n \in \mathbb{N}} V_n$ is provided by the map

$$v \mapsto (E^{K_n}(v) - E^{K_{n-1}}(v))_{n \in \mathbb{N}}, \tag{10.26}$$

a fact, whose simple proof we leave to the reader. Now, let U be a subrepresentation of V. Then, since $V = V^{\infty_{\mathbb{R}}}$, cf. (10.21), and given our description of the seminorms on $V^{\infty_{\mathbb{R}}}$ as the $p_{\alpha,D}$, we get

$$U = U^{\infty_{\mathbb{R}}} \tag{10.27}$$

as seminormed spaces. Moreover, as every vector $v \in V$ – and hence *a fortiori* every $u \in U$ – is fixed by an open compact subgroup of $G(\mathbb{A}_f)$, we get

$$U = \bigcup_{n\in\mathbb{N}} U^{K_n} \cong \bigcup_{n\in\mathbb{N}} \bigoplus_{i\leq n} U_i = \bigoplus_{n\in\mathbb{N}} U_n \tag{10.28}$$

as vector spaces, where the second isomorphism follows from reading (10.24) algebraically. However, since U_n is clearly a topological vector subspace of V_n for all $n \in \mathbb{N}$, [Bou03], II, §4, Prop. 8, shows that $\bigoplus_{n\in\mathbb{N}} U_n$ with the subspace topology from $\bigoplus_{n\in\mathbb{N}} V_n$ must be isomorphic to $\bigoplus_{n\in\mathbb{N}} U_n$ (with its intrinsic topology coming from the one of the summands, cf. §1.2.2). But as $\bigoplus_{n\in\mathbb{N}} V_n \cong V$ as seminormed spaces, cf. (10.25), this observation together with (10.28) just implies that U (with the subspace topology from V) is isomorphic to $\bigoplus_{n\in\mathbb{N}} U_n$ as seminormed spaces by the map described in (10.26). Hence, in summary

$$U \cong \bigoplus_{n\in\mathbb{N}} U_n \cong \varinjlim \bigoplus_{i\leq n} U_i \overset{(10.24)}{\cong} \varinjlim U^{K_n} \overset{(10.27)}{=} \varinjlim (U^{\infty_\mathbb{R}})^{K_n} = U^{\infty_\mathbb{A}}.$$

As the composition of the isomorphisms in this sequence is the identity map, this shows that U is smooth. Whence, the first assertion follows.

Now, let us suppose that V/U is a $G(\mathbb{A})$-representation. Recall that we have just shown that U is a smooth subrepresentation. Therefore, using [Bou03], II, §4, Sect. 5, Prop. 8.(ii), we get

$$\begin{aligned} V/U &\overset{(10.25)}{\cong} \Big(\bigoplus_{n\in\mathbb{N}} V_n\Big)/\Big(\bigoplus_{n\in\mathbb{N}} U_n\Big) \cong \bigoplus_{n\in\mathbb{N}} V_n/U_n \cong \varinjlim \bigoplus_{i\leq n} V_i/U_i \\ &\cong \varinjlim \Big(\bigoplus_{i\leq n} V_i\Big)/\Big(\bigoplus_{i\leq n} U_i\Big) \overset{(10.24)}{\cong} \varinjlim V^{K_n}/U^{K_n}, \end{aligned} \tag{10.29}$$

However, as one easily verifies,

$$\begin{aligned} (V/U)^{K_n} &\to V^{K_n}/U^{K_n} \\ v + U &\mapsto E^{K_n}(v) + U^{K_n} \end{aligned} \tag{10.30}$$

is an isomorphism of seminormed spaces. Hence, composing its inverse with all the isomorphisms in (10.29), we finally get an equality of seminormed spaces

$$V/U = \varinjlim (V/U)^{K_n}. \tag{10.31}$$

Recalling (10.21), we still need to prove that $V/U = (V/U)^{\infty_\mathbb{R}}$. To this end, we observe that applying the functor of taking G_∞-smooth vectors to the exact sequence of $G(\mathbb{A})$-representations $0 \to U \hookrightarrow V \twoheadrightarrow V/U \to 0$, and recalling (10.21) once more, we obtain a sequence of continuous, G_∞-equivariant linear maps $0 \to U \hookrightarrow V \to (V/U)^{\infty_\mathbb{R}} \to 0$. As $(V/U)^{\infty_\mathbb{R}} \subseteq V/U$, the homomorphism $V \to (V/U)^{\infty_\mathbb{R}}$ must remain surjective, whence this sequence is exact, too. Thus, the identity map defines a continuous linear bijection $V/U \to (V/U)^{\infty_\mathbb{R}}$. Since its inverse is continuous by Lem. 10.19, $V/U = (V/U)^{\infty_\mathbb{R}}$ as seminormed spaces. As we already explained, this shows the last claim of the proposition. □

10.3 Admissible representations

Recall our fixed maximal compact subgroup $K_\mathbb{A}$ of $G(\mathbb{A})$ and let (ρ, W) be an irreducible representation of $K_\mathbb{A}$ as defined at the end of §10.1. If (π, V) is a representation of $G(\mathbb{A})$, we will denote by $V(\rho)$ the *ρ-isotypical component in* V, i.e., the image of the map

$$\begin{aligned} \mathrm{Hom}_{K_\mathbb{A}}(W, V) \otimes W &\to V \\ \varphi \otimes w &\mapsto \varphi(w). \end{aligned}$$

Here, mimicing the local case, $\mathrm{Hom}_{K_\mathbb{A}}(W, V)$ denotes the space of all $K_\mathbb{A}$-equivariant linear maps from W to V. Combining our Thm. 1.8 with the following fundamental result, they are all automatically continuous.

Proposition 10.32 (see [Joh76], Thm. 3.9; [Bor72], Cor. 3.6). *If a representation (ρ, W) of $K_\mathbb{A}$ is irreducible, then it is finite-dimensional and unitary.*

As fundamental as the previous result is the following definition.

Definition 10.33. A representation (π, V) of $G(\mathbb{A})$ is called *admissible*, if for all irreducible representations (ρ, W) of $K_\mathbb{A}$, the ρ-isotypical component $V(\rho)$ is finite-dimensional.

Obviously, if (π, V) is admissible, then so is $(\pi^{\infty_\mathbb{A}}, V^{\infty_\mathbb{A}})$.

It will be useful to have the following reformulation of admissibility: Recall that K_∞ is a maximal compact subgroup of G_∞, and that G_∞ is a local archimedean group. For an irreducible (and hence automatically finite-dimensional, see Cor. 3.9) representation (ρ_∞, W_∞) of K_∞, and an open compact subgroup K_f of $G(\mathbb{A}_f)$, let $\mathrm{Hom}_{K_\infty}(W_\infty, V^{K_f})$ be the analogue of (2.31), i.e., the space of all K_∞-equivariant linear maps $f : W_\infty \to V^{K_f}$ between the K_∞-stable spaces W_∞ and V^{K_f}. We let $V^{K_f}(\rho_\infty)$ be the image of the natural map

$$\begin{aligned} \mathrm{Hom}_{K_\infty}(W_\infty, V^{K_f}) \otimes W_\infty &\to V^{K_f} \\ \psi \otimes w_\infty &\mapsto \psi(w_\infty). \end{aligned}$$

We get the following.

Lemma 10.34. *A representation (π, V) of $G(\mathbb{A})$ is admissible if and only if $V^{K_f}(\rho_\infty)$ is finite-dimensional for all open compact subgroups K_f of $G(\mathbb{A}_f)$ and all irreducible representations (ρ_∞, W_∞) of K_∞.*

Sketch of a proof. Suppose $V^{K_f}(\rho_\infty)$ is finite-dimensional for all open compact subgroups K_f of $G(\mathbb{A}_f)$ and all irreducible representations (ρ_∞, W_∞) of K_∞. Let (ρ, W) be an irreducible representation of $K_{\mathbb{A}}$. Then, combining Prop. 10.32 (finite-dimensionality and unitarity of (ρ, W)), [Dei10], Thm. 7.5.29 (decomposition of ρ along the decomposition $K_{\mathbb{A}} = K_\infty \times K_{\mathbb{A}_f}$), Thm. 1.8 (finite-dimensional seminormed spaces are all Hilbert spaces), and the proof of Lem. 4.35 (revealing, that every vector in W is fixed by an open compact subgroup of $K_{\mathbb{A}_f}$), there exists an irreducible representation (ρ_∞, W_∞) of K_∞ and an open compact subgroup K_f of $K_{\mathbb{A}_f}$, such that

$$V(\rho) \subseteq V^{K_f}(\rho_\infty).$$

Hence, as K_f is also open and compact in $G(\mathbb{A}_f)$, $\dim_{\mathbb{C}} V(\rho) \leq \dim_{\mathbb{C}} V^{K_f}(\rho_\infty) < \infty$, and so the representation (π, V) is admissible.

Now suppose that (π, V) is admissible and fix in a first step an arbitrary open compact subgroup K_f of $K_{\mathbb{A}_f}$ and an arbitrary irreducible representation (ρ_∞, W_∞) of K_∞. We let $\mathcal{R} = \mathcal{R}(\rho_\infty, K_f)$ be the set[1] of all equivalence classes of irreducible $K_{\mathbb{A}}$-representations (ρ, W), which admit a non-zero K_f-invariant vector and whose restriction $\rho|_{K_\infty}$ to K_∞ is isomorphic to a direct sum of copies of ρ_∞. Observe that, since K_f has finite index in $K_{\mathbb{A}_f}$, $\mathcal{R}$ is a finite set by Frobenius reciprocity, cf. [Bor72], §2.5. Now, again combining Prop. 10.32, [Dei10], Thm. 7.5.29, and Thm. 1.8, together with [Bor72], Prop. 3.6, one shows that

$$V^{K_f}(\rho_\infty) \subseteq \bigoplus_{[(\rho, W)] \in \mathcal{R}} V(\rho). \tag{10.35}$$

Hence, as $\mathcal{R}$ is finite and each $V(\rho)$ is finite-dimensional by admissibility of (π, V), the space $V^{K_f}(\rho_\infty)$ is finite-dimensional for all open compact

[1] The very purists among the readers may want to recall our Prop. 10.32 and then consult the footnote on p. 193 in [Dei10] here.

subgroups K_f of $K_{\mathbb{A}_f}$ and all irreducible representations (ρ_∞, W_∞) of K_∞. For a general open compact subgroup K_f of $G(\mathbb{A}_f)$ the corresponding claim follows from this, as $K_f \cap K_{\mathbb{A}_f}$ is an open compact subgroup of $K_{\mathbb{A}_f}$ and obviously $V^{K_f} \subseteq V^{K_f \cap K_{\mathbb{A}_f}}$. □

Corollary 10.36. *If a representation (π, V) of $G(\mathbb{A})$ satisfies the conditions of Def. 10.33 with respect to the given maximal compact subgroup $K_{\mathbb{A}}$, then it satisfies them with respect to all maximal compact subgroups of $G(\mathbb{A})$. In other words, the notion of admissibility is independent of the choice of $K_{\mathbb{A}}$.*

Proof. Let $K'_{\mathbb{A}} = K'_\infty \times K'_{\mathbb{A}_f}$ be another maximal compact subgroup of $G(\mathbb{A})$. By Lem. 10.34, a representation (π, V) of $G(\mathbb{A})$ is admissible (with respect to $K_{\mathbb{A}}$) if and only if $V^{K_f}(\rho_\infty)$ is finite-dimensional for all open compact subgroups K_f of $G(\mathbb{A}_f)$ and all irreducible representations (ρ_∞, W_∞) of K_∞. But then, as K_∞ is conjugate to K'_∞ by an element in G_∞, cf. [PlaRap94], Thm. 3.13, also $V^{K_f}(\rho'_\infty)$ is finite-dimensional for all open compact subgroups K_f of $G(\mathbb{A}_f)$ and all irreducible representations $(\rho'_\infty, W'_\infty)$ of K'_∞. Hence, $V(\rho')$ is finite-dimensional for all irreducible representations (ρ', W') of $K'_{\mathbb{A}}$ by Lem. 10.34. □

Having elaborated on questions of independence and equivalence of our notion of admissibility, in the next result we shall encounter an important class of admissible representations of $G(\mathbb{A})$. This theorem should be seen as a global analogue of Thm. 3.33.

Theorem 10.37. *Every irreducible unitary representation (π, V) of $G(\mathbb{A})$ is admissible.*

Proof. The reader will find this as Thm. 4.(2) in [Fla79], where it is attributed to Gel'fand–Graev–Piatetski-Shapiro [GGPS69] and Bernstein [Ber74], but without further comment or explanation. We will at the very least sketch here of how to derive a proof of this result from the literature.

Firstly, we recall that all the local groups G_v are tame (a notion which we have already mentioned in Thm. 3.37, but did not discuss it further), or, following [Dix77], §13.9.4, "of type I". This can be seen as follows: Let first v be archimedean. Then, it was proved in [Dix57], corollary on p. 328, that G_v is "of type I". See also [Wal92], Thm. 14.6.10. Let now v be non-archimedean and fix an arbitrary $f \in C_c^\infty(G_\mathsf{v})$, i.e., a smooth function $f : G_\mathsf{v} \to \mathbb{C}$ with compact support. Then, by the sheer definition of smoothness and compact support, there must be open compact subgroups C_1 and C_2 of G_v such that $f(c_1 g_\mathsf{v}) = f(g_\mathsf{v}) = f(g_\mathsf{v} c_2)$ for all $g_\mathsf{v} \in G_\mathsf{v}$, $c_i \in C_i$, $i = 1, 2$. We set $C_\mathsf{v} := C_1 \cap C_2$, which is an open compact subgroup of G_v, and obtain that f hence defines a function on $C_\mathsf{v} \backslash G_\mathsf{v} / C_\mathsf{v}$. Next, we fix an arbitrary topologically irreducible unitary representation $(\pi_\mathsf{v}, V_\mathsf{v})$ of G_v. Then, a combination of Thm. 3.35 and Exc. 2.36, or, equivalently, the reference [Ber74], Thm. 1, mentioned above, shows that $\dim_\mathbb{C} V_\mathsf{v}^{C_\mathsf{v}} < \infty$. As a consequence, the convolution

$$\pi_\mathsf{v}(f)v := \int_G f(g_\mathsf{v}) \cdot \pi_\mathsf{v}(g_\mathsf{v})v \ dg_\mathsf{v} \qquad\qquad \forall v \in V_\mathsf{v},$$

must be of finite rank and hence a compact linear operator. We claim that this implies that $\pi_\mathsf{v}(f)$ is a compact linear operator for all $f \in C^*(G_\mathsf{v})$, i.e., for all functions in the C*-algebra associated with G_v, cf. [Dix77], §13.9.1 for the very definition, which we will not repeat here. Indeed, as $C_c^\infty(G_\mathsf{v})$ is dense in the Banach space $L^1(G_\mathsf{v})$ of all (equivalence classes of) integrable measurable functions on G_v (for this combine the Stone-Weierstraß-theorem with the density of $C_c(G_\mathsf{v})$ in $L^1(G_\mathsf{v})$) and since also the assignment $L^1(G_\mathsf{v}) \to \mathrm{Aut}_\mathbb{C}^{ct}(V_\mathsf{v})$ is continuous, $\pi_\mathsf{v}(f)$ is a compact linear operator for all $[f] \in L^1(G_\mathsf{v})$. It now follows from the definition of $C^*(G_\mathsf{v})$, cf. [Dix77], §13.9.1 again, that $\pi_\mathsf{v}(f)$ is a compact linear operator for all $f \in C^*(G_\mathsf{v})$. Therefore, combining Prop. 2.7.4, Prop. 13.3.1, Prop. 13.3.4 and §13.3.5 of [Dix77], G_v is "liminal" in the sense of [Dix77], Def. 4.2.1. It finally follows from §4.3.1 and Thm. 5.5.2, *ibidem*, that G_v is "of type I".

Hence, in summary, all the local groups G_v, may v be archimedean or non-archimedean, are "of type I".

We observe that this implies that the groups $G_p := \mathrm{Res}_{F/\mathbb{Q}}(\mathbf{G})(\mathbb{Q}_p) = \prod_{\mathsf{v}|p} G_\mathsf{v}$, cf. [Pla-Rap94] (5.6) and (5.8), where p is either a rational prime or the archimedean place of $\mathbb{Q}$, are "of type I", as it was shown in [GMac53], corollary on p. 200 (together with [Dix77], §13.9.4). Now, recall the finite set T_0 from §9.1, outside of which $\mathbf{G}/F_\mathsf{v}$ is unramified. Our Prop. 7.20.(ii) together with Thm. 3.35 finally shows that the groups G_p satisfy the condition mentioned in [GGPS69], p. 274.

Therefore, the main theorem of [GGPS69], Chp. 3, §3.3, may be applied to π as viewed as a representation of $\mathrm{Res}_{F/\mathbb{Q}}(\mathbf{G})(\mathbb{A}_\mathbb{Q})$ and so we obtain an isomorphism of $\mathrm{Res}_{F/\mathbb{Q}}(\mathbf{G})(\mathbb{A}_\mathbb{Q})$-representations $\pi \cong \widehat{\bigotimes}'_p \pi_p$, where the right hand side denotes the restricted Hilbert space tensor product, cf. §8.2.2, of topologically irreducible unitary representations π_p of G_p, see Rem. 10.39 below for the precise definition of the action of $\mathrm{Res}_{F/\mathbb{Q}}(\mathbf{G})(\mathbb{A}_\mathbb{Q})$. For the moment, we confine ourselves to noting that at almost all p, the irreducible admissible representation on the space of smooth vectors in π_p (cf. Thm. 3.35) is unramified. Applying [GGPS69], Lem. 1 in §3.3, to the individual π_p, we finally end up with an isomorphism of $G(\mathbb{A})$-representations

$$\pi \cong \widehat{\bigotimes}'_{\mathsf{v}\in S} \pi_\mathsf{v}, \tag{10.38}$$

where $(\pi_\mathsf{v}, V_\mathsf{v})$ denotes a topologically irreducible unitary representation of G_v, whose irreducible admissible representation on its space of smooth vectors is unramified for almost all v.

Let now (ρ, W) be an irreducible (and hence finite-dimensional, cf. Prop. 10.32) representation of $K_\mathbb{A}$. Then, it is easy to see, cf. [Bum98], Lem. 3.3.1 (whose proof, although given only for $\mathbf{G} = \mathbf{GL}_n/F$, perfectly works in complete generality) that ρ decomposes as a $K_\mathbb{A}$-representation as a restricted Hilbert space tensor product

$$\rho \cong \widehat{\bigotimes}'_{\mathsf{v}\in S} \rho_\mathsf{v}$$

of irreducible representations $(\rho_\mathsf{v}, W_\mathsf{v})$ of K_v, which are equal to the trivial one-dimensional representation $\rho_\mathsf{v} = \mathbf{1}_\mathsf{v}$ on $W_\mathsf{v} = \mathbb{C}$ for almost all v. Hence, we are left with a finite set $S_{\pi,\rho}$ of places, where either the irreducible admissible G_v-representation on the space of smooth vectors in π_v is not unramified, or $\rho_\mathsf{v} \neq \mathbf{1}_\mathsf{v}$. In other words, for $\mathsf{v} \notin S_{\pi,\rho}$, $\dim_\mathbb{C} V_\mathsf{v}(\rho_\mathsf{v}) = \dim_\mathbb{C} V_\mathsf{v}^{K_\mathsf{v}} = 1$ by Prop. 7.20.(ii) and the simple observation that each K_v-invariant vector in π_v must obviously be smooth. We get isomorphisms of vector spaces

$$\begin{aligned} V(\rho) &\cong \widehat{\bigotimes}'_{\mathsf{v}\in S} V_\mathsf{v}(\rho_\mathsf{v}) \\ &\cong \widehat{\bigotimes}_{\mathsf{v}\in S_{\pi,\rho}} V_\mathsf{v}(\rho_\mathsf{v}) \mathbin{\widehat{\otimes}} \widehat{\bigotimes}'_{\mathsf{v}\notin S_{\pi,\rho}} V_\mathsf{v}(\rho_\mathsf{v}) \\ &\cong \widehat{\bigotimes}_{\mathsf{v}\in S_{\pi,\rho}} V_\mathsf{v}(\rho_\mathsf{v}) \mathbin{\widehat{\otimes}} \widehat{\bigotimes}'_{\mathsf{v}\notin S_{\pi,\rho}} V_\mathsf{v}(\mathbf{1}_\mathsf{v}) \\ &\cong \widehat{\bigotimes}_{\mathsf{v}\in S_{\pi,\rho}} V_\mathsf{v}(\rho_\mathsf{v}) \mathbin{\widehat{\otimes}} \widehat{\bigotimes}'_{\mathsf{v}\notin S_{\pi,\rho}} V_\mathsf{v}^{K_\mathsf{v}} \\ &\cong \widehat{\bigotimes}_{\mathsf{v}\in S_{\pi,\rho}} V_\mathsf{v}(\rho_\mathsf{v}), \end{aligned}$$

where the last space must be finite-dimensional by Thm. 3.33 and the finiteness of $S_{\pi,\rho}$. This completes the (sketch of the) proof. □

Remark 10.39 (Representations on restricted Hilbert space tensor products). A remark is in order as to how the right-hand side of (10.38) can be made a representation of $G(\mathbb{A})$. To this end, we identify $\mathbb{N}$ with the set of places S of F and consider a sequence $((\pi_\mathsf{v}, H_\mathsf{v}))_{\mathsf{v}\in S}$ of topologically irreducible unitary representations of G_v, subject to the condition that outside a finite set of places $S_0 \supseteq S_\infty$, the irreducible admissible unitary G_v-representation on the space of smooth vectors of V_v is unramified. (We recall that by Thm. 3.35 and Prop. 9.2 this is a well-posed requirement.)

Recall our fixed choice of a maximal compact subgroup $K_\mathbb{A} = \prod_{\mathsf{v}\in S} K_\mathsf{v}$ of $G(\mathbb{A})$ from §9.2. Prop. 7.20.(ii) together with the simple observation that a $\pi_\mathsf{v}(K_\mathsf{v})$-invariant vector in V_v must obviously be smooth, shows that $V_\mathsf{v}^{K_\mathsf{v}}$ is one-dimensional for every $\mathsf{v} \notin S_0$. For each such place we fix a unit vector $v_\mathsf{v}^\circ \in V_\mathsf{v}^{K_\mathsf{v}}$ and consider the restricted Hilbert space tensor product,

$$V := \widehat{\bigotimes}'_{\mathsf{v}\in S} V_\mathsf{v},$$

cf. §8.2.2, with respect to this choice.

Let $g \in G(\mathbb{A})$. As $g = (g_\mathsf{v})_{\mathsf{v}\in S}$, where $g_\mathsf{v} \in K_\mathsf{v}$ for all but finitely many places, g defines a linear map $\pi(g)$ on the algebraic direct limit of the Hilbert space tensor products $\widehat{\bigotimes}_{\mathsf{v}\in T} V_\mathsf{v}$, $T \supseteq S_0$ a finite set of places, see §8.2.2, which on pure tensors reads as

$$\pi(g)(\otimes_{\mathsf{v}\in S}\, v_\mathsf{v}) := \otimes_{\mathsf{v}\in S}\, \pi_\mathsf{v}(g_\mathsf{v})v_\mathsf{v}, \tag{10.40}$$

(since $v_\mathsf{v}^\circ \in V_\mathsf{v}^{K_\mathsf{v}}$ for almost all places, this is well-defined). Obviously, $\pi(g)$ preserves the Hermitian form (8.13). It hence extends to a continuous linear map on the Hilbert space completion $V = \widehat{\bigotimes}'_{\mathsf{v}\in S} V_\mathsf{v}$ and we obtain a well-defined group homomorphism

$$\pi : G(\mathbb{A}) \to \mathrm{Aut}_\mathbb{C}(V)$$

$$g \mapsto \pi(g).$$

It is easy to see that for each fixed $v \in V$, the map $g \mapsto \pi(g)v$ is continuous. Therefore, (invoking the simple, short argument on the top of p. 11 in [Kna86] for instance) (π, V) is a representation of $G(\mathbb{A})$. The reader is also referred to the original source, [GGPS69], Chp. 3, §3.2, p. 273, which treats the case of reductive groups over $\mathbb{Q}$.

Let us finally point out that the very choice of the unit vectors v_{v}° is not a crucial one. Indeed, for any other choice of unit vectors $v_{\mathsf{v}}^{\flat} \in V_{\mathsf{v}}^{K_{\mathsf{v}}}$, $\mathsf{v} \notin S_0$, there is a unique $\mu_{\mathsf{v}} \in S^1$ such that $v_{\mathsf{v}}^{\flat} = \mu_v \cdot v_{\mathsf{v}}^{\circ}$ by the one-dimensionality of $V_{\mathsf{v}}^{K_{\mathsf{v}}}$. Therefore, for every finite set of places $T \supseteq S_0$, multiplication by $\prod_{\mathsf{v}\in T\setminus S_0} \mu_{\mathsf{v}}$ defines a unitary automorphism of $\widehat{\bigotimes}_{\mathsf{v}\in T} V_{\mathsf{v}}$. This implies that $V := \widehat{\bigotimes}'_{\mathsf{v}\in S} V_{\mathsf{v}}$ is in fact unique up to canonical isomorphisms of Hilbert spaces.

10.4 $(\mathfrak{g}_\infty, K_\infty, G(\mathbb{A}_f))$-modules

Similar to the case of local archimedean groups, where the notion of a $(\mathfrak{g}, K)$-module provided a purely algebraic "approximation" of a G-representation, we will need a certain algebraification of the concept of an (admissible) representation of $G(\mathbb{A})$. To this end, recall once more that G_∞ is an archimedean local group, denote by $\mathfrak{g}_\infty$ its Lie algebra and recall our fixed maximal compact subgroup K_∞ of G_∞.

Definition 10.41. (1) A $(\mathfrak{g}_\infty, K_\infty, G(\mathbb{A}_f))$-*module* is a $(\mathfrak{g}_\infty, K_\infty)$-module, see Def. 3.2, (π_0, V_0), together with a commuting smooth linear action of $G(\mathbb{A}_f)$, i.e., a group homomorphism $\pi_f : G(\mathbb{A}_f) \to \mathrm{Aut}_{\mathbb{C}}(V_0)$, such that $G(\mathbb{A}_f) \to V_0$, $g_f \mapsto \pi_f(g_f)v$ is locally constant for each $v \in V_0$, and such that $\pi_0(X)$ and $\pi_0(k_\infty)$ commute with $\pi_f(g_f)$ for all $X \in \mathfrak{g}_\infty$, $k_\infty \in K_\infty$ and $g_f \in G(\mathbb{A}_f)$. For notational simplicity, we will suppress π_f[2] in the notation of a $(\mathfrak{g}_\infty, K_\infty, G(\mathbb{A}_f))$-module and just write (π_0, V_0).

(2) A $(\mathfrak{g}_\infty, K_\infty, G(\mathbb{A}_f))$-module (π_0, V_0) is *irreducible*, if $V_0 \neq \{0\}$ and V_0 does not admit any proper linear subspace, which is invariant under the action of $\mathfrak{g}_\infty$, K_∞ and $G(\mathbb{A}_f)$.

(3) Two $(\mathfrak{g}_\infty, K_\infty, G(\mathbb{A}_f))$-module are *isomorphic*, or *equivalent*, if there exists a linear, $\mathfrak{g}_\infty$-, K_∞- and $G(\mathbb{A}_f)$-equivariant bijection between them.

(4) A $(\mathfrak{g}_\infty, K_\infty, G(\mathbb{A}_f))$-module (π_0, V_0) is called *admissible*, if for all irreducible representations (ρ, W) of $K_{\mathbb{A}}$, the ρ-isotypical component $V_0(\rho)$, i.e., the image of the natural map

$$\begin{aligned}\mathrm{Hom}_{K_{\mathbb{A}}}(W, V_0) \otimes W &\to V_0\\ \varphi_0 \otimes w &\mapsto \varphi_0(w).\end{aligned}$$

is finite-dimensional.

Again, for an irreducible representation (ρ_∞, W_∞) of K_∞ and an open compact subgroup K_f of $G(\mathbb{A}_f)$, let $V_0^{K_f}(\rho_\infty)$ be the image of the natural

[2] As far as the needs of this book are concerned, this does not mean a big lack of precision, as the actions π_0 as well as π_f will be of the same, self-explanative origin.

map

$$\begin{aligned}\mathrm{Hom}_{K_\infty}(W_\infty, V_0^{K_f}) \otimes W_\infty &\to V_0^{K_f} \\ \psi_0 \otimes w_\infty &\mapsto \psi_0(w_\infty).\end{aligned}$$

We get the following.

Lemma 10.42. *A $(\mathfrak{g}_\infty, K_\infty, G(\mathbb{A}_f))$-module (π_0, V_0) is admissible, if and only if $V_0^{K_f}(\rho_\infty)$ is finite-dimensional for all open compact subgroups K_f of $G(\mathbb{A}_f)$ and all irreducible representations (ρ_∞, W_∞) of K_∞.*

Proof. This follows essentially in the same way as Lem. 10.34. Only the use of [Bor72], Prop. 3.6 is not legitimate any more, as V_0 is not supposed to be a seminormed space. However, our assumption of smoothness of the action of $G(\mathbb{A}_f)$ serves as a mean to repair this defect: Indeed, combining [Ren10], Thm. IV.1.1, our Thm. 1.8 and [BouII04], VIII, §2, Prop. 1, with [Bus-Hen06], Prop. 2.3, [Wal89], Lem. 3.3.3, and [Dei10], Thm. 7.5.29 one again obtains an inclusion

$$V_0^{K_f}(\rho_\infty) \subseteq \bigoplus_{[(\rho,W)]\in\mathcal{R}} V_0(\rho),$$

i.e., the analogue of (10.35). From here the proof may be finished verbatim. □

As a corollary we get that our notion of an admissible $(\mathfrak{g}_\infty, K_\infty, G(\mathbb{A}_f))$-module is equivalent to the identical notion defined in [Bor-Wal00], XII.2.2. Moreover, it is obviously independent of our choice of $K_{\mathbb{A}_f}$.

As in (3.5), for a $G(\mathbb{A})$-representation (π, V) we now define

$$V_{(K_\infty)} := \{v \in V \mid \dim_{\mathbb{C}}(\langle \pi(K_\infty)v \rangle) < \infty\}$$

to be the space of K_∞-finite vectors in V. It is dense in V by [Var77], Lem. II.7.10. In view of our Lem. 10.34 and 10.42, the following analogue of Lem. 3.6 is then clear by the explanations in Rem. 3.7 and the ones right after (3.11) and provides the above mentioned algebraification of (admissible) $G(\mathbb{A})$-representations in terms of (admissible) $(\mathfrak{g}_\infty, K_\infty, G(\mathbb{A}_f))$-modules:

Lemma 10.43. *The assignment $V \mapsto V_0 := V_{(K_\infty)}^{\infty_{\mathbb{A}}}$ defines a functor from the category of $G(\mathbb{A})$-representations (and continuous $G(\mathbb{A})$-equivariant linear maps) to the category of $(\mathfrak{g}_\infty, K_\infty, G(\mathbb{A}_f))$-modules (and linear $\mathfrak{g}_\infty$-, K_∞- and $G(\mathbb{A}_f)$-equivariant maps). If V is admissible, then V_0 is admissible. Hence, a smooth representation of $G(\mathbb{A})$ is admissible, if and only if its underlying $(\mathfrak{g}_\infty, K_\infty, G(\mathbb{A}_f))$-module $V_{(K_\infty)}^{\infty_{\mathbb{A}}} = V_{(K_\infty)}$ is.*

While admissibility passes on easily, it is more difficult to prove.

Proposition 10.44. *Let (π, V) be an irreducible unitary $G(\mathbb{A})$-representation. Then, its underlying $(\mathfrak{g}_\infty, K_\infty, G(\mathbb{A}_f))$-module $V^{\infty_\mathbb{A}}_{(K_\infty)}$ is irreducible.*

Sketch of a proof. Recall from (10.38) that

$$V \cong \widehat{\bigotimes}'_{\mathsf{v}\in S} V_\mathsf{v},$$

where $(\pi_\mathsf{v}, V_\mathsf{v})$ denotes a topologically irreducible unitary representation of G_v, whose space of smooth vectors is unramified for almost all v. Let us write $V_\infty := \widehat{\bigotimes}_{\mathsf{v}\in S_\infty} V_\mathsf{v}$ and $V_f := \widehat{\bigotimes}'_{\mathsf{v}\in S_f} V_\mathsf{v}$. Then, by construction,

$$V \cong V_\infty \widehat{\otimes} V_f \tag{10.45}$$

as $G(\mathbb{A})$-representations. Let K_n be any open compact subgroup in our fixed sequence $(K_n)_{n\in\mathbb{N}}$. As (π, V) is admissible by Thm. 10.37, $V^{K_n}(\rho_\infty)$ must be finite-dimensional for each irreducible representation (ρ_∞, W_∞) of K_∞ by Prop. 10.34. The decomposition $V \cong V_\infty \widehat{\otimes} V_f$ induces an injection of vector spaces $V_\infty(\rho_\infty) \otimes V_f^{K_n} \hookrightarrow V^{K_n}(\rho_\infty)$, whence $V_f^{K_n}$ must be finite-dimensional. This implies that for each K_n as above, we have identifications of $(\mathfrak{g}_\infty, K_\infty)$-modules

$$\begin{aligned} V^{K_n}_{(K_\infty)} &\cong (V_\infty \widehat{\otimes} V_f)^{K_n}_{(K_\infty)} \\ &= (V_\infty \otimes V_f^{K_n})_{(K_\infty)} \\ &= (V_\infty)_{(K_\infty)} \otimes V_f^{K_n}, \end{aligned} \tag{10.46}$$

where the second equality follows in by the same argument leading to [Bor-Wal00], X.6.2.(2). In particular, we have an isomorphism of $(\mathfrak{g}_\infty, K_\infty, G(\mathbb{A}_f))$-modules

$$V^{\infty_\mathbb{A}}_{(K_\infty)} \cong \bigcup_{n\in\mathbb{N}} \Big((V_\infty)_{(K_\infty)} \otimes V_f^{K_n} \Big) \cong (V_\infty)_{(K_\infty)} \otimes \bigcup_{n\in\mathbb{N}} V_f^{K_n} \cong (V_\infty)_{(K_\infty)} \otimes V_f^{\infty_f}.$$

(The very attentive reader will have noticed that here we viewed V_f as a representation of $G(\mathbb{A})$, as we may, in order to give meaning to the symbol $V_f^{\infty_f}$.) We observe that V_∞ is an irreducible unitary representation of G_∞ by Thm. 3.37, hence it is admissible by Thm. 3.33, and so Thm. 3.16 implies that $(V_\infty)_{(K_\infty)}$ is an irreducible $(\mathfrak{g}_\infty, K_\infty)$-module. By construction, $V_f = \widehat{\bigotimes}'_{\mathsf{v}\in S_f} V_\mathsf{v}$, whence the fact that the space of smooth vectors in almost all V_v is unramified (and therefore $\dim_\mathbb{C} V_\mathsf{v}^{K_\mathsf{v}} = 1$ (cf. Prop. 7.20) for almost all places $\mathsf{v} \in S$), implies that $V_f^{\infty_f}$ allows a $G(\mathbb{A}_f)$-equivariant linear bijection with the algebraic restricted tensor product (cf. §12.3.1 below for the precise definition) of the irreducible admissible G_v-representations on the spaces of smooth vectors inside V_v, $\mathsf{v} \in S_f$. In summary, the $(\mathfrak{g}_\infty, K_\infty, G(\mathbb{A}_f))$-module $V_{(K_\infty)}^{\infty_\mathbb{A}}$ is irreducible by [Bum98], Prop. 3.4.4, Prop. 3.4.8 and Thm. 3.4.4 *ibidem.* □

Example 10.47. Let (π, V) be an irreducible unitary $G(\mathbb{A})$-representation. Then, we have just seen in the argument leading to Prop. 10.44 that for every $n \in \mathbb{N}$ the space $V_f^{K_n}$ is finite-dimensional. Hence, for every $n \in \mathbb{N}$, the identity map provides a continuous embedding $\imath_n : V_f^{K_n} \hookrightarrow V_f$ with closed image, cf. Thm. 1.8. Taking the strict inductive limit we obtain a continuous embedding $\imath : \varinjlim V_f^{K_n} \hookrightarrow \varinjlim V_f = V_f$ by definition of the inductive limit topology. Clearly, $\varinjlim V_f^{K_n} = V_f^{\infty_f}$ equipped with the finest locally convex topology, cf. Cor. 8.3. However, amplifying the argument of the proof of Lem. 4.35 shows that $\imath(V_f^{\infty_f})$ is dense in V_f (Exercise!), whence, in general not closed, although the image of each step $\imath_n(V_f^{K_n})$ is closed in V_f.

The following proposition is taken from [Gro-Žun23]:

Proposition 10.48. *Let (π, V) be a smooth $G(\mathbb{A})$-representation. Let V_1 be an admissible $(\mathfrak{g}_\infty, K_\infty, G(\mathbb{A}_f))$-submodule of $V_{(K_\infty)}$ that is dense in V. Then, $V_{(K_\infty)} = V_1$ and (π, V) is admissible.*

Proof. Let (ρ_∞, W_∞) be an irreducible representation of K_∞. Then, for each of our open compact subgroups $K_n \subseteq G(\mathbb{A}_f)$ the (normalized) linear operator $E_{\rho_\infty}^{K_n} : V \to V$, defined by

$$E_{\rho_\infty}^{K_n}(v) := \frac{\dim_\mathbb{C}(W_\infty)}{\mathrm{vol}_{dk_f}(K_n)} \int_{K_\infty} \int_{K_n} \mathrm{tr}(\rho_\infty(k_\infty)^{-1})\ \pi(k_\infty \cdot k_f)v\ dk_f\ dk_\infty$$

is a continuous projection onto $V^{K_n}(\rho_\infty)$ and restricts to a projection of V_1 onto $V_1^{K_n}(\rho_\infty)$ (where we gave the obvious meaning to the latter space), see [Mui-Žun20], Lem. 19. Therefore, denoting by "Cl_V" taking the topological closure within V,

$$\begin{aligned} V^{K_n}(\rho_\infty) &= E^{K_n}_{\rho_\infty}(V) \\ &= E^{K_n}_{\rho_\infty}(\mathrm{Cl}_V(V_1)) \\ &\subseteq \mathrm{Cl}_V(E^{K_n}_{\rho_\infty}(V_1)) \\ &= \mathrm{Cl}_V(V_1^{K_n}(\rho_\infty)) \\ &= V_1^{K_n}(\rho_\infty), \end{aligned} \tag{10.49}$$

where the inclusion (10.49) holds by continuity of $E^{K_n}_{\rho_\infty}$, and the last equality holds as $V_1^{K_n}(\rho_\infty)$ is finite-dimensional by Lem. 10.42 and hence closed in V as a consequence of Thm. 1.8. Since trivially $V^{K_n}(\rho_\infty) \supseteq V_1^{K_n}(\rho_\infty)$, this shows the equality $V^{K_n}(\rho_\infty) = V_1^{K_n}(\rho_\infty)$. Consequently, applying [Wal89], Lem. 3.3.3 twice, namely once to the $(\mathfrak{g}_\infty, K_\infty)$-modules $V^{K_n}_{(K_\infty)}$, whose union $\bigcup_{n\in\mathbb{N}} V^{K_n}_{(K_\infty)}$ equals $V^{\infty_\mathbb{A}}_{(K_\infty)} = V_{(K_\infty)}$, and another time to the $(\mathfrak{g}_\infty, K_\infty)$-modules $V_1^{K_n}$, whose union $\bigcup_{n\in\mathbb{N}} V_1^{K_n}$ must equal V_1, as every vector in $V_1 \subseteq V$ is fixed by an open compact subgroup of $G(\mathbb{A}_f)$, we get

$$V_{(K_\infty)} = \bigcup_{n\in\mathbb{N}} \bigoplus_{[(\rho_\infty, W_\infty)]} V^{K_n}(\rho_\infty) = \bigcup_{n\in\mathbb{N}} \bigoplus_{[(\rho_\infty, W_\infty)]} V_1^{K_n}(\rho_\infty) = V_1,$$

and so the first claim of the proposition follows.

In order to see the second claim, just observe that, as V is smooth, Lem. 10.43 shows, that V it is admissible, if and only if its underlying $(\mathfrak{g}_\infty, K_\infty, G(\mathbb{A}_f))$-module $V^{\infty_\mathbb{A}}_{(K_\infty)} = V_{(K_\infty)}$ is. However, by what we have just proved, $V_{(K_\infty)} = V_1$, and so admissibility of V follows from the assumption that V_1 is admissible. □

Among the admissible representations, there is a subclass of representations, for which the passage to their underlying $(\mathfrak{g}_\infty, K_\infty, G(\mathbb{A}_f))$-modules works most efficiently.

Definition 10.50. A smooth representation (π, V) of $G(\mathbb{A})$ is called a *Casselman–Wallach representation* of $G(\mathbb{A})$, if for every open compact subgroup K_f of $G(\mathbb{A}_f)$, the G_∞-stable closed subspace V^{K_f} of V is a Casselman–Wallach representation of G_∞, cf. Def. 3.25.

We record that by Lem. 10.34, every Casselman–Wallach representation (π, V) of $G(\mathbb{A})$ is admissible and, recalling (10.21), that $V = \varinjlim V^{K_n}$ is necessarily an LF-space. Moreover, we get, cf. [Gro-Žun23]:

Proposition 10.51.

(1) Let (π, V) be a Casselman–Wallach representation of $G(\mathbb{A})$. If U is a subrepresentation of V, then U and V/U are Casselman–Wallach representations of $G(\mathbb{A})$.

(2) Two Casselman–Wallach representations (π, V) and (π', V') of $G(\mathbb{A})$ are isomorphic if and only if the underlying $(\mathfrak{g}_\infty, K_\infty, G(\mathbb{A}_f))$-modules $V_{(K_\infty)}$ and $V'_{(K_\infty)}$ are isomorphic.

Proof. (1) We first observe that V/U is indeed a representation of $G(\mathbb{A})$: In fact, one directly verifies that (10.29) still holds in the present situation, i.e., that we have $V/U \cong \varinjlim V^{K_n}/U^{K_n}$ as seminormed spaces. As by assumption V^{K_n} and U^{K_n} are both Fréchet spaces, so is the their quotient, cf. Lem. 1.11, hence V/U is complete and barreled by Lem. 8.2 and Lem. 1.10. Consequently, V/U is a representation of $G(\mathbb{A})$, as explained in the end of §10.1.

Knowing that V/U is in a representation of $G(\mathbb{A})$, assertion (1) follows directly using Prop. 10.23 and Lem. 3.27 and the fact that the isomorphism of seminormed spaces (10.30) is obviously G_∞-equivariant.

(2) Clearly, it is enough to prove that a $(\mathfrak{g}_\infty, K_\infty, G(\mathbb{A}_f))$-module isomorphism $f_0 : V_{(K_\infty)} \xrightarrow{\sim} W_{(K_\infty)}$ extends to an isomorphism of $G(\mathbb{A})$-representations $f : V \xrightarrow{\sim} W$. By [Wal92], Thm. 11.6.7 & Cor. 11.6.8, for every $n \in \mathbb{N}$ we can extend the $(\mathfrak{g}_\infty, K_\infty)$-module isomorphism $f_0|_{V^{K_n}_{(K_\infty)}} : V^{K_n}_{(K_\infty)} \xrightarrow{\sim} W^{K_n}_{(K_\infty)}$ in a unique way to an isomorphism of G_∞-representations $f_n : V^{K_n} \xrightarrow{\sim} W^{K_n}$. See also Thm. 3.26. The uniqueness of this extension guarantees that f_{n+1} extends f_n. Hence, passing to the direct limit, we obtain a well-defined isomorphism of G_∞-representations $f : V \to W$, which extends f_0. As f_0 is $G(\mathbb{A}_f)$-equivariant, so is f by continuity, and therefore f is an isomorphism of $G(\mathbb{A})$-representations as desired. □

Chapter 11

Automorphic Forms and Smooth-Automorphic Forms

11.1 Spaces of smooth functions with predetermined growth

Let $C^\infty(G(F)\backslash G(\mathbb{A}))$ be the vector subspace of $C^\infty(G(\mathbb{A}))$, consisting of all smooth functions which are invariant under multiplication by $G(F)$ from the left, i.e.,

$$C^\infty(G(F)\backslash G(\mathbb{A})) := \{\varphi \in C^\infty(G(\mathbb{A})) \mid \varphi(\gamma g) = \varphi(g) \quad \forall \gamma \in G(F), g \in G(\mathbb{A})\}.$$

For $n \in \mathbb{N}$ and $d \in \mathbb{Z}_{>0}$, we define

$$C^\infty_{umg,d}(G(F)\backslash G(\mathbb{A}))^{K_n}$$

to be the vector space of all $\varphi \in C^\infty(G(F)\backslash G(\mathbb{A}))$, which are invariant under K_n by right-translation and which are furthermore of *uniform moderate growth* with respect to d, i.e., which satisfy that for every $D \in \mathcal{U}(\mathfrak{g}_\infty)$,

$$p_{d,D}(\varphi) := \sup_{g \in G(\mathbb{A})} |(D\varphi)(g)| \, \|g\|^{-d} < \infty. \tag{11.1}$$

Here, $D \in \mathcal{U}(\mathfrak{g}_\infty)$ acts on φ by the extension of the differentiation of right-translation R_∞ of G_∞, i.e., if $D = X_1 \cdot \ldots \cdot X_k \in \mathcal{U}(\mathfrak{g}_\infty)$, then

$$(D\varphi)(g) := \left.\frac{\mathrm{d}}{\mathrm{d}t_1 \cdot \ldots \cdot \mathrm{d}t_k}\right|_{t_1=\ldots=t_k=0} \varphi(g \exp(t_1 X_1) \ldots \exp(t_k X_k)),$$

and this action is extended linearly to all $D \in \mathcal{U}(\mathfrak{g}_\infty)$.

Exercise 11.2. Let $n \in \mathbb{N}$ and $d \in \mathbb{Z}_{>0}$.
(1) Show that the functions $p_{d,D}$, D running though $\mathcal{U}(\mathfrak{g}_\infty)$, define a point-separating family of seminorms on $C^\infty_{umg,d}(G(F)\backslash G(\mathbb{A}))^{K_n}$.
(2) Show moreover that there is an inclusion of vector spaces

$$C^\infty_{umg,d}(G(F)\backslash G(\mathbb{A}))^{K_n} \subseteq C^\infty_{umg,d+1}(G(F)\backslash G(\mathbb{A}))^{K_n}, \tag{11.3}$$

Hint: Look up (9.8).

In order to make $C^\infty_{umg,d}(G(F)\backslash G(\mathbb{A}))^{K_n}$ into a seminormed space, we equip it with the locally convex topology, defined by the seminorms $p_{d,D}$, where d is fixed and $D \in \mathcal{U}(\mathfrak{g}_\infty)$.

We would like to show that with this natural topology $C^\infty_{umg,d}(G(F)\backslash G(\mathbb{A}))^{K_n}$ is a Fréchet space. To this end, it will be useful to be able to relate the space $C^\infty_{umg,d}(G(F)\backslash G(\mathbb{A}))^{K_n}$ to spaces of smooth functions on G_∞: Therefore, let us recall that for each $n \in \mathbb{N}$, there exists a *finite* subset $C_n \subseteq G(\mathbb{A}_f)$ such that $G(\mathbb{A})$ can be written as a disjoint union of sets

$$G(\mathbb{A}) = \bigsqcup_{c\in C_n} G(F)(G_\infty \cdot c \cdot K_n), \tag{11.4}$$

see [Bor63], Thm. 5.1. Moreover, for every $c \in C_n$, the congruence subgroup

$$\Gamma_{c,K_n} := cK_nc^{-1} \cap G(F)$$

of $G(F)$ embeds into G_∞ as a discrete subgroup of finite covolume modulo $A^{\mathbb{R}}_G$, which is a consequence of Prop. 9.11 and (11.4). Now, let $C^\infty_{umg,d}(\Gamma_{c,K_n}\backslash G_\infty)$ denote the space of all smooth, left Γ_{c,K_n}-invariant functions $f : G_\infty \to \mathbb{C}$ such that

$$p_{\infty,d,D}(f) := \sup_{g_\infty\in G_\infty} |(Df)(g_\infty)| \, \|g_\infty\|^{-d} < \infty, \qquad D \in \mathcal{U}(\mathfrak{g}_\infty).$$

As in the global case, the functions $p_{\infty,d,D}$, $D \in \mathcal{U}(\mathfrak{g}_\infty)$, define a point-separating family of seminorms on $C^\infty_{umg,d}(\Gamma_{c,K_n}\backslash G_\infty)$ and we equip $C^\infty_{umg,d}(\Gamma_{c,K_n}\backslash G_\infty)$ with the locally convex topology defined by it. The following result is fundamental, see [Gro-Žun23]:

Proposition 11.5. *For each $n \in \mathbb{N}$ and $d \in \mathbb{Z}_{>0}$, the assignment $\varphi \mapsto (\varphi(\_ \cdot c))_{c\in C_n}$ defines an isomorphism of Fréchet spaces*

$$C^\infty_{umg,d}(G(F)\backslash G(\mathbb{A}))^{K_n} \cong \bigoplus_{c\in C_n} C^\infty_{umg,d}(\Gamma_{c,K_n}\backslash G_\infty),$$

which restricts to an isomorphism of the corresponding closed (and hence Fréchet) subspaces of $A^{\mathbb{R}}_G$-invariant functions

$$C^\infty_{umg,d}([G])^{K_n} \cong \bigoplus_{c\in C_n} C^\infty_{umg,d}(A^{\mathbb{R}}_G\Gamma_{c,K_n}\backslash G_\infty).$$

Proof. Firstly, we note that as a direct consequence of [Wal94], Lem. 2.7, the spaces $C^\infty_{umg,d}(\Gamma_{c,K_n}\backslash G_\infty)$ are Fréchet, whence so is their finite direct sum, cf. Lem. 1.12. As a next step, the reader will easily convince him-/herself that the map $\varphi \mapsto (\varphi(\_\_ \cdot c))_{c\in C_n}$ is a continuous linear bijection, whose inverse may be constructed as follows: For a $(f_c)_{c\in C_n} \in \bigoplus_{c\in C_n} C^\infty_{umg,d}(\Gamma_{c,K_n}\backslash G_\infty)$, define

$$\varphi(\delta g_\infty c\ell) := f_c(g_\infty), \qquad \delta \in G(F),\ g_\infty \in G_\infty,\ c \in C_n,\ \ell \in K_n.$$

Then, φ is a well-defined element of $C^\infty_{umg,d}(G(F)\backslash G(\mathbb{A}))^{K_n}$. In order to show that the resulting inverse map $(f_c)_{c\in C_n} \mapsto \varphi$ of $\varphi \mapsto (\varphi(\_\_ \cdot c))_{c\in C_n}$ is indeed continuous (only from which it will finally follow that $C^\infty_{umg,d}(G(F)\backslash G(\mathbb{A}))^{K_n}$ with its above defined structure as a semi-normed space is also a Fréchet space), we fix a Siegel set $\mathfrak{S}$ in $G(\mathbb{A})$ as in [Mœ-Wal95], p. 20, with the property that $G(\mathbb{A}) = G(F)\mathfrak{S}$. That such a choice of a Siegel set always exists – a fact which is stated without proof in [Mœ-Wal95], p. 20 – is a consequence of Thm. F in [Spr94]. (See also [Bor69], Thm. 13.1 for a more classical result in this direction.) Therefore, left-invariance of each φ by $G(F)$ and [Mœ-Wal95], I.2.2(vii) show that there exists $r \in \mathbb{R}_{>0}$ such that for every $D \in \mathcal{U}(\mathfrak{g}_\infty)$,

$$\begin{aligned} p_{d,D}(\varphi) &= \sup_{\gamma g \in G(F)\mathfrak{S}} |(D\varphi)(\gamma g)| \|\gamma g\|^{-d} \\ &\leq r \cdot \sup_{g\in\mathfrak{S}} |(D\varphi)(g)| \|g\|^{-d}. \end{aligned} \tag{11.6}$$

Following the explanations of the proof of Thm. 5.2 in [Pla-Rap94], the image of the projection of $\mathfrak{S}$ to $G(\mathbb{A}_f)$ is compact, hence, (11.4) implies that for each $n \in \mathbb{N}$, there must be a finite set $\Delta_n \subseteq G(F)$ such that $\mathfrak{S} \subseteq \Delta_n G_\infty C_n K_n$. So, (11.6) is furthermore bounded above by

$$r \cdot \sup_{\substack{\delta\in\Delta_n,\, g_\infty\in G_\infty \\ c\in C_n,\, \ell\in K_n}} |(D\varphi)(\delta g_\infty c\ell)|\, \|\delta g_\infty c\ell\|^{-d}. \tag{11.7}$$

Next, invoking (9.8), we obtain the bound

$$\begin{aligned} \|g_\infty\| &= \left\|\delta^{-1}\delta g_\infty c\ell\ell^{-1}c^{-1}\right\| \\ &\leq C_0^3 \cdot \left\|\delta^{-1}\right\| \cdot \|\delta g_\infty c\ell\| \cdot \left\|\ell^{-1}\right\| \cdot \left\|c^{-1}\right\| \\ &= C_0^3 \cdot \|\delta\| \cdot \|\delta g_\infty c\ell\| \cdot \|\ell\| \cdot \|c\|, \end{aligned}$$

where the last equation follows as $\|x\| = \left\|x^{-1}\right\|$ by the definition of the adelic group norm, cf. §9.1. In particular,

$$\|\delta g_\infty c\ell\|^{-1} \leq C_0^3 \cdot \|\delta\| \cdot \|\ell\| \cdot \|c\| \cdot \|g_\infty\|^{-1},$$

which, inserted into (11.7) gives us the bounds

$$\begin{aligned} p_{d,D}(\varphi) &\leq r \cdot \sup_{\substack{\delta\in\Delta_n,\, g_\infty\in G_\infty \\ c\in C_n,\, \ell\in K_n}} |(D\varphi)(\delta g_\infty c\ell)|\, C_0^{3d}\, \|\delta\|^d \, \|c\|^d \, \|\ell\|^d \, \|g_\infty\|^{-d} \\ &= r\, C_0^{3d} \cdot \sup_{\substack{\delta\in\Delta_n,\, g_\infty\in G_\infty \\ c\in C_n,\, \ell\in K_n}} |(Df_c)(g_\infty)|\, \|\delta\|^d \, \|c\|^d \, \|\ell\|^d \, \|g_\infty\|^{-d} \\ &\leq \; r\, C_0^{3d} \left(\max_{\delta\in\Delta_n} \|\delta\|^d\right) \left(\max_{\ell\in K_n} \|\ell\|^d\right) \left(\max_{c\in C_n} \|c\|^d p_{\infty,d,D}(f_c)\right). \end{aligned}$$

Consequently, $(f_c)_{c\in C_n} \mapsto \varphi$ is continuous, whence $\varphi \mapsto (\varphi(\_\!\_ \cdot c))_{c\in C_n}$ is an isomorphism of seminormed spaces and so it also follows that $C^\infty_{umg,d}(G(F)\backslash G(\mathbb{A}))^{K_n}$ is a Fréchet space as claimed.

The remaining claim about the subspaces of $A_G^{\mathbb{R}}$-invariant functions follows from this, using again the arguments of [Wal94], Lem. 2.7, which imply that $C^\infty_{umg,d}(\Gamma_{c,K_n}\backslash G_\infty)$ is a G_∞-representation by right-translation R_∞. As the isomorphism of seminormed spaces $(f_c)_{c\in C_n} \mapsto \varphi$ is obviously $R_\infty(G_\infty)$-equivariant, $C^\infty_{umg,d}(G(F)\backslash G(\mathbb{A}))^{K_n}$ is also a G_∞-representation by right-translation by Exc. 2.11. It follows that the intersection of the kernels of the continuous linear operators $R_\infty(a) - id_\infty$, $a \in A_G^{\mathbb{R}}$, is closed in $C^\infty_{umg,d}(G(F)\backslash G(\mathbb{A}))^{K_n}$ as well as in $\bigoplus_{c\in C_n} C^\infty_{umg,d}(\Gamma_{c,K_n}\backslash G_\infty)$. Therefore – these intersections being $C^\infty_{umg,d}([G])^{K_n}$, respectively, $\bigoplus_{c\in C_n} C^\infty_{umg,d}(A_G^{\mathbb{R}}\Gamma_{c,K_n}\backslash G_\infty)$ (recall that $A_G^{\mathbb{R}}$ is contained in the center of $G(\mathbb{A})$) – each of these spaces is Fréchet by Lem. 1.11, whence the G_∞-equivariant isomorphism $\varphi \mapsto (\varphi(\_\!\_ \cdot c))_{c\in C_n}$ restricts to an isomorphism of Fréchet spaces

$$C^\infty_{umg,d}([G])^{K_n} \cong \bigoplus_{c\in C_n} C^\infty_{umg,d}(A_G^{\mathbb{R}}\Gamma_{c,K_n}\backslash G_\infty)$$

as claimed. □

We fix the notation

$$C^\infty_{umg,d}(G(F)\backslash G(\mathbb{A})) := \bigcup_{n\in\mathbb{N}} C^\infty_{umg,d}(G(F)\backslash G(\mathbb{A}))^{K_n}$$

and

$$C^\infty_{umg}(G(F)\backslash G(\mathbb{A})) := \bigcup_{d\in\mathbb{Z}_{>0}} C^\infty_{umg,d}(G(F)\backslash G(\mathbb{A})).$$

Consistently with our previous use of the notion of uniform moderate growth, a smooth left $G(F)$-invariant function φ shall henceforth be called of *uniform moderate growth (with respect to d)*, if $\varphi \in C^{\infty}_{umg}(G(F)\backslash G(\mathbb{A}))$ (respectively, if $\varphi \in C^{\infty}_{umg,d}(G(F)\backslash G(\mathbb{A}))$). The reader shall be warned that – although each space $C^{\infty}_{umg,d}(G(F)\backslash G(\mathbb{A}))^{K_n}$ is Fréchet by Prop. 11.5 – neither of the two spaces $C^{\infty}_{umg,d}(G(F)\backslash G(\mathbb{A}))$ and $C^{\infty}_{umg}(G(F)\backslash G(\mathbb{A}))$ is complete when equipped with the natural locally convex topology defined by the seminorms $p_{d,D}$ (where D runs through $\mathcal{U}(\mathfrak{g}_\infty)$ in the first case, while also d runs through $\mathbb{Z}_{>0}$ in the second case), as it is implied by the following exercise.

Exercise 11.8. For a subset $Y \subseteq G(\mathbb{A})$ let char_Y be the characteristic function on Y. Show that the pointwise limit of the series $\sum_{n\in\mathbb{N}} \frac{1}{2^n}\mathrm{char}_{G(F)G_\infty K_n}$ is not invariant under any open compact subgroup of $G(\mathbb{A}_f)$. Conclude that $\sum_{n\in\mathbb{N}} \frac{1}{2^n}\mathrm{char}_{G(F)G_\infty K_n}$ does not converge to an element in $C^{\infty}_{umg,d}(G(F)\backslash G(\mathbb{A}))$, if the latter space carries the locally convex topology defined by the seminorms $p_{d,D}$, $D \in \mathcal{U}(\mathfrak{g}_\infty)$.

We will now choose an arbitrary, but fixed ideal $\mathcal{J} \lhd \mathcal{Z}(\mathfrak{g}_\infty)$ of finite codimension. For $n \in \mathbb{N}$ and $d \in \mathbb{Z}_{>0}$, we let

$$\mathcal{A}^{\infty}_d(G)^{K_n,\mathcal{J}^n} := \{\varphi \in C^{\infty}_{umg,d}(G(F)\backslash G(\mathbb{A}))^{K_n} \mid \mathcal{J}^n\varphi = \{0\}\}.$$

We obtain the following.

Proposition 11.9. *For all $n \in \mathbb{N}$ and $d \in \mathbb{Z}_{>0}$, the spaces $\mathcal{A}^{\infty}_d(G)^{K_n,\mathcal{J}^n}$ are Casselman–Wallach representations of G_∞ acted upon by right-translation R_∞.*

Proof. This may be found as Prop. 2.5 in [Gro-Žun23], whose proof we explain here: Firstly, we observe that the arguments of [Wal94], §2.5 and Lem. 2.7 imply that $C^{\infty}_{umg,d}(\Gamma_{c,K_n}\backslash G_\infty)$ is a smooth G_∞-representation by right-translation R_∞. Hence, so is $C^{\infty}_{umg,d}(G(F)\backslash G(\mathbb{A}))^{K_n}$ as it follows from Prop. 11.5 (in particular from the end of its proof, recalling Exc. 2.22). Therefore, the subspace $\mathcal{A}^{\infty}_d(G)^{K_n,\mathcal{J}^n}$ being the intersection of the kernels of the continuous linear operators $R_\infty(J)$, $J \in \mathcal{J}^n$, is closed in $C^{\infty}_{umg,d}(G(F)\backslash G(\mathbb{A}))^{K_n}$ and hence itself Fréchet by Lem. 1.11. As it is closed under $R_\infty(G_\infty)$, $\mathcal{A}^{\infty}_d(G)^{K_n,\mathcal{J}^n}$ is a smooth G_∞-representation under right-translation by Cor. 2.21.

As explained in the proof of [Wal94], Lem. 2.7, smoothness of $(R_\infty, C^{\infty}_{umg,d}(\Gamma_{c,K_n}\backslash G_\infty))$ already implies that this representation is of moderate growth. Hence, so is $C^{\infty}_{umg,d}(G(F)\backslash G(\mathbb{A}))^{K_n}$ by Prop. 11.5 and so is the smooth subrepresentation $\mathcal{A}^{\infty}_d(G)^{K_n,\mathcal{J}^n}$, cf. Lem. 3.24.

As a final ingredient, we note that the $(\mathfrak{g}_\infty, K_\infty)$-module $\mathcal{A}^\infty_d(G)^{K_n,\mathcal{J}^n}_{(K_\infty)}$ underlying $\mathcal{A}^\infty_d(G)^{K_n,\mathcal{J}^n}$ is admissible by [Bor-Jac79], §4.3.(i). Since it is also $\mathcal{Z}(\mathfrak{g}_\infty)$-finite by its very definition, Thm. 3.14 shows that $\mathcal{A}^\infty_d(G)^{K_n,\mathcal{J}^n}_{(K_\infty)}$ is finitely generated. Hence, $\left(R_\infty, \mathcal{A}^\infty_d(G)^{K_n,\mathcal{J}^n}\right)$ is a Casselman–Wallach representations of G_∞, which completes the proof. □

11.2 The LF-space of smooth-automorphic forms

After all our preparatory work, we are now ready to give the following fundamental definition.

Definition 11.10. A *smooth-automorphic form* is an element of one of the spaces

$$\mathcal{A}^\infty_{\mathcal{J}}(G) := \bigcup_{n,d} \mathcal{A}^\infty_d(G)^{K_n,\mathcal{J}^n},$$

where $\mathcal{J}$ runs through the ideals of finite codimension in $\mathcal{Z}(\mathfrak{g}_\infty)$.

The more experienced reader will have noticed that our definition of a smooth-automorphic form is at first sight different to what one usually finds in the (rather sparse) literature on the topic (where, as far as we know, [Wal94], §6.1, respectively, [Cog04], §2.3, are the first instances, where the concept of smooth-automorphic forms was introduced (in the former reference in the more classical context of real reductive groups and their arithmetic subgroups, respectively, in the latter, in the adelic setting, but only for $\mathbf{G} = \mathbf{GL}_n$)). Let us convince here shortly that all approaches lead to the same class of objects.

Lemma 11.11. *A smooth function $\varphi : G(\mathbb{A}) \to \mathbb{C}$ is a smooth-automorphic form, if and only if it satisfies the following list of conditions:*

(1) φ is left $G(F)$-invariant,
(2) φ is right $K_{\mathbb{A}_f}$-finite,
(3) φ is annihilated by an ideal $\mathcal{J} \lhd \mathcal{Z}(\mathfrak{g}_\infty)$ of finite codimension,
(4) φ is of uniform moderate growth.

Proof. This is a triviality once we recall from the end of §10.1 that every smooth function $\varphi : G(\mathbb{A}) \to \mathbb{C}$ is automatically right-$K_{\mathbb{A}_f}$-finite, whence condition (2) could just have been removed from the above list[1]. □

[1]... and in fact it is only stated there, in order to match formulations, which the reader will quite often find in the literature, cf. [Cog04], §2.3, or, [Bor-Jac79], §4.2.

The notion of a smooth-automorphic form hence only depends on the group scheme $\mathbf{G}/F^2$.

Either using the previous lemma, but also directly from Def. 11.10, it is easy to see that $\mathcal{A}^\infty_{\mathcal{J}}(G)$ is stable under right-translation by $G(\mathbb{A})$, i.e., that for all $\varphi \in \mathcal{A}^\infty_{\mathcal{J}}(G)$ and $g \in G(\mathbb{A})$, also $R(g)\varphi := \varphi(\_ \cdot g)$ is in $\mathcal{A}^\infty_{\mathcal{J}}(G)$. However, as we want to make $\mathcal{A}^\infty_{\mathcal{J}}(G)$ into a *representation* of $G(\mathbb{A})$ under right-translation R, we will have to specify a locally convex topology on $\mathcal{A}^\infty_{\mathcal{J}}(G)$, which makes it into a complete seminormed space, such that

$$\begin{aligned} G(\mathbb{A}) \times \mathcal{A}^\infty_{\mathcal{J}}(G) &\to \mathcal{A}^\infty_{\mathcal{J}}(G) \\ (g, \varphi) &\mapsto R(g)\varphi \end{aligned}$$

is continuous.

Recalling the warning example, Exc. 11.8, of how *not* to do this, another try to achieve this could be to equip $\mathcal{A}^\infty_{\mathcal{J}}(G)$ with the inductive limit topology given by the natural inclusions $\mathcal{A}^\infty_d(G)^{K_n,\mathcal{J}^n} \hookrightarrow \mathcal{A}^\infty_{\mathcal{J}}(G)$, $(n, d) \in \mathbb{N} \times \mathbb{Z}_{>0}$.

However, this approach bears another problem, namely that now the abstract theory of inductive limits does not provide a simple key of how to show that the resulting locally convex topology on $\mathcal{A}^\infty_{\mathcal{J}}(G)$ is Hausdorff at all, and even less so, of how to show that the resulting space is complete. For instance, employing the usual estimates of the seminorms at hand, it remains unclear whether the Fréchet topology on $\mathcal{A}^\infty_{d+1}(G)^{K_{n+1},\mathcal{J}^{n+1}}$ (which is defined by the seminorms $p_{d+1,D}$, $D \in \mathcal{U}(\mathfrak{g}_\infty)$) induces the original Fréchet topology on the subspace $\mathcal{A}^\infty_d(G)^{K_n,\mathcal{J}^n}$ (which is defined by the seminorms $p_{d,D}$, $D \in \mathcal{U}(\mathfrak{g})$). In particular it is not clear, whether $\mathcal{A}^\infty_d(G)^{K_n,\mathcal{J}^n}$ is closed in $\mathcal{A}^\infty_{d+1}(G)^{K_{n+1},\mathcal{J}^{n+1}}$. To cut a long story short, making $\mathcal{A}^\infty_{\mathcal{J}}(G)$ into a representation of $G(\mathbb{A})$ by right-translation, needs another ingredient, which is tuned to fit the concrete space $\mathcal{A}^\infty_{\mathcal{J}}(G)$.

Fortunately, this at first sight tricky situation may be remedied by the following general result.

Proposition 11.12. *Let $\mathcal{J} \lhd \mathcal{Z}(\mathfrak{g}_\infty)$ be an arbitrary, but fixed ideal of finite codimension. Then, there exists an integer $d \in \mathbb{Z}_{>0}$ such that*

$$\mathcal{A}^\infty_{\mathcal{J}}(G) \subseteq C^\infty_{umg,d}(G(F)\backslash G(\mathbb{A})).$$

[2]... and our choice of hyperspecial maximal compact subgroups K_v used in order to define $G(\mathbb{A})$.

That is, having fixed $\mathcal{J}$, there is an exponent $d = d(\mathcal{J})$ of growth, such that all smooth-automorphic forms in $\mathcal{A}^\infty_\mathcal{J}(G)$ satisfy (11.42) *for the same d.*

Proof. The (very technical) proof is given in all details in [Gro-Žun23], Prop. 2.8, to which we refer. □

Prop. 11.12, implies that there exists a smallest $d = d_0 \in \mathbb{Z}_{>0}$ such that $\mathcal{A}^\infty_\mathcal{J}(G) \subset C^\infty_{umg,d}(G(F)\backslash G(\mathbb{A}))$. Let us fix such an exponent for a moment. (We shall soon see that the actual choice of d is irrelevant, cf. Lem. 11.14 below.) It is easy to see that for each $n \in \mathbb{N}$, the identity map $\mathcal{A}^\infty_{d_0}(G)^{K_n,\mathcal{J}^n} \hookrightarrow \mathcal{A}^\infty_{d_0}(G)^{K_{n+1},\mathcal{J}^{n+1}}$ has closed image. Indeed, the seminorms on $\mathcal{A}^\infty_{d_0}(G)^{K_n,\mathcal{J}^n}$ are independent of n and so the subspace-topology on $\mathcal{A}^\infty_{d_0}(G)^{K_n,\mathcal{J}^n}$ inherited from $\mathcal{A}^\infty_{d_0}(G)^{K_{n+1},\mathcal{J}^{n+1}}$ is nothing but the original topology of $\mathcal{A}^\infty_{d_0}(G)^{K_n,\mathcal{J}^n}$, in which it is complete, cf. Prop. 11.9, whence closed in the ambient space $\mathcal{A}^\infty_{d_0}(G)^{K_{n+1},\mathcal{J}^{n+1}}$.

We may hence equip the space of smooth-automorphic forms $\mathcal{A}^\infty_\mathcal{J}(G)$ with its natural LF-space topology with defining sequence $V_n := \mathcal{A}^\infty_{d_0}(G)^{K_n,\mathcal{J}^n}$:

$$\mathcal{A}^\infty_\mathcal{J}(G) = \varinjlim \mathcal{A}^\infty_{d_0}(G)^{K_n,\mathcal{J}^n}. \tag{11.13}$$

As explained in §8.1.2, this way $\mathcal{A}^\infty_\mathcal{J}(G)$ becomes a complete, bornological, barreled seminormed space.

The next result shows that this approach does not only work for our particular (minimal) choice d_0, but universally for whatever choice of d, which satisfies the conditions of Prop. 11.12.

Lemma 11.14. *The LF-space topology on $\mathcal{A}^\infty_\mathcal{J}(G)$ is independent of the choice of the exponent of growth. More precisely, for every $d \geq d_0$ we have an equality of seminormed spaces*

$$\varinjlim \mathcal{A}^\infty_d(G)^{K_n,\mathcal{J}^n} = \varinjlim \mathcal{A}^\infty_{d_0}(G)^{K_n,\mathcal{J}^n}.$$

Proof. Let $d \geq d_0$. Obviously, it is enough to show that for each $n \in \mathbb{N}$,

$$\mathcal{A}^\infty_{d_0}(G)^{K_n,\mathcal{J}^n} = \mathcal{A}^\infty_d(G)^{K_n,\mathcal{J}^n} \tag{11.15}$$

as seminormed spaces. Clearly, by Prop. 11.12, (11.15) holds as vector spaces, i.e., the identity map

$$\mathrm{id} : \mathcal{A}^\infty_{d_0}(G)^{K_n,\mathcal{J}^n} \to \mathcal{A}^\infty_d(G)^{K_n,\mathcal{J}^n}$$

is a linear bijection. Moreover, as for every $D \in \mathcal{U}(\mathfrak{g}_\infty)$

$$\begin{aligned} p_{d,D}(\varphi) &= \sup_{g \in G(\mathbb{A})} |(D\varphi)(g)| \|g\|^{-d} \\ &\overset{(9.8)}{\leq} c_0^{-(d-d_0)} \sup_{g \in G(\mathbb{A})} |(D\varphi)(g)| \|g\|^{-d_0} \\ &= c_0^{d_0-d} \cdot p_{d_0,D}(\varphi) \end{aligned}$$

the identity map id is continuous. Hence, id is an isomorphism of Fréchet spaces by Prop. 11.9 and Thm. 1.7. This shows the claim. □

Exercise 11.16. Show that if $(\varphi_i)_{i\in I}$ is a convergent net in $\mathcal{A}^\infty_\mathcal{J}(G)$ with limit φ, then the net of complex numbers $((D\varphi_i)(g))_{i\in I}$ converges in $\mathbb{C}$ to $(D\varphi)(g)$ for every $D \in \mathcal{U}(\mathfrak{g}_\infty)$ and $g \in G(\mathbb{A})$.

11.3 Smooth-automorphic forms as representation of $G(\mathbb{A})$

The next result – fundamental for all that will follow – builds a bridge between the LF-space $\mathcal{A}^\infty_\mathcal{J}(G)$ and the representation theory of $G(\mathbb{A})$.

Theorem 11.17. *Acted upon by right-translation R, the LF-space $\mathcal{A}^\infty_\mathcal{J}(G)$ is a smooth representation of $G(\mathbb{A})$.*

Proof. We will divide the proof into two steps: In the first step, we will show that $(R, \mathcal{A}^\infty_\mathcal{J}(G))$ is a representation of $G(\mathbb{A})$, while in the second step we will verify that this representation is indeed smooth.

Step 1: We have already noticed that $\mathcal{A}^\infty_\mathcal{J}(G)$ is stable under right-translation by elements in $G(\mathbb{A})$ and – being an LF-space – a complete seminormed space, cf. §11.2. Hence, to prove that $(R, \mathcal{A}^\infty_\mathcal{J}(G))$ is a representation of $G(\mathbb{A})$, it remains to show that

$$\begin{aligned} G(\mathbb{A}) \times \mathcal{A}^\infty_\mathcal{J}(G) &\to \mathcal{A}^\infty_\mathcal{J}(G) \\ (g, \varphi) &\mapsto R(g)\varphi \end{aligned} \tag{11.18}$$

is continuous. As we explained in §11.2 above, the space $\mathcal{A}^\infty_\mathcal{J}(G)$ is also barreled. Therefore, in order to show that (11.18) is continuous, it is enough to verify that the same map is *separately* continuous, cf. [BouII04], VIII, §2, Prop. 1.

Let us first show continuity of (11.18) in the second variable, whence we assume to have chosen and fixed an arbitrary $g \in G(\mathbb{A})$. As implied by Lem. 8.2, we need to prove continuity of

$$R(g) : \mathcal{A}_d^\infty(G)^{K_n,\mathcal{J}^n} \to \mathcal{A}_\mathcal{J}^\infty(G)$$

for every $n \in \mathbb{N}$. Let us write $g = (g_\infty, g_f) \in G_\infty \times G(\mathbb{A}_f) = G(\mathbb{A})$. Recalling Prop. 11.9, $R((g_\infty, id_f)) = R_\infty(g_\infty)$ is an automorphism of the Fréchet space $\mathcal{A}_d^\infty(G)^{K_n,\mathcal{J}^n}$ and hence it is also continuous as a map

$$R((g_\infty, id_f)) : \mathcal{A}_d^\infty(G)^{K_n,\mathcal{J}^n} \to \mathcal{A}_\mathcal{J}^\infty(G),$$

by construction of the topology on $\mathcal{A}_\mathcal{J}^\infty(G)$, cf. Lem. 8.2. On the other hand, the image of the linear map

$$R((id_\infty, g_f)) : \mathcal{A}_d^\infty(G)^{K_n,\mathcal{J}^n} \to \mathcal{A}_\mathcal{J}^\infty(G),$$

obviously lies inside (any) $\mathcal{A}_d^\infty(G)^{K_m,\mathcal{J}^m}$, with $m \gg n$ such that $K_m \subseteq g_f K_n g_f^{-1} \cap K_n$: Here, as in the proof of Prop. 10.10, we observe that such an m exists, because $(K_n)_{n\in\mathbb{N}}$ defines a base of neighborhoods of the identity in $G(\mathbb{A}_f)$. Therefore, for every $g_f \in G(\mathbb{A}_f)$ we obtain an $m \gg n$ and a linear map given by right-translation by g_f

$$R_m(g_f) : \mathcal{A}_d^\infty(G)^{K_n,\mathcal{J}^n} \to \mathcal{A}_d^\infty(G)^{K_m,\mathcal{J}^m}.$$

Moreover, recalling (9.8), we get for every $g \in G(\mathbb{A})$,

$$\|g\|^{-1} \leq C_0 \cdot \|gg_f\|^{-1}\|g_f\|.$$

Consequently, for every $D \in \mathcal{U}(\mathfrak{g}_\infty)$ and $\varphi \in \mathcal{A}_d^\infty(G)^{K_n,\mathcal{J}^n}$

$$\begin{aligned} p_{d,D}(R_m(g_f)\varphi) &= \sup_{g\in G(\mathbb{A})} |(D\varphi)(gg_f)|\|g\|^{-d} \\ &\leq C_0^d \sup_{g\in G(\mathbb{A})} |(D\varphi)(gg_f)|\|gg_f\|^{-d}\|g_f\|^d \\ &= C_0^d\|g_f\|^d p_{d,D}(\varphi). \end{aligned}$$

Therefore, for every $g_f \in G(\mathbb{A}_f)$, the linear map $R_m(g_f)$ is continuous. Again, recalling Lem. 8.2, this shows that $R((id_\infty, g_f))$ is continuous. In summary, $R(g) = R((id_\infty, g_f)) \circ R((g_\infty, id_f))$ being the composition of continuous maps is itself a continuous map $\mathcal{A}_d^\infty(G)^{K_n,\mathcal{J}^n} \to \mathcal{A}_\mathcal{J}^\infty(G)$ for each $n \in \mathbb{N}$. Consequently, as explained above, $R(g) : \mathcal{A}_\mathcal{J}^\infty(G) \to \mathcal{A}_\mathcal{J}^\infty(G)$ is continuous for all $g \in G(\mathbb{A})$.

We will now prove continuity of (11.18) in the first variable. So, let $\varphi \in \mathcal{A}_\mathcal{J}^\infty(G)$ be an arbitrary, but fixed smooth-automorphic form and consider its orbit map c_φ. We may fix an $n \in \mathbb{N}$ such that $\varphi \in \mathcal{A}_d^\infty(G)^{K_n,\mathcal{J}^n}$.

Prop. 11.9 now implies that c_φ restricts to a smooth, and therefore continuous map, $c_\varphi : G_\infty \to \mathcal{A}_d^\infty(G)^{K_n,\mathcal{J}^n}$. By construction of the topology on $\mathcal{A}_\mathcal{J}^\infty(G)$, cf. Lem. 8.2, the corresponding map $c_\varphi : G_\infty \to \mathcal{A}_\mathcal{J}^\infty(G)$ is also continuous. On the other hand, φ being right-invariant under the open compact subgroup K_n, we also obtain a continuous restriction $c_\varphi : G(\mathbb{A}_f) \to \mathcal{A}_\mathcal{J}^\infty(G)$. Hence, by the barreledness of $\mathcal{A}_\mathcal{J}^\infty(G)$ and the just verified continuity of $R((id_\infty, g_f)) : \mathcal{A}_\mathcal{J}^\infty(G) \to \mathcal{A}_\mathcal{J}^\infty(G)$, the map $G(\mathbb{A}_f) \times \mathcal{A}_\mathcal{J}^\infty(G) \to \mathcal{A}_\mathcal{J}^\infty(G)$, $(g_f, \phi) \mapsto R(g_f)\phi$ is jointly continuous. Therefore, c_φ, which factors as

$$\begin{array}{ccccccc} G(\mathbb{A}) & \xrightarrow{\sim} & G_\infty \times G(\mathbb{A}_f) & \longrightarrow & \mathcal{A}_\mathcal{J}^\infty(G) \times G(\mathbb{A}_f) & \longrightarrow & \mathcal{A}_\mathcal{J}^\infty(G) \\ g & \longmapsto & (g_\infty, g_f) & \longmapsto & (c_\varphi(g_\infty), g_f) & \longmapsto & R(g_f)c_\varphi(g_\infty) = c_\varphi(g) \end{array}$$

is continuous as it is a composition of continuous maps.

In summary, we have just shown that (11.18) is separately continuous, whence jointly continuous, by our above explanations. It follows that $(R, \mathcal{A}_\mathcal{J}^\infty(G))$ is a representation of $G(\mathbb{A})$.

Step 2: Knowing that $(R, \mathcal{A}_\mathcal{J}^\infty(G))$ is a representation of $G(\mathbb{A})$, we may now prove its smoothness. To this end, we will use our alternative description of what it means for a representation of $G(\mathbb{A})$ to be smooth, given by (10.21), i.e., we will show that we have the following two equalities of seminormed spaces

$$\mathcal{A}_\mathcal{J}^\infty(G) = \mathcal{A}_\mathcal{J}^\infty(G)^{\infty_\mathbb{R}} = \varinjlim \mathcal{A}_\mathcal{J}^\infty(G)^{K_n}. \tag{11.19}$$

The first equality may be seen as follows. Firstly, we observe that each step $\mathcal{A}_d^\infty(G)^{K_n,\mathcal{J}^n}$ in the strict inductive limit $\mathcal{A}_\mathcal{J}^\infty(G) = \varinjlim \mathcal{A}_d^\infty(G)^{K_n,\mathcal{J}^n}$ is a smooth representation of G_∞ under right-translation by Prop. 11.9. As each step $\mathcal{A}_d^\infty(G)^{K_n,\mathcal{J}^n}$ inherits from $\mathcal{A}_\mathcal{J}^\infty(G)$ its original topology, cf. Lem. 8.2, for every $\varphi \in \mathcal{A}_\mathcal{J}^\infty(G)$ the restricted orbit map $c_\varphi|_{G_\infty} : G_\infty \to \mathcal{A}_\mathcal{J}^\infty(G)$ (whose image lands inside $\mathcal{A}_d^\infty(G)^{K_n,\mathcal{J}^n}$) is also smooth, i.e., we obtain an equality of vector spaces $\mathcal{A}_\mathcal{J}^\infty(G) = \mathcal{A}_\mathcal{J}^\infty(G)^{\infty_\mathbb{R}}$. Recalling our description of the seminormed topology on $\mathcal{A}_\mathcal{J}^\infty(G)^{\infty_\mathbb{R}}$ from §10.2, we are hence left to prove that for every continuous seminorm p_α on $\mathcal{A}_\mathcal{J}^\infty(G)$ and $D \in \mathcal{U}(\mathfrak{g}_\infty)$, the seminorm $p_{\alpha,D}$ is continuous, i.e., by Lem. 8.2, that the restrictions $p_{\alpha,D}\big|_{\mathcal{A}_d^\infty(G)^{K_n,\mathcal{J}^n}}$ are continuous. However, this holds by the smoothness of $(R_\infty, \mathcal{A}_d^\infty(G)^{K_n,\mathcal{J}^n})$, see §2.3. Hence, $\mathcal{A}_\mathcal{J}^\infty(G) = \mathcal{A}_\mathcal{J}^\infty(G)^{\infty_\mathbb{R}}$ as seminormed spaces, which shows the first equality in (11.19).

In order to see the second equality, i.e., that $\mathcal{A}^\infty_{\mathcal{J}}(G) = \varinjlim \mathcal{A}^\infty_{\mathcal{J}}(G)^{K_n}$ holds topologically, we need to show that that the linear isomorphism provided by the identity map

$$\mathcal{A}^\infty_{\mathcal{J}}(G) = \varinjlim \mathcal{A}^\infty_{d}(G)^{K_n,\mathcal{J}^n} \xrightarrow{\text{id}} \varinjlim \mathcal{A}^\infty_{\mathcal{J}}(G)^{K_n}$$

is bicontinuous. By Lem. 8.2, this is equivalent to showing that for all $m \in \mathbb{N}$ the natural inclusions

$$\mathcal{A}^\infty_{d}(G)^{K_m,\mathcal{J}^m} \hookrightarrow \varinjlim \mathcal{A}^\infty_{\mathcal{J}}(G)^{K_n} \quad \text{and} \quad \mathcal{A}^\infty_{\mathcal{J}}(G)^{K_m} \hookrightarrow \mathcal{A}^\infty_{\mathcal{J}}(G) \tag{11.20}$$

are continuous. It is obvious that the second inclusion is continuous (as $\mathcal{A}^\infty_{\mathcal{J}}(G)^{K_m}$ was given the subspace topology from $\mathcal{A}^\infty_{\mathcal{J}}(G)$) and equally obviously the inclusion $\mathcal{A}^\infty_{d}(G)^{K_m,\mathcal{J}^m} \hookrightarrow \mathcal{A}^\infty_{\mathcal{J}}(G)$ is continuous by construction of the strict inductive limit topology on $\mathcal{A}^\infty_{\mathcal{J}}(G)$. However, since the image of the latter map lands inside $\mathcal{A}^\infty_{\mathcal{J}}(G)^{K_m} \subset \mathcal{A}^\infty_{\mathcal{J}}(G)$, which embeds continuously into $\varinjlim \mathcal{A}^\infty_{\mathcal{J}}(G)^{K_n}$ by definition of the strict inductive limit topology on $\varinjlim \mathcal{A}^\infty_{\mathcal{J}}(G)^{K_n}$, also the first map in (11.20) is continuous. This completes the proof. □

11.4 Automorphic forms

Having shown that $\mathcal{A}^\infty_{\mathcal{J}}(G)$ is a smooth representation of $G(\mathbb{A})$ by right-translation, the definition of an *automorphic form* comes quite effortless:

Definition 11.21. An *automorphic form* (with respect to K_∞) is an element of one of the spaces

$$\mathcal{A}_{\mathcal{J}}(G) := \mathcal{A}^\infty_{\mathcal{J}}(G)_{(K_\infty)},$$

where $\mathcal{J}$ runs through the ideals of finite codimension in $\mathcal{Z}(\mathfrak{g}_\infty)$.

Recall from Lem. 10.43 that each space $\mathcal{A}_{\mathcal{J}}(G)$ is a $(\mathfrak{g}_\infty, K_\infty, G(\mathbb{A}_f))$-module. We remark that, as opposed to the definition of a smooth-automorphic form, which depended only on the group-scheme $\mathbf{G}/F$, the notion of an automorphic form obviously depends on the choice of the maximal compact subgroup K_∞ of G_∞.

Similarly to the case of smooth-automorphic forms, our above definition is not quite what one finds in the literature. In order to explain, why our approach is equivalent to the usual one, recall that a smooth function $\varphi : G(\mathbb{A}) \to \mathbb{C}$ is of *moderate growth*, if there is a $d \in \mathbb{Z}_{>0}$ such that

$$\sup_{g \in G(\mathbb{A})} |\varphi(g)| \, \|g\|^{-d} < \infty.$$

The following lemma proves that our definition matches what one finds in the literature (cf. [Bor-Jac79], Def. 4.2):

Lemma 11.22. *A smooth function $\varphi : G(\mathbb{A}) \to \mathbb{C}$ is an automorphic form (with respect to K_∞), if and only if it satisfies the following list of conditions:*

(1) φ is left $G(F)$-invariant,
(2) φ is right $K_{\mathbb{A}}$-finite,
(3) φ is annihilated by an ideal $\mathcal{J} \lhd \mathcal{Z}(\mathfrak{g}_\infty)$ of finite codimension,
(4) φ is of moderate growth.

Sketch of a proof. Recalling Lem. 11.11, the only thing that remains to be shown is that a smooth function satisfying the conditions (1)–(4) is of uniform moderate growth. Reading the proof of Prop. 11.5 once again, it amounts to an easy standard check to see that for each $n \in \mathbb{N}$ the assignment $\varphi \mapsto (\varphi(\_\cdot c))_{c\in C_n}$ induces an isomorphism of vector spaces

$$C^\infty_{mg,d}(G(F)\backslash G(\mathbb{A}))^{K_n} \xrightarrow{\sim} \bigoplus_{c\in C_n} C^\infty_{mg,d}(\Gamma_{c,K_n}\backslash G_\infty)$$

between the vector subspace $C^\infty_{mg,d}(G(F)\backslash G(\mathbb{A}))^{K_n}$ of $C^\infty(G(F)\backslash G(\mathbb{A}))^{K_n}$ of smooth functions of moderate growth with respect to d and the direct sum $\bigoplus_{c\in C_n} C^\infty_{mg,d}(\Gamma_{c,K_n}\backslash G_\infty)$, where $C^\infty_{mg,d}(\Gamma_{c,K_n}\backslash G_\infty)$ denotes those smooth functions $f \in C^\infty(\Gamma_{c,K_n}\backslash G_\infty)$, for which $\sup_{g_\infty\in G_\infty} |f(g_\infty)| \, \|g_\infty\|^{-d} < \infty$. Obviously, $\varphi \mapsto (\varphi(\_\cdot c))_{c\in C_n}$ is $R_\infty(G_\infty)$-equivariant, hence it maps the K_∞-finite and $\mathcal{Z}(\mathfrak{g}_\infty)$-finite functions in $C^\infty(G(F)\backslash G(\mathbb{A}))^{K_n}$ onto the K_∞-finite and $\mathcal{Z}(\mathfrak{g}_\infty)$-finite functions in $\bigoplus_{c\in C_n} C^\infty_{mg,d}(\Gamma_{c,K_n}\backslash G_\infty)$. However, every such function f in $C^\infty_{mg,d}(\Gamma_{c,K_n}\backslash G_\infty)$ is in fact of *uniform* moderate growth (with respect to the same exponent d) as it follows by a straight forward calculation from a general theorem of Harish-Chandra (partly also due to Jacquet and Borel), cf. [HCh66], §8, Thm. 1.[3] Therefore, every function φ satisfying (1)–(4) is of uniform moderate growth by Prop. 11.5. □

Exercise 11.23. Complete the argument of the above proof that [HCh66], §8, Thm. 1 implies that a K_∞-finite and $\mathcal{Z}(\mathfrak{g}_\infty)$-finite f in $C^\infty_{mg,d}(\Gamma_{c,K_n}\backslash G_\infty)$ is in $C^\infty_{umg,d}(\Gamma_{c,K_n}\backslash G_\infty)$.

As mentioned in §10.4, $\mathcal{A}_{\mathcal{J}}(G) = \mathcal{A}^\infty_{\mathcal{J}}(G)_{(K_\infty)}$ is a dense subspace of $\mathcal{A}^\infty_{\mathcal{J}}(G)$, i.e., every smooth-automorphic form is topologically "arbitrarily close" to a classical automorphic form. However, from the perspective of vector spaces $\mathcal{A}_{\mathcal{J}}(G)$ is "very far" from $\mathcal{A}^\infty_{\mathcal{J}}(G)$: Indeed, $\mathcal{A}_{\mathcal{J}}(G)$ is

[3] Harish-Chandra's theorem is for connected groups G_∞ only, but the proof generalizes easily.

of countable dimension as it follows from a theorem of Harish-Chandra – see [HCh68], Thm. 1 for the classical reference and [Bor-Jac79], Thm. 1.7 and 4.3.(i) for its adelic version – whereas $\mathcal{A}^\infty_\mathcal{J}(G)$ is of uncountable dimension as it contains non-finite-dimensional Fréchet spaces $\mathcal{A}^\infty_d(G)^{K_n,\mathcal{J}^n}$, cf. §1.1.

The next lemma, which is also taken from [Gro-Žun23], will be an important technical tool later.

Lemma 11.24. *Let $\{V_{0,i}\}_{i\in I}$ be an arbitrary family of $(\mathfrak{g}_\infty, K_\infty)$-submodules of the space $\mathcal{A}_d(G)^{K_n,\mathcal{J}^n} := \mathcal{A}^\infty_d(G)^{K_n,\mathcal{J}^n}_{(K_\infty)}$. If the sum of the $V_{0,i}$ is direct, then, the sum of G_∞-subrepresentations $\mathrm{Cl}_{\mathcal{A}^\infty_d(G)^{K_n,\mathcal{J}^n}}(V_{0,i})$ of $\mathcal{A}^\infty_d(G)^{K_n,\mathcal{J}^n}$ is also direct.*

Proof. Let us abbreviate $\overline{V_{0,i}} := \mathrm{Cl}_{\mathcal{A}^\infty_d(G)^{K_n,\mathcal{J}^n}}(V_{0,i})$. Then, for every $i \in I$, combining Prop. 11.9 with Thm. 3.16 implies that $\overline{V_{0,i}}$ is indeed a G_∞-subrepresentation of $\mathcal{A}^\infty_d(G)^{K_n,\mathcal{J}^n}$ and it satisfies $\left(\overline{V_{0,i}}\right)_{(K_\infty)} = V_{0,i}$. Now, suppose that $\sum_{i\in I}\phi_i = 0$, for $\phi_i \in \overline{V_{0,i}}$, where $\phi_i = 0$ for all but finitely many $i \in I$. We have to show that indeed $\phi_i = 0$ for all $i \in I$. To this end, consider an arbitrary irreducible representation (ρ_∞, W_∞) of K_∞. Since, $\sum_{i\in I}\phi_i = 0$ and this sum is finite, we have $\sum_{i\in I} E_{\rho_\infty}(\phi_i) = 0$. As $E_{\rho_\infty}(\phi_i) \in \left(\overline{V_{0,i}}\right)_{(K_\infty)} = V_{0,i}$, it follows that $E_{\rho_\infty}(\phi_i) = 0$ for all (ρ_∞, W_∞) and all $i \in I$. However, since $\mathcal{A}^\infty_d(G)^{K_n,\mathcal{J}^n}$ is a smooth representation of G_∞ (and hence of K_∞) by Prop. 11.9, [HCh66], Lem. 5, implies that

$$\phi_i = \sum_{[(\rho_\infty, W_\infty)]} E_{\rho_\infty}(\phi_i) = 0$$

for every $i \in I$. □

11.5 Remarks on the full space of all smooth-automorphic forms

In this section we will consider the space of *all* smooth-automorphic forms

$$\mathcal{A}^\infty(G) := \bigcup_{\mathcal{J}} \mathcal{A}^\infty_\mathcal{J}(G),$$

the union ranging over all ideals $\mathcal{J}$ of finite codimension in $\mathcal{Z}(\mathfrak{g}_\infty)$, and we will see how it compares to the individual spaces $\mathcal{A}^\infty_\mathcal{J}(G)$. In doing so, we shall also see why we give preference to studying the spaces $\mathcal{A}^\infty_\mathcal{J}(G)$ rather than the full space $\mathcal{A}^\infty(G)$, although it may seem to be the more natural object of interest.

The goal of this section is to prove the following theorem.

Theorem 11.25. *The natural map given by summation of functions induces an isomorphism of vector spaces*

$$\mathcal{A}^{\infty}(G) \cong \bigoplus_{\mathfrak{m}} \mathcal{A}^{\infty}_{\mathfrak{m}}(G),$$

the sum ranging over the family of all maximal ideals $\mathfrak{m}$ in $\mathcal{Z}(\mathfrak{g}_\infty)$. This sum is uncountable (unless in the trivial case that $\mathfrak{g}_\infty = \{0\}$).

This result does not seem to be (too) well-known, at least the only reference in the literature, known to the author of this book, is [Beu-Cha-Zyd21], §2.7.1, where the statement may be found without proof. In the proof below, we will follow closely some detailed explanations, which were communicated to us by Sonja Žunar. We start off with some analysis of ideals in polynomial algebras. For the rest of this section, we shall hence abbreviate $R := \mathbb{C}[X_1, \ldots, X_n]$ for some arbitrary, but fixed $n \in \mathbb{Z}_{>0}$.

Lemma 11.26. *Let $\mathcal{I}$ be an ideal of R of finite codimension $r \in \mathbb{Z}_{>0}$. Then, the number of maximal ideals of R containing $\mathcal{I}$ is at most r.*

Proof. Assume that there exist $r+1$ mutually distinct maximal ideals $\mathfrak{m}_1, \ldots, \mathfrak{m}_{r+1}$ of R containing $\mathcal{I}$. Note that for every j, $1 \leq j \leq r$, we have proper inclusions

$$\bigcap_{k=1}^{j} \mathfrak{m}_k \supset \bigcap_{k=1}^{j+1} \mathfrak{m}_k. \tag{11.27}$$

Indeed, if we had equality, then

$$\prod_{k=1}^{j} \mathfrak{m}_k \subseteq \bigcap_{k=1}^{j} \mathfrak{m}_k = \bigcap_{k=1}^{j+1} \mathfrak{m}_k \subseteq \mathfrak{m}_{j+1},$$

which, as $\mathfrak{m}_{j+1}$ is prime, implies that $\mathfrak{m}_k \subseteq \mathfrak{m}_{j+1}$ for some $k \in \{1, \ldots, j\}$, an obvious contradiction. Therefore,

$$\begin{aligned} r &= \dim_{\mathbb{C}} R/\mathcal{I} \\ &= \underbrace{\dim_{\mathbb{C}} R/\mathfrak{m}_1}_{\geq 1} + \sum_{j=1}^{r} \underbrace{\dim_{\mathbb{C}} \left(\cap_{k=1}^{j} \mathfrak{m}_k\right) / \left(\cap_{k=1}^{j+1} \mathfrak{m}_k\right)}_{\geq 1 \text{ by } (11.27)} \\ &\quad + \underbrace{\dim_{\mathbb{C}} \left(\cap_{k=1}^{r+1} \mathfrak{m}_k\right)/\mathcal{I}}_{\geq 0} \\ &\geq r+1, \end{aligned}$$

which is impossible. Hence, there are at most r maximal ideals of R containing $\mathcal{I}$. □

Lemma 11.28. *Following Lem. 11.26, let $\mathfrak{m}_1, \dots, \mathfrak{m}_d$ be the finitely many mutually distinct maximal ideals of R containing $\mathcal{I}$. Then, there exists an $N \in \mathbb{Z}_{>0}$ such that*

$$\mathcal{I} \supseteq \prod_{j=1}^{d} \mathfrak{m}_j^N.$$

Proof. As a consequence of Hilbert's Nullstellensatz, the radical of $\mathcal{I}$ may be written as the intersection

$$\operatorname{rad} \mathcal{I} = \bigcap_{j=1}^{d} \mathfrak{m}_j, \tag{11.29}$$

cf. [Rei95], Thm. 5.6 and Cor. 1.12 *ibidem*. Moreover, we claim that there exists an $N \in \mathbb{Z}_{>0}$ such that $\mathcal{I} \supseteq (\operatorname{rad} \mathcal{I})^N$. Indeed, in a noetherian ring, such as R, cf. [Rei95], Thm. 3.6, every ideal is finitely generated, cf. [Rei95], Prop.-Def. 3.2. So, let $\operatorname{rad} \mathcal{I}$ be generated by elements $x_1, ..., x_\ell$. Then, there are positive integers m_i, $1 \leq i \leq \ell$, such that $x_i^{m_i} \in \mathcal{I}$. It follows that every summand in the expansion of an expression of the form $\left(\sum_{i=1}^{\ell} \lambda_i x_i\right)^N$ with $\lambda_i \in R$ and $N := m_1 + ... + m_\ell$ is a multiple of some $x_i^{m_i}$, hence is in $\mathcal{I}$. Therefore, $\mathcal{I} \supseteq (\operatorname{rad} \mathcal{I})^N$. Combing this simple fact with (11.29), we obtain

$$\mathcal{I} \supseteq (\operatorname{rad} \mathcal{I})^N = \left(\bigcap_{j=1}^{d} \mathfrak{m}_j\right)^N \supseteq \left(\prod_{j=1}^{d} \mathfrak{m}_j\right)^N = \prod_{j=1}^{d} \mathfrak{m}_j^N.$$

□

Lemma 11.30. *Let A be a commutative ring with unity $e \in A$, let V be an A-module and let $v \in V$. If there exist mutually distinct maximal ideals $\mathfrak{m}_1, \dots, \mathfrak{m}_k, \mathfrak{m}'_1, \dots, \mathfrak{m}'_\ell$ of A and integers $a_1, \dots, a_k, b_1, \dots, b_\ell \in \mathbb{Z}_{>0}$ such that*

$$\mathfrak{m}_1^{a_1} \cdots \mathfrak{m}_k^{a_k}.v = (\mathfrak{m}'_1)^{b_1} \cdots (\mathfrak{m}'_\ell)^{b_\ell}.v = 0,$$

then $v = 0$.

Proof. Let v be as in the statement of the lemma. We consider the annihilating ideal $\operatorname{Ann}(v) := \{a \in A \mid a.v = 0\}$. Obviously it suffices to prove that $\operatorname{Ann}(v) = A$, since then $e \in \operatorname{Ann}(v)$, hence $e.v = 0$, i.e., $v = 0$.

Arguing by contradiction, suppose that $\mathrm{Ann}(v)$ was a proper ideal of A. Then, there exists a maximal ideal $\mathfrak{m}$ of A such that $\mathrm{Ann}(v) \subseteq \mathfrak{m}$. However, since $\mathfrak{m}$ is prime, the inclusions

$$\mathfrak{m}_1^{a_1} \cdots \mathfrak{m}_k^{a_k} \subseteq \mathrm{Ann}(v) \subseteq \mathfrak{m} \qquad \text{and} \qquad (\mathfrak{m}_1')^{b_1} \cdots (\mathfrak{m}_\ell')^{b_\ell} \subseteq \mathrm{Ann}(v) \subseteq \mathfrak{m}$$

imply that $\mathfrak{m}_j \subseteq \mathfrak{m}$ for some $j \in \{1, \dots, k\}$ and $\mathfrak{m}_t' \subseteq \mathfrak{m}$ for some $t \in \{1, \dots, \ell\}$, i.e., by the maximality of these ideals, that $\mathfrak{m}_j = \mathfrak{m} = \mathfrak{m}_t'$. But this contradicts the assumption that the maximal ideals $\mathfrak{m}_j$ and $\mathfrak{m}_t'$ are mutually distinct. Thus, $\mathrm{Ann}(v) = A$. □

For an ideal $\mathcal{I}$ of R and an R-module V, we will write

$$V^{\mathcal{I}} := \{v \in V \mid \mathcal{I}.v = 0\} \quad \text{and} \quad V_{\mathcal{I}} := \left\{v \in V \mid \mathcal{I}^j.v = 0 \text{ for some } j \in \mathbb{Z}_{>0}\right\}.$$

Proposition 11.31. *Let V be an R-module such that for every $v \in V$, the annihilating ideal $\mathrm{Ann}(v) = \{a \in R \mid a.v = 0\}$ is of finite codimension in R. Then, summation of elements in V induces an isomorphism*

$$V \cong \bigoplus_{\mathfrak{m}} V_{\mathfrak{m}},$$

the sum ranging over all maximal ideals $\mathfrak{m}$ of R.

Proof. We will first show that the sum of R-submodules $V_{\mathfrak{m}}$ on the right hand side is direct: Suppose we are given mutually distinct maximal ideals $\mathfrak{m}_1, \dots, \mathfrak{m}_s$ of R and vectors $v_j \in V_{\mathfrak{m}_j}$, $1 \leq j \leq s$, such that

$$v_1 + \ldots + v_s = 0. \tag{11.32}$$

By the definition of the R-modules $V_{\mathfrak{m}_j}$, there exists an $N \in \mathbb{Z}_{>0}$ such that

$$\mathfrak{m}_j^N.v_j = 0 \qquad \text{for all } j \in \{1, ..., s\}. \tag{11.33}$$

Moreover, for every $j \in \{1, ..., s\}$ we have

$$\left(\prod_{k \neq j} \mathfrak{m}_k^N\right).v_j \overset{(11.32)}{=} \left(\prod_{k \neq j} \mathfrak{m}_k^N\right).\left(-\sum_{l \neq j} v_l\right) = -\sum_{l \neq j} \underbrace{\left(\prod_{k \neq j} \mathfrak{m}_k^N\right).v_l}_{\overset{(11.33)}{=} 0} = 0. \tag{11.34}$$

Applying our general Lem. 11.30, it follows that $v_j = 0$ for all $j \in \{1, ..., s\}$. Hence, the sum of the R-submodules $V_{\mathfrak{m}}$, $\mathfrak{m}$ ranging over all maximal ideals $\mathfrak{m}$ of R, is indeed direct.

It remains to prove that for any $v \in V$, there exist finitely many maximal ideals $\mathfrak{m}_1, \ldots, \mathfrak{m}_s$ of R such that $v \in \sum_{j=1}^{s} V_{\mathfrak{m}_j}$. Since $\mathrm{Ann}(v)$ is an ideal of finite codimension in R, it follows from Lem. 11.26 that there exists an integer $d \in \mathbb{Z}_{>0}$ and mutually distinct maximal ideals $\mathfrak{m}_1, \ldots, \mathfrak{m}_d$ of R and integers $N_1, \ldots, N_d \in \mathbb{Z}_{>0}$ such that

$$\mathcal{J} := \prod_{j=1}^{d} \mathfrak{m}_j^{N_j} \subseteq \mathrm{Ann}(v).$$

Note that the ideal $\mathcal{J}$ is of finite codimension in R, hence the $\mathbb{C}$-algebra $\overline{R} := R/\mathcal{J}$ is a finite-dimensional complex vector space. Let us introduce the following notation: For $r \in R$, let

$$\overline{r} := r + \mathcal{J} \in R/\mathcal{J},$$

and, more generally, for a subset $S \subseteq R$, let

$$\overline{S} := (S + \mathcal{J})/\mathcal{J} \subseteq R/\mathcal{J}.$$

We observe that with this notation for every $j \in \{1, \ldots, d\}$ we have

$$\overline{\mathfrak{m}_j}^{N_j} = \overline{\mathfrak{m}_j}^{N_j+1}. \tag{11.35}$$

Indeed,

$$\begin{aligned}
\overline{\mathfrak{m}_j}^{N_j+1} &= \mathfrak{m}_j^{N_j+1} + \prod_k \mathfrak{m}_k^{N_k} \\
&= \mathfrak{m}_j^{N_j}\left(\mathfrak{m}_j + \prod_{k \neq j} \mathfrak{m}_k^{N_k}\right) \\
&= \mathfrak{m}_j^{N_j} R \\
&= \mathfrak{m}_j^{N_j} \\
&= \mathfrak{m}_j^{N_j} + \prod_k \mathfrak{m}_k^{N_k} \\
&= \overline{\mathfrak{m}_j}^{N_j},
\end{aligned}$$

where we remark that the third equality holds since, if $\mathfrak{m}_j + \prod_{k \neq j} \mathfrak{m}_k^{N_k}$ were a proper ideal of R, then a maximal ideal $\mathfrak{m}$ containing it would contain both $\mathfrak{m}_j$ and $\prod_{k \neq j} \mathfrak{m}_k^{N_k}$ and hence by its primeness both $\mathfrak{m}_j$ and $\mathfrak{m}_k$ for some $j \neq k$, which is impossible.

We now apply Nakayama's lemma, in the precise form to be found in [SLan02], Lem. X.4.1, to the localized ring $A = \left(\overline{R} \setminus \overline{\mathfrak{m}_j}\right)^{-1}\overline{R}$, the ideal

$\mathfrak{a} = \left(\overline{R} \setminus \overline{\mathfrak{m}_j}\right)^{-1}\overline{\mathfrak{m}_j}$ and the A-module $E = \left(\overline{R} \setminus \overline{\mathfrak{m}_j}\right)^{-1}\overline{\mathfrak{m}_j}^{N_j}$ and obtain that $\left(\overline{R} \setminus \overline{\mathfrak{m}_j}\right)^{-1}\overline{\mathfrak{m}_j}^{N_j+1} = 0$. Hence, recalling our observation (11.35) just made above, $\left(\overline{R} \setminus \overline{\mathfrak{m}_j}\right)^{-1}\overline{\mathfrak{m}_j}^{N_j} = 0$.

Fixing a basis $\overline{m_{j,1}}, \ldots, \overline{m_{j,d_j}}$ of the complex vector space $\overline{\mathfrak{m}_j}^{N_j}$, it follows that for every $k \in \{1, \ldots, d_j\}$ we have

$$\frac{\overline{m_{j,k}}}{1} = \frac{0}{1} \qquad \text{in } \left(\overline{R} \setminus \overline{\mathfrak{m}_j}\right)^{-1}\overline{R},$$

i.e., there exists $r_{j,k} \in R \setminus \mathfrak{m}_j$ such that $\overline{r_{j,k}}\,\overline{m_{j,k}} = 0$. But this implies that

$$\left(\prod_{k=1}^{d_j} \overline{r_{j,k}}\right)\overline{\mathfrak{m}_j}^{N_j} = 0.$$

Let $r_j := \prod_{k=1}^{d_j} r_{j,k}$. We have

$$\overline{r_j} \in \overline{R} \setminus \overline{\mathfrak{m}_j} \qquad \text{and} \qquad \overline{r_j}\,\overline{\mathfrak{m}_j}^{N_j} = 0. \tag{11.36}$$

We now put

$$r := \sum_{j=1}^{d} r_j \tag{11.37}$$

and claim that $\overline{r} \in \overline{R}^{\times}$. To this end, arguing by contradiction, suppose that $\overline{r} \notin \overline{R}^{\times}$. Then, $\overline{r}$ is contained in a maximal ideal of $\overline{R}$, i.e., in $\overline{\mathfrak{m}_j}$ for some $j \in \{1, \ldots, d\}$: Here we recall that the maximal ideals of $\overline{R}$ are exactly the ideals $\overline{\mathfrak{m}_1}, \ldots, \overline{\mathfrak{m}_d}$. Indeed, the maximal ideals of $\overline{R}$ are obviously the ideals of the form $\overline{\mathfrak{m}}$, where $\mathfrak{m}$ is a maximal ideal of R containing $\mathcal{J} = \mathfrak{m}_1^{N_1} \cdots \mathfrak{m}_d^{N_d}$, and the only such maximal ideals $\mathfrak{m}$ are the ideals $\mathfrak{m}_1, \ldots, \mathfrak{m}_d$, since $\mathfrak{m}$ being prime, the inclusion $\mathfrak{m}_1^{N_1} \cdots \mathfrak{m}_d^{N_d} \subseteq \mathfrak{m}$ implies that $\mathfrak{m}_j \subseteq \mathfrak{m}$ for some j. Having noted this, without loss of generality, we may suppose that

$$\overline{r} \in \overline{\mathfrak{m}_1}. \tag{11.38}$$

Since $\overline{r_1}\,\overline{\mathfrak{m}_1}^{N_1} = 0$ by (11.36),

$$\overline{r_1}\,\overline{r}^{N_1} = 0. \tag{11.39}$$

Next, note that if $j \neq k$, then by (11.36)

$$\overline{\mathfrak{m}_j}^{N_j}\overline{r_j}\,\overline{r_k} = \overline{\mathfrak{m}_k}^{N_k}\overline{r_j}\,\overline{r_k} = 0,$$

which by Lem. 11.30 implies that

$$\overline{r_j}\,\overline{r_k} = 0 \qquad \text{for } j \neq k. \tag{11.40}$$

Taking the N_1-th power of (11.37) and applying (11.40) we obtain that

$$\overline{r}^{N_1} = \sum_{j=1}^{d} \overline{r_j}^{N_1}, \tag{11.41}$$

which, multiplying (11.41) by $\overline{r_1}$ and using (11.39) and (11.40), yields

$$0 = \overline{r}_1^{N_1+1}. \tag{11.42}$$

Now we have

$$\overline{\mathfrak{m}_1} \overset{(11.38)}{\ni} \overline{r}^{N_1+1} \underset{(11.40)}{\overset{(11.37)}{=}} \sum_{j=1}^{d} \overline{r_j}^{N_1+1} \overset{(11.42)}{=} \sum_{j=2}^{d} \overline{r_j}^{N_1+1} \overset{(11.40)}{=} \left(\sum_{j=2}^{d} \overline{r_j} \right)^{N_1+1},$$

which, as $\overline{\mathfrak{m}_1}$ is prime, implies that

$$\sum_{j=2}^{d} \overline{r_j} \in \overline{\mathfrak{m}_1}, \tag{11.43}$$

hence

$$\overline{r_1} \overset{(11.37)}{=} \underbrace{\overline{r}}_{\in \overline{\mathfrak{m}_1} \text{ by } (11.38)} - \underbrace{\sum_{j=2}^{d} \overline{r_j}}_{\in \overline{\mathfrak{m}_1} \text{ by } (11.43)} \in \overline{\mathfrak{m}_1},$$

which is in contradiction with (11.36). This proves that $\overline{r} \in \overline{R}^{\times}$.

Knowing this, there hence exists $t \in R$ such that $\overline{t}\,\overline{r} = 1$ in $\overline{R}$, i.e., $1 = tr + a$ for some $a \in \mathcal{J}$. Therefore, coming back to our fixed $v \in V$, we have

$$v = 1.v = (tr + a).v = tr.v \overset{(11.37)}{=} \sum_{j=1}^{d} tr_j.v.$$

We observe that for every j, we have $tr_j.v \in V_{\mathfrak{m}_j}$: Indeed,

$$\mathfrak{m}_j^{N_j}.tr_j.v = tr_j\mathfrak{m}_j^{N_j}.v \subseteq \mathcal{J}.v = 0,$$

where the inclusion holds since $r_j\mathfrak{m}_j^{N_j} \subseteq \mathcal{J}$ by (11.36). Thus, $v \in \sum_{j=1}^{d} V_{\mathfrak{m}_j}$, which completes the proof. □

We may now give the:

Proof of Thm. 11.25: If $\mathfrak{g}_\infty = \{0\}$, then the result is trivial. So let us assume that $\mathfrak{g}_\infty \neq \{0\}$. Then, recalling that the algebra $\mathcal{Z}(\mathfrak{g}_\infty)$ is isomorphic to $\mathbb{C}[X_1, \dots, X_n]$ for some $n \in \mathbb{Z}_{>0}$, cf. [Bor07], §1.3, the direct sum decomposition $\mathcal{A}^\infty(G) \cong \bigoplus_{\mathfrak{m}} \mathcal{A}^\infty_{\mathfrak{m}}(G)$ is an immediate corollary of Prop. 11.31. Using the isomorphism of algebras $\mathcal{Z}(\mathfrak{g}_\infty) \cong \mathbb{C}[X_1, \dots, X_n]$ once more and the Weak Nullstellensatz, cf. [Rei95], Cor. 5.2, it is also clear that there are uncountable many maximal ideals in $\mathcal{Z}(\mathfrak{g}_\infty)$. □

Thm. 11.25 gives us an ad hoc method to topologize the space of all smooth-automorphic forms, simply by letting

$$\mathcal{A}^\infty(G) = \bigoplus_{\mathfrak{m}} \mathcal{A}^\infty_{\mathfrak{m}}(G), \tag{11.44}$$

i.e., we equip $\mathcal{A}^\infty(G)$ with the direct sum topology, cf. §1.2.2, coming from the LF-summands $\mathcal{A}^\infty_{\mathfrak{m}}(G)$. Clearly, as the sum is uncountable (unless $\mathfrak{g}_\infty = \{0\}$), $\mathcal{A}^\infty(G)$ is not an LF-space itself, though it is a complete seminormed space by Lem. 1.12. It is this lack of an LF-structure (and hence the absence of the Open Mapping Theorem), which made us give preference to working with the spaces $\mathcal{A}^\infty_{\mathcal{J}}(G)$ rather than with the full space $\mathcal{A}^\infty(G)$.

To make full circle, let us now compare the seminormed space $\mathcal{A}^\infty(G)$ with its smaller siblings $\mathcal{A}^\infty_{\mathcal{J}}(G)$. Our result in this regard is as follows:

Proposition 11.45. *Suppose that $\mathcal{J} \lhd \mathcal{Z}(\mathfrak{g}_\infty)$ is an ideal of finite codimension. We let $\mathfrak{m}_1, \dots, \mathfrak{m}_d$ and $N \in \mathbb{Z}_{>0}$ be as in the statement of Lem. 11.28. Then,*

$$\mathcal{A}^\infty_{\mathcal{J}}(G) \cong \bigoplus_{j=1}^{d} \mathcal{A}^\infty_{\mathfrak{m}_j}(G).$$

In particular, the LF-space $\mathcal{A}^\infty_{\mathcal{J}}(G)$ is a closed topological vector subspace of $\mathcal{A}^\infty(G)$.

Proof. Firstly, we observe that by the very proof of Prop. 11.31, we know that if V is any R-module and $\mathfrak{m}_1, \dots, \mathfrak{m}_k$ are mutually distinct maximal ideals of R and $M \in \mathbb{Z}_{>0}$, then,

$$V^{\mathfrak{m}_1^M \cdots \mathfrak{m}_k^M} = \bigoplus_{j=1}^{k} V^{\mathfrak{m}_j^M}. \tag{11.46}$$

Now, let $\mathcal{J}$ be an arbitrary, but fixed ideal of $\mathcal{Z}(\mathfrak{g}_\infty)$ of finite codimension and let $\mathfrak{m}_1$,..., $\mathfrak{m}_d$ and $N \in \mathbb{Z}_{>0}$ be as in Lem. 11.28. Then,

$$\prod_{j=1}^{d} \mathfrak{m}_j^N \subseteq \mathcal{J} \subseteq \bigcap_{j=1}^{d} \mathfrak{m}_j = \prod_{j=1}^{d} \mathfrak{m}_j.$$

It follows that for every $n \in \mathbb{N}$ the identity map defines continuous inclusions

$$\mathcal{A}^{\infty}_{d_0}(G)^{K_n,\mathfrak{m}_1^n\cdots\mathfrak{m}_d^n} \hookrightarrow \mathcal{A}^{\infty}_{d_0}(G)^{K_n,\mathcal{J}^n} \hookrightarrow \mathcal{A}^{\infty}_{d_0}(G)^{K_n,\mathfrak{m}_1^{nN}\cdots\mathfrak{m}_d^{nN}},$$

where we used d_0 as we did right below Prop. 11.12. Whence, Cor. 8.4 shows that $\mathcal{A}^{\infty}_{\mathcal{J}}(G) = \varinjlim \mathcal{A}^{\infty}_{d_0}(G)^{K_n,\mathcal{J}^n}$ is isomorphic to $\varinjlim \mathcal{A}^{\infty}_{d_0}(G)^{K_n,\mathfrak{m}_1^n\cdots\mathfrak{m}_d^n}$ as seminormed spaces. However, for the latter space we obtain

$$\begin{aligned} \varinjlim \mathcal{A}^{\infty}_{d_0}(G)^{K_n,\mathfrak{m}_1^n\cdots\mathfrak{m}_d^n} &\overset{(11.46)}{=} \varinjlim \bigoplus_{j=1}^{d} \mathcal{A}^{\infty}_{d_0}(G)^{K_n,\mathfrak{m}_j^n} \\ &= \bigoplus_{j=1}^{d} \varinjlim \mathcal{A}^{\infty}_{d_0}(G)^{K_n,\mathfrak{m}_j^n} \\ &= \bigoplus_{j=1}^{d} \mathcal{A}^{\infty}_{\mathfrak{m}_j}(G). \end{aligned}$$

This shows the claim. □

Chapter 12

Automorphic Representations and Smooth-Automorphic Representations

12.1 First facts and definitions

Definition 12.1. A $G(\mathbb{A})$-representation (π, V) is called a *smooth-automorphic representation*, if $V = U/W$, where $W \subseteq U$ are $G(\mathbb{A})$-subrepresentations of $(R, \mathcal{A}^\infty_{\mathcal{J}}(G))$ for some ideal $\mathcal{J} \lhd \mathcal{Z}(\mathfrak{g}_\infty)$ of finite codimension. If $W = \{0\}$, then (π, V) is called a *smooth-automorphic subrepresentation.*

Obviously, by Thm. 11.17 and Prop. 10.23 every smooth-automorphic representation is smooth, hence there is no hidden conflict of terminology in our use of the word "smooth". On the other hand, it should be noted carefully that by assuming that V is a representation, we explicitly *suppose* that the quotient $U/W = V$ is complete and that the induced action is continuous, as this is generally wrong for representations of $G(\mathbb{A})$ as pointed out after Def. 10.1. However, in most practical situations, it follows automatically that a pair $W \subseteq U$ of $G(\mathbb{A})$-subrepresentations of $\mathcal{A}^\infty_{\mathcal{J}}(G)$ gives rise to a representation on the quotient U/W: This holds, for instance, if the quotients U^{K_n}/W^{K_n} are complete and barreled for $n \gg 0$, because then so is U/W by (10.29) and Lem. 8.2, whence $(R, U/W)$ is a representation of $G(\mathbb{A})$ by [BouII04], VIII, §2, Prop. 1. More specific and very useful in practice is the following.

Lemma 12.2. *If a smooth-automorphic subrepresentation U is annihilated by a power of $\mathcal{J}$, it is a Casselman–Wallach representation of $G(\mathbb{A})$ and so is every quotient U/W for a subrepresentation $W \subseteq U$. In particular, this holds, if U is finitely generated as a $G(\mathbb{A})$-representation.*

Proof. Suppose that there is a $k \in \mathbb{N}$ such that $\mathcal{J}^k \cdot U = \{0\}$. Then, (recalling once more that the actions of G_∞ and $G(\mathbb{A}_f)$ commute) U^{K_n} is a G_∞-subrepresentation of $\mathcal{A}^\infty_d(G)^{K_n,\mathcal{J}^n} \subseteq \mathcal{A}^\infty_\mathcal{J}(G)$ for all $n \geq k$. Now, as $\mathcal{A}^\infty_d(G)^{K_n,\mathcal{J}^n}$ inherits from $\mathcal{A}^\infty_\mathcal{J}(G)$ its original Fréchet space topology by Lem. 8.2 and Prop. 11.9, U^{K_n} is a Casselman–Wallach representation of G_∞ for $n \geq k$ by Prop. 11.9 and Lem. 3.27 and hence, again using Lem. 3.27, U^{K_f} is a Casselman–Wallach representation of G_∞ for every open compact subgroup K_f of $G(\mathbb{A}_f)$. Therefore, recalling Prop. 10.23 and Thm. 11.17, U is a Casselman–Wallach representation of $G(\mathbb{A})$ by definition, cf. Def. 10.50, and consequently so is the quotient U/W for every subrepresentation W of U by Prop. 10.51. The remaining assertion follows from this as every of the finitely many generators φ_i of U must lie in some $\mathcal{A}^\infty_d(G)^{K_{n_i},\mathcal{J}^{n_i}}$, which implies that U is annihilated by the power $\mathcal{J}^{\max_i n_i}$. □

As observed right after Def. 10.50, if the conditions of Lem. 12.2 are satisfied, U and all of its quotients U/W are LF-spaces and in fact admissible $G(\mathbb{A})$-representations by Lem. 10.34. In particular we obtain the following.

Corollary 12.3. *Let $\varphi \in \mathcal{A}^\infty_\mathcal{J}(G)$. Then the smooth-automorphic subrepresentation $V_\varphi := \mathrm{Cl}_{\mathcal{A}^\infty_\mathcal{J}(G)}(\langle R(G(\mathbb{A}))\varphi\rangle)$ spanned by φ is admissible.*

Remark 12.4. The reader will have noticed that admissibility of V_φ does not follow from a combination of the sheer fact that this representation is generated by one vector and some abstract considerations, but rather needed that every single smooth-automorphic form is $\mathcal{Z}(\mathfrak{g}_\infty)$-finite. Indeed, it may fail that an abstract $G(\mathbb{A})$-representation, which is generated by one vector, is admissible.

Remark 12.5. In §11.5 we argued, why we focused on the spaces $\mathcal{A}^\infty_\mathcal{J}(G)$, rather than on the full space of all smooth-automorphic forms $\mathcal{A}^\infty(G)$ – its natural topology involves an uncountable limit and is therefore too little convenient. Here, Lem. 12.2 provides us a reason for why we do not concentrate on the smaller space $\mathcal{A}^\infty(G)^\mathcal{J}$ of all smooth-automorphic forms, which are annihilated by a fixed ideal $\mathcal{J} \lhd \mathcal{Z}(\mathfrak{g}_\infty)$ (rather than by an arbitrary power of such an ideal): Its representation theory would be "too convenient" as then all smooth-automorphic representations – irreducible or not – would be admissible, which the author of this book considers unnatural.

We continue with the fundamental definition of an automorphic representation.

Definition 12.6. A $(\mathfrak{g}_\infty, K_\infty, G(\mathbb{A}_f))$-module (π_0, V_0) is called an *automorphic representation*, if $V_0 = U_0/W_0$, where $W_0 \subseteq U_0$ are

$(\mathfrak{g}_\infty, K_\infty, G(\mathbb{A}_f))$-submodules of $(R_0, \mathcal{A}_\mathcal{J}(G))$ for some ideal $\mathcal{J} \lhd \mathcal{Z}(\mathfrak{g}_\infty)$ of finite codimension. If $W_0 = \{0\}$, then (π_0, V_0) is called an *automorphic subrepresentation.*

Of course, unlike the expression "automorphic representation" might naively suggest, an automorphic representation *is not* a representation of $G(\mathbb{A})$, but only a $(\mathfrak{g}_\infty, K_\infty, G(\mathbb{A}_f))$-module. However, as this is the way the reader will find the notion of an automorphic representation in the literature, we will surrender to this abuse of terminology. To some excuse, the results of our next section – taken from [Gro-Žun23], §3 – serve as a way to partly legitimate this notional faux pas.

12.2 A dictionary between smooth-automorphic representations and automorphic representations

Proposition 12.7. *Let V_0 be an automorphic subrepresentation of $\mathcal{A}_\mathcal{J}(G)$. Its topological closure in $\mathcal{A}^\infty_\mathcal{J}(G)$, $V := \mathrm{Cl}_{\mathcal{A}^\infty_\mathcal{J}(G)}(V_0)$, is then a smooth-automorphic subrepresentation.*

Proof. Obviously, it suffices to prove that $R(G(\mathbb{A}))V \subseteq V$, i.e., that V is $G(\mathbb{A})$-invariant under right-translation. To this end, let

$$U := \mathrm{Cl}_{\mathcal{A}^\infty_\mathcal{J}(G)}(\langle R(G(\mathbb{A}))V_0 \rangle) \tag{12.8}$$

be the topological closure in $\mathcal{A}^\infty_\mathcal{J}(G)$ of the vector space spanned by all $R(g)v_0$, $g \in G(\mathbb{A})$ and $v_0 \in V_0$. By construction, U is the smallest smooth-automorphic subrepresentation containing V_0. Moreover, clearly U contains V, whence it is enough to show that $U = V$. Arguing by contradiction, suppose that $U \supsetneq V$.

As V is closed in U, we may apply the Hahn-Banach theorem to the pair $V \subsetneq U$, cf. [Jar81], Sect. 7.2, Cor. 2.(a), and hence obtain a non-zero continuous linear functional $u : U \to \mathbb{C}$, which vanishes constantly on V. Given $\phi \in V_0$ and $g \in G(\mathbb{A})$, let $\varphi_\phi(g) := u(R(g)\phi)$. We claim that $\varphi_\phi \in C^\infty(G(\mathbb{A}))$. Indeed, we have $\varphi_\phi = u \circ c_\phi$, and ϕ being in $V_0 \subseteq \mathcal{A}^\infty_\mathcal{J}(G)$, the orbit map c_ϕ is a smooth function $G(\mathbb{A}) \to U \subseteq \mathcal{A}^\infty_\mathcal{J}(G)$ by Thm. 11.17. Hence, as moreover $u : U \to \mathbb{C}$ is continuous and so bounded, $\varphi_\phi(\_ \cdot g_f) : G_\infty \to \mathbb{C}$ maps smooth curves to smooth curves by [Kri-Mic97], Cor. 2.11, and therefore is a smooth function by Cor. 3.14 *ibidem*, for every fixed $g_f \in G(\mathbb{A}_f)$. Again by Thm. 11.17, φ_ϕ is uniformly locally constant, hence, in summary, φ_ϕ is smooth in the sense of Def. 10.3, i.e., in $C^\infty(G(\mathbb{A}))$.

We now claim that φ_ϕ is also $\mathcal{Z}(\mathfrak{g}_\infty)$-finite. In order to see this, observe that by the just verified smoothness of φ_ϕ, for every $X \in \mathfrak{g}_\infty$ and $g \in G(\mathbb{A})$, we have the (well-defined) equalities

$$(X \cdot \varphi_\phi)(g) = \frac{\mathrm{d}}{\mathrm{d}t}\bigg|_{t=0} u(R(g\exp(tX))\phi) = u(R(g)R(X)\phi) = \varphi_{X\cdot\phi}(g),$$

and obviously these extend linearly to all $D \in \mathcal{U}(\mathfrak{g}_\infty)$, i.e., we finally obtain

$$(D \cdot \varphi_\phi)(g) = \varphi_{D\cdot\phi}(g) = u(R(g)R(D)\phi) \tag{12.9}$$

for all $D \in \mathcal{U}(\mathfrak{g}_\infty)$ and $g \in G(\mathbb{A})$. But this shows that φ_ϕ is $\mathcal{Z}(\mathfrak{g}_\infty)$-finite, as claimed above: Indeed, by assumption so is ϕ, i.e., slightly more precisely, there is an $n \in \mathbb{N}$ such that $\mathcal{J}^n \cdot \phi = 0$. By (12.9), also $\mathcal{J}^n \cdot \varphi_\phi = 0$, hence φ_ϕ is indeed $\mathcal{Z}(\mathfrak{g}_\infty)$-finite as asserted.

Next, by an obvious calculation, the reader may verify that φ_ϕ is $K_\mathbb{A}$-finite on the right. Consequently, for every $g_f \in G(\mathbb{A}_f)$, the function

$$\varphi_\phi(\_\_ \cdot g_f) : G^\circ_\infty \to \mathbb{C}$$

on the connected component of the identity $id_\infty \in G_\infty$ is smooth, K_∞-finite and $\mathcal{Z}(\mathfrak{g}_\infty)$-finite. This implies that $\varphi_\phi(\_\_ \cdot g_f)$ is a real analytic function: In fact, the arguments presented in [Bor07], §3.15, transfer verbatim to the situation at hand (Exercise!), showing that $\varphi_\phi(\_\_ \cdot g_f)$ is annihilated by a so-called elliptic differential operator. Analyticity of $\varphi_\phi(\_\_ \cdot g_f)$ now follows from a general result in partial differential equations, cf. [LBer-Sch71], Appendix to §4. As a consequence, we may suppose that $\varphi_\phi(\_\_ \cdot g_f)$ is represented by a multi-variable Taylor series around the identity $id_\infty \in G^\circ_\infty$.

However, as

$$(D \cdot \varphi_\phi(\_\_ \cdot g_f))(id_\infty) = (D \cdot \varphi_\phi)(g_f) \overset{(12.9)}{=} u(R(g_f)R(D)\phi) = 0,$$

for all $D \in \mathcal{U}(\mathfrak{g}_\infty)$ (since $R(g_f)R(D)\phi \in V$, on which u is supposed to be identically zero), all partial derivatives of $\varphi_\phi(\_\_ \cdot g_f)$ vanish at $id_\infty \in G^\circ_\infty$. But this shows that the above Taylor series of $\varphi_\phi(\_\_ \cdot g_f)$ is constantly zero, whence so is $\varphi_\phi(\_\_ \cdot g_f)$, i.e., $\varphi_\phi(g_\infty g_f) = 0$, for all $g_\infty \in G^\circ_\infty$ and $g_f \in G(\mathbb{A}_f)$. Recalling that $\phi \in V_0$ in the definition of φ_ϕ was arbitrary, we obtain that our continuous linear form u must be zero on the whole space

$$\begin{aligned}\mathrm{Cl}_{\mathcal{A}^\infty_{\mathcal{J}}(G)}(\langle R(G^\circ_\infty \times G(\mathbb{A}_f))V_0\rangle) &= \mathrm{Cl}_{\mathcal{A}^\infty_{\mathcal{J}}(G)}(\langle R(G^\circ_\infty \times G(\mathbb{A}_f))R(K_\infty)V_0\rangle)\\ &= \mathrm{Cl}_{\mathcal{A}^\infty_{\mathcal{J}}(G)}(\langle R(G^\circ_\infty K_\infty \times G(\mathbb{A}_f))V_0\rangle)\\ &= \mathrm{Cl}_{\mathcal{A}^\infty_{\mathcal{J}}(G)}(\langle R(G(\mathbb{A}))V_0\rangle)\\ &\overset{(12.8)}{=} U,\end{aligned}$$

where the first equality follows from V_0 being by assumption K_∞-invariant under right-translation, the second equality holds as the action of K_∞ and $G(\mathbb{A}_f)$ commute and the third equality follows from [Kna02], Prop. 7.19.(b). However, this implies that u is identically zero – a contradiction. Hence, $V = U$ and therefore V is $G(\mathbb{A})$-invariant under right-translation. □

Theorem 12.10. *The assignments $V \mapsto V_{(K_\infty)}$ and $V_0 \mapsto \mathrm{Cl}_{\mathcal{A}^\infty_{\mathcal{J}}(G)}(V_0)$ set up pairwise inverse bijections between the family of all admissible smooth-automorphic subrepresentations V of $\mathcal{A}^\infty_{\mathcal{J}}(G)$ and the family of all admissible automorphic subrepresentations V_0 of $\mathcal{A}_{\mathcal{J}}(G)$. This correspondence respects irreducibility.*

Proof. We repeat a variant of the proof of [Gro-Žun23], Thm. 3.7. To start off, we recall that by [Bor-Jac79], Prop. 4.5.(4)[1], every irreducible automorphic subrepresentation V_0 is admissible. Therefore, if we are able to show that the underlying $(\mathfrak{g}_\infty, K_\infty, G(\mathbb{A}_f))$-module of an irreducible smooth-automorphic subrepresentation V of $\mathcal{A}^\infty_{\mathcal{J}}(G)$ remains irreducible, then – V being a smooth $G(\mathbb{A})$-representation by Prop. 11.17 and Prop. 10.23 – V is admissible by Lem. 10.43. It is therefore enough to prove the following four claims:

(1) Let V_0 be an admissible $(\mathfrak{g}_\infty, K_\infty, G(\mathbb{A}_f))$-submodule of $\mathcal{A}_{\mathcal{J}}(G)$. Then, $V = \mathrm{Cl}_{\mathcal{A}^\infty_{\mathcal{J}}(G)}(V_0)$ is an admissible $G(\mathbb{A})$-representation, and we have $V_{(K_\infty)} = V_0$.
(2) Let V be an admissible smooth-automorphic subrepresentation. Then, $V_0 = V_{(K_\infty)}$ is an admissible $(\mathfrak{g}_\infty, K_\infty, G(\mathbb{A}_f))$-module, and we have $\mathrm{Cl}_{\mathcal{A}^\infty_{\mathcal{J}}(G)}(V_0) = V$.
(3) Let V_0 be an irreducible $(\mathfrak{g}_\infty, K_\infty, G(\mathbb{A}_f))$-submodule of $\mathcal{A}_{\mathcal{J}}(G)$. Then, $V = \mathrm{Cl}_{\mathcal{A}^\infty_{\mathcal{J}}(G)}(V_0)$ is an irreducible $G(\mathbb{A})$-representation.

[1]The careful reader will also want to use [Bum98], Thm. 3.4.2, Prop. 3.4.4 and Prop. 3.4.8 together with our Lem. 10.42, in order to see why [Bor-Jac79], Prop. 4.5.(4) implies that an irreducible automorphic subrepresentation is an admissible $(\mathfrak{g}_\infty, K_\infty, G(\mathbb{A}_f))$-module.

(4) Let V be an irreducible smooth-automorphic subrepresentation. Then, the $(\mathfrak{g}_\infty, K_\infty, G(\mathbb{A}_f))$-module $V_0 = V_{(K_\infty)}$ is irreducible.

Proof of (1): By Prop. 12.7, V is a $G(\mathbb{A})$-subrepresentation of $\mathcal{A}^\infty_{\mathcal{J}}(G)$, and hence smooth by Prop. 11.17 and Prop. 10.23. By our very construction, V_0 is a $(\mathfrak{g}_\infty, K_\infty, G(\mathbb{A}_f))$-submodule of $V_{(K_\infty)}$ and dense in V. However, as V_0 is assumed to be admissible, Prop. 10.48 implies that $V_{(K_\infty)} = V_0$ and so that the smooth $G(\mathbb{A})$-representation V is admissible.

Proof of (2): Admissibility of V_0 is a trivial consequence of its definition, knowing that V is admissible. The remaining assertion follows from $V_{(K_\infty)}$ being dense in V, cf. [Var77], Lem. II.7.10.

Proof of (3): As observed in the beginning of our proof, V_0 is admissible, so by (1) above, V is a $G(\mathbb{A})$-subrepresentation of $\mathcal{A}^\infty_{\mathcal{J}}(G)$ satisfying $V_{(K_\infty)} = V_0$. Let $U \neq \{0\}$ be a $G(\mathbb{A})$-subrepresentation of V. Since $U_{(K_\infty)}$ is dense in U, see again [Var77], Lem. II.7.10, $U_{(K_\infty)} \neq \{0\}$, which by the irreducibility of V_0 implies that $U_{(K_\infty)} = V_0$. Therefore, $U = \mathrm{Cl}_{\mathcal{A}^\infty_{\mathcal{J}}(G)}(V_0) = V$ and so V is irreducible.

Proof of (4): As $V_{(K_\infty)}$ is dense in V, we must have $V_{(K_\infty)} \neq \{0\}$. Thus, it is enough to show that for every $0 \neq \phi \in V_{(K_\infty)}$, the smallest $(\mathfrak{g}_\infty, K_\infty, G(\mathbb{A}_f))$-submodule $V_{0,\phi} := \langle \phi \rangle_{(\mathfrak{g}_\infty, K_\infty, G(\mathbb{A}_f))}$ of $\mathcal{A}_{\mathcal{J}}(G)$ containing ϕ satisfies

$$V_{0,\phi} = V_{(K_\infty)}.$$

Combining Prop. 12.7 and the irreducibility of V, we get $\mathrm{Cl}_{\mathcal{A}^\infty_{\mathcal{J}}(G)}(V_{0,\phi}) = V$. Hence, applying (1) to $V_{0,\phi}$ – which, being spanned by one single automorphic form, is admissible by [Bor-Jac79], Prop. 4.5.(4) – yields $V_{(K_\infty)} = V_{0,\phi}$. □

The next theorem shows that the above dictionary between admissible smooth-automorphic subrepresentations and admissible automophic subrepresentations extends naturally to general subquotients, if these are assumed irreducible.

Theorem 12.11. *If V is an irreducible smooth-automorphic representation, then $V_{(K_\infty)}$ is an irreducible automorphic representation. In particular, every irreducible smooth-automorphic representation is admissible. Conversely, let V_0 be an irreducible automorphic representation. Then,* *there exists an irreducible smooth-automorphic representation V such that $V_{(K_\infty)} \cong V_0$.*

Proof. "$\Rightarrow$": Let $V = U/W$ be an irreducible smooth-automorphic $G(\mathbb{A})$-representation. In order to show that the $(\mathfrak{g}_\infty, K_\infty, G(\mathbb{A}_f))$-module $V_{(K_\infty)}$ is irreducible, let us denote by $\underline{\varphi}$ (respectively, $\underline{S}$) the image of a $\varphi \in \mathcal{A}^\infty_{\mathcal{J}}(G)$ (respectively, of a subset $S \subseteq \mathcal{A}^\infty_{\mathcal{J}}(G)$) under the canonical projection

$$\mathcal{A}^\infty_{\mathcal{J}}(G) \twoheadrightarrow \mathcal{A}^\infty_{\mathcal{J}}(G)/W.$$

Once more we recall that $U_{(K_\infty)}$ is dense in U by [Var77], Lem. II.7.10, whence we must have $U_{(K_\infty)} \setminus W \neq \emptyset$. It is therefore enough to prove that for every $\phi \in U_{(K_\infty)}$, which is not in W, the $(\mathfrak{g}_\infty, K_\infty, G(\mathbb{A}_f))$-submodule

$$V_{0,\phi} = \langle \underline{\phi} \rangle_{(\mathfrak{g}_\infty, K_\infty, G(\mathbb{A}_f))} = \underline{\langle \phi \rangle_{(\mathfrak{g}_\infty, K_\infty, G(\mathbb{A}_f))}}$$

of $V_{(K_\infty)}$, which is spanned by ϕ, equals $V_{(K_\infty)}$.

In order to see this, we note that by Lem. 12.7,

$$\mathrm{Cl}_U\left(\langle \phi \rangle_{(\mathfrak{g}_\infty, K_\infty, G(\mathbb{A}_f))} + W_{(K_\infty)}\right) = \mathrm{Cl}_{\mathcal{A}^\infty_{\mathcal{J}}(G)}\left(\langle \phi \rangle_{(\mathfrak{g}_\infty, K_\infty, G(\mathbb{A}_f))} + W_{(K_\infty)}\right)$$

is a $G(\mathbb{A})$-subrepresentation of U. Clearly, it contains W as a proper subspace, so, by the irreducibility of V, it equals U. It follows that $U_{0,\phi} := \langle \phi \rangle_{(\mathfrak{g}_\infty, K_\infty, G(\mathbb{A}_f))} + W_{(K_\infty)}$ is dense in U, hence $\underline{U_{0,\phi}} = V_{0,\phi}$ is dense in $\underline{U} = V$. Recall once more that $V_{0,\phi}$ is an admissible $(\mathfrak{g}_\infty, K_\infty, G(\mathbb{A}_f))$-submodule of $V_{(K_\infty)}$ by [Bor-Jac79], Prop. 4.5.(4). Hence, as V is also a smooth $G(\mathbb{A})$-representation by Prop. 11.17 and Prop. 10.23, our Prop. 10.48 finally implies that $V_{0,\phi} = V_{(K_\infty)}$, as desired. Hence, $V_{(K_\infty)}$ is irreducible as claimed.

Admissibility of V now follows from Prop. 10.48.

"$\Leftarrow$": Let $W_0 \subseteq U_0$ be $(\mathfrak{g}_\infty, K_\infty, G(\mathbb{A}_f))$-submodules of $\mathcal{A}_{\mathcal{J}}(G)$ such that $U_0/W_0 = V_0$. Let us fix a $\phi_0 \in U_0$, which is not in W_0 and denote by $U_1 := \langle \phi_0 \rangle_{(\mathfrak{g}_\infty, K_\infty, G(\mathbb{A}_f))}$ the $(\mathfrak{g}_\infty, K_\infty, G(\mathbb{A}_f))$-submodule of U_0, which is spanned by ϕ. By the irreducibility of U_0/W_0, the canonical $(\mathfrak{g}_\infty, K_\infty, G(\mathbb{A}_f))$-equivariant linear map $U_1 \to U_0/W_0$ is surjective, hence, denoting its kernel by W_1, we have isomorphisms of $(\mathfrak{g}_\infty, K_\infty, G(\mathbb{A}_f))$-modules:

$$U_1/W_1 \cong U_0/W_0 = V_0. \tag{12.12}$$

Next, we observe that [Bor-Jac79], Prop. 4.5.(4) implies that the $(\mathfrak{g}_\infty, K_\infty, G(\mathbb{A}_f))$-submodules $W_1 \subseteq U_1$ of $\mathcal{A}_{\mathcal{J}}(G)$ are admissible, whence by Thm. 12.10 their topological closures $\overline{W}_1 := \mathrm{Cl}_{\mathcal{A}^\infty_{\mathcal{J}}(G)}(W_1)$ and $\overline{U}_1 := \mathrm{Cl}_{\mathcal{A}^\infty_{\mathcal{J}}(G)}(U_1)$ in $\mathcal{A}^\infty_{\mathcal{J}}(G)$ are admissible smooth-automorphic subrepresentations satisfying

$$(\overline{W}_1)_{(K_\infty)} = W_1 \qquad \text{and} \qquad (\overline{U}_1)_{(K_\infty)} = U_1. \tag{12.13}$$

Let us now fix a $k \in \mathbb{Z}_{>0}$ such that $\mathcal{J}^k \cdot \phi_0 = 0$. Since the action of $\mathcal{J}^k$ is continuous and commutes with the action of $G(\mathbb{A})$, the whole $G(\mathbb{A})$-representation $\overline{U}_1$ is annihilated by $\mathcal{J}^k$, whence Lem. 12.2 shows that the quotient $V := \overline{U}_1/\overline{W}_1$ is indeed a representation of $G(\mathbb{A})$ (in fact, even a Casselman–Wallach representation).

We claim that $V_{(K_\infty)} \cong V_0$. In order to see this, we convince ourselves that for every isomorphism class $[(\rho_\infty, W_\infty)]$ of irreducible K_∞-representations the definition of the projector E_{ρ_∞} defined in (2.32) carries over verbatim from the local archimedean case to the global representation V, whence [Wal89], Lem. 3.3.3 implies that we have an isomorphism of $(\mathfrak{g}_\infty, K_\infty, G(\mathbb{A}_f))$-modules:

$$V_{(K_\infty)} \cong \bigoplus_{[(\rho_\infty, W_\infty)]} E_{\rho_\infty}(\overline{U}_1/\overline{W}_1).$$

But this shows that as $(\mathfrak{g}_\infty, K_\infty, G(\mathbb{A}_f))$-modules

$$\begin{aligned} V_{(K_\infty)} &\cong \bigoplus_{[(\rho_\infty, W_\infty)]} (E_{\rho_\infty}(\overline{U}_1) + \overline{W}_1)/\overline{W}_1 \\ &\overset{\text{[Wal89], Lem. 3.3.3}}{\cong} ((\overline{U}_1)_{(K_\infty)} + \overline{W}_1)/\overline{W}_1 \\ &\overset{(12.13)}{=} (U_1 + \overline{W}_1)/\overline{W}_1. \end{aligned}$$

Observe that the canonical map $U_1/W_1 \to (U_1 + \overline{W}_1)/\overline{W}_1$ is an isomorphism: Indeed, surjectivity being obvious, we easily see that the map is also injective as $U_1 \cap \overline{W}_1 = (\overline{W}_1)_{(K_\infty)} \overset{(12.13)}{=} W_1$. Hence, finally,

$$V_{(K_\infty)} \cong U_1/W_1 \overset{(12.12)}{\cong} V_0.$$

We still need to argue why V is irreducible. However, knowing that $V_{(K_\infty)} \cong V_0$, this follows exactly as in the proof of claim (3) in Thm. 12.10. $\square$

Corollary 12.14. *The underlying* $(\mathfrak{g}_\infty, K_\infty, G(\mathbb{A}_f))$*-module* $V_{(K_\infty)}$ *of an irreducible smooth-automorphic representation* V *allows a completion as an irreducible smooth-automorphic Casselman–Wallach representation of* $G(\mathbb{A})$*. This completion is unique up to isomorphism of* $G(\mathbb{A})$*-representations.*

Proof. Going once more through the proof of Thm. 12.11 shows the first assertion. The second assertion is a direct consequence of Prop. 10.51.(2). □

12.3 The restricted tensor product theorem

12.3.1 *The algebraic restricted tensor product*

In order to deal with smooth-automorphic representations, we will need an algebraic version of the ideas studied in §8.2.2 and Rem. 10.39.

Let $(V_i)_{i\in I}$ be a family of vector spaces, indexed by a set I, such that for all but finitely many i, we are given a fixed choice of non-zero vector $x_i^\circ \in V_i$. Formulated slightly more rigorously, we assume to be given a set I, a finite subset $I_0 \subseteq I$, a family of vector spaces $(V_i)_{i\in I}$ and an element $(x_i^\circ)_{i\in I\setminus I_0} \in \prod_{i\in I\setminus I_0}(V_i \setminus \{0\})$. Then, the family,

$$\mathcal{I} := \{J \subseteq I : I_0 \subseteq J \text{ and } J \text{ is finite}\}$$

is a directed set with respect to inclusion. If $J, J' \in \mathcal{I}$ with $J \subseteq J'$, we define a linear map

$$\lambda_{J,J'} : \bigotimes_{i\in J} V_i \to \bigotimes_{i\in J'} V_i$$

as the linear extension of the assignment

$$\bigotimes_{i\in J} x_i \mapsto \bigotimes_{i\in J} x_i \otimes \bigotimes_{i\in J'\setminus J} x_i^\circ.$$

Then, $\mathcal{I}$ together with the maps $\lambda_{J,J'}$ forms a direct system, and we may hence consider its algebraic direct limit $\varinjlim_{J\in\mathcal{I}}(\bigotimes_{i\in J} V_i)$, i.e., the formal disjoint union $\coprod_{J\in\mathcal{I}} \bigotimes_{i\in J} V_i$ modulo the equivalence relation that two elements $v_J \in \bigotimes_{i\in J} V_i$ and $w_{J'} \in \bigotimes_{i\in J'} V_i$ are to be identified, if and only if there is a $J'' \supseteq J \cup J'$ such that $\lambda_{J,J''}(v_J) = \lambda_{J',J''}(w_{J'})$. We denote this direct limit by

$$\bigotimes_{i\in I}{}' V_i := \varinjlim_{J\in\mathcal{I}}\left(\bigotimes_{i\in J} V_i\right)$$

and call it the *restricted tensor product of the V_i's* (with respect to $(x_i^\circ)_{i\in I\setminus I_0}$). We may think of $\bigotimes'_{i\in I} V_i$ as the linear span of all the tensors $\bigotimes_{i\in I} x_i$, where $x_i = x_i^\circ$ for all but finitely many i's.[2]

Now, put $I = S_f$, the set of non-archimedean places of our global field F and assume that for each $\mathsf{v} \in S_f$ we are given a representation $(\pi_\mathsf{v}, V_\mathsf{v})$ of the local non-archimedean group G_v. We assume moreover that for all but finitely places $\mathsf{v} \in S_f$ these representations are unramified (which, as we recall, in view of Prop. 9.2 is an unproblematic assumption for the underlying group scheme $\mathbf{G}/F$). Recall from Prop. 7.20.(ii) that for every such place, $V_\mathsf{v}^{K_\mathsf{v}}$ is one-dimensional and so we may choose a nonzero vector $v_\mathsf{v}^\circ \in V_\mathsf{v}^{K_\mathsf{v}}$, which is in fact unique up to rescaling by elements of $\mathbb{C}^\times$, and let

$$V_f := \bigotimes_{\mathsf{v}\in S_f}{}' V_\mathsf{v}$$

be the restricted tensor product of the family $(V_\mathsf{v})_{\mathsf{v}\in S_f}$ with respect to the collection of the v_v°'s. As all v_v° are K_v-invariant by assumption, we obtain a well-defined group homomorphism $\pi_f : G(\mathbb{A}_f) \to \mathrm{Aut}_\mathbb{C}(V_f)$ given by (the linear extension of) the map

$$\pi_f(g_f)(\otimes'_{\mathsf{v}\in S_f} v_\mathsf{v}) := \otimes'_{\mathsf{v}\in S_f} \pi_\mathsf{v}(g_\mathsf{v})v_\mathsf{v}. \tag{12.15}$$

Here, clearly, $g_f = (g_\mathsf{v})_{\mathsf{v}\in S_f} \in G(\mathbb{A}_f) = \prod'_{\mathsf{v}\in S_f} G_\mathsf{v}$. The pair (π_f, V_f) is called the *restricted tensor product of the representations* $(\pi_\mathsf{v}, V_\mathsf{v})$ (with respect to the choice of v_v°'s). We will also write $\pi_f = \bigotimes'_{\mathsf{v}\in S_f} \pi_\mathsf{v}$.

12.3.2 *The restricted tensor product theorem for irreducible smooth-automorphic representations*

As we have seen in Cor. 12.14, the underlying $(\mathfrak{g}_\infty, K_\infty, G(\mathbb{A}_f))$-module $V_{(K_\infty)}$ of an irreducible smooth-automorphic representation V allows a completion as an irreducible smooth-automorphic Casselman–Wallach representation of $G(\mathbb{A})$, which is unique up to isomorphism. Our next result provides a fundamental local-global principle for all these representations, showing that they factor in a unique way over the places $\mathsf{v} \in S$ of F. Recall the concept of the (completed) inductive tensor product from §8.2.1, respectively, Def. 8.10.

[2]This formulation is of course a little problematic, as it entails the subtlety that a human, who is given a (pure) tensor in $\bigotimes_{i\in I} V_i$ is supposed to be able to read off the individual factors from it, as otherwise, s/he could not tell whether or not $x_i = x_i^\circ$ for all but finitely many i (leave alone the question of how s/he could ever manage to check potentially uncountably many i's in finite time). Therefore, the reader, who wants to be on the safe side, should use our definition as a direct limit.

Theorem 12.16. *Let (π, V) be an irreducible smooth-automorphic Casselman–Wallach representation of $G(\mathbb{A})$. Then, for each place $\mathsf{v} \in S$, there is an irreducible admissible representation $(\pi_\mathsf{v}, V_\mathsf{v})$ of G_v, which is a Casselman–Wallach representation, if $\mathsf{v} \in S_\infty$, and which are unramified outside a finite set $S_0 \subset S_f$, such that for every $(v_\mathsf{v}^\circ)_{\mathsf{v} \in S_f \setminus S_0} \in \prod_{\mathsf{v} \in S_f \setminus S_0}(V_\mathsf{v}^{K_\mathsf{v}} \setminus \{0\})$, we obtain an isomorphism of $G(\mathbb{A})$-representations*

$$\pi \cong \overline{\bigotimes_{\mathsf{in}}}_{\mathsf{v} \in S_\infty} \pi_\mathsf{v} \ \overline{\otimes}_{\mathsf{in}} \ \bigotimes_{\mathsf{v} \in S_f}{}' \pi_\mathsf{v}, \tag{12.17}$$

where the underlying vector space of the restricted tensor product representation $\bigotimes'_{\mathsf{v} \in S_f} \pi_\mathsf{v}$ (with respect to $(v_\mathsf{v}^\circ)_{\mathsf{v} \in S_f \setminus S_0}$) is equipped with the finest locally convex topology. Among all local representations with the aforementioned properties, the π_v's are unique up to isomorphism.

Proof. Let (π, V) be as in the statement of the theorem. Then, Thm. 12.11 shows that $V_{(K_\infty)}$ is an irreducible automorphic representation and hence it is in particular an irreducible admissible $(\mathfrak{g}_\infty, K_\infty, G(\mathbb{A}_f))$-module (see the explanation in the beginning of the proof of Thm. 12.10, or again [Bor-Jac79], Prop. 4.5.(4)). We may hence apply [Fla79], Thm. 3 (the reader may find a detailed exposition of the same result as the very contents of [Bum98], §3.4, to which we also refer) to $V_{(K_\infty)}$: We therefore get for each $\mathsf{v} \in S_\infty$ an irreducible $(\mathfrak{g}_\mathsf{v}, K_\mathsf{v})$-module $(\pi_{0,\mathsf{v}}, V_{0,\mathsf{v}})$, and for each $\mathsf{v} \in S_f$ an irreducible admissible G_v-representation $(\pi_\mathsf{v}, V_\mathsf{v})$, which are unramified outside a finite set $S_0 \subset S_f$, such that for every $(v_\mathsf{v}^\circ)_{\mathsf{v} \in S_f \setminus S_0} \in \prod_{\mathsf{v} \in S_f \setminus S_0}(V_\mathsf{v}^{K_\mathsf{v}} \setminus \{0\})$, we obtain an isomorphism of $(\mathfrak{g}_\infty, K_\infty, G(\mathbb{A}_f))$-modules

$$V_{(K_\infty)} \cong \bigotimes_{\mathsf{v} \in S_\infty} V_{0,\mathsf{v}} \ \otimes \ \bigotimes_{\mathsf{v} \in S_f}{}' V_\mathsf{v}. \tag{12.18}$$

Moreover, all these $(\mathfrak{g}_\mathsf{v}, K_\mathsf{v})$-modules $(\pi_{0,\mathsf{v}}, V_{0,\mathsf{v}})$ and G_v-representations $(\pi_\mathsf{v}, V_\mathsf{v})$ are unique up to isomorphism. We observe that $\bigotimes'_{\mathsf{v} \in S_f} V_\mathsf{v}^{K_{n,\mathsf{v}}}$ is finite-dimensional for every $n \in \mathbb{N}$: Indeed, this is a direct consequence of the admissibility of the local non-archimedean representations $(\pi_\mathsf{v}, V_\mathsf{v})$ of G_v, which, as we recall, are unramified at almost all places, whence $\dim_\mathbb{C}(V_\mathsf{v}^{K_{n,\mathsf{v}}}) = \dim_\mathbb{C}(V_\mathsf{v}^{K_\mathsf{v}}) = 1$, by Prop. 7.20, at all but finitely many $\mathsf{v} \in S_f$. Therefore, [Bum98], Prop. 3.4.4 together with Thm. 3.4.2, *ibidem*, shows that the $(\mathfrak{g}_\infty, K_\infty)$-module $\bigotimes_{\mathsf{v} \in S_\infty} V_{0,\mathsf{v}} \otimes \bigotimes'_{\mathsf{v} \notin S_\infty} V_\mathsf{v}^{K_{n,\mathsf{v}}}$ is admissible and finitely generated. By (12.18) and (12.15) it is isomorphic to $V_{(K_\infty)}^{K_n}$. Hence, Thm. 3.26 implies that there must be an isomorphism of

Casselman–Wallach completions

$$\overline{V^{K_n}_{(K_\infty)}}^{\mathsf{CW}} \cong \overline{\left(\bigotimes_{\mathsf{v}\in S_\infty} V_{0,\mathsf{v}} \otimes \bigotimes_{\mathsf{v}\in S_f}{}' V_{\mathsf{v}}^{K_{n,\mathsf{v}}}\right)}^{\mathsf{CW}}.$$

Since V^{K_n} is a Casselman–Wallach representation of G_∞ by assumption, applying Thm. 3.26 once more together with Lem. 1.15, we obtain an isomorphism of G_∞-representations

$$V^{K_n} \cong \overline{\left(\bigotimes_{\mathsf{v}\in S_\infty} V_{0,\mathsf{v}}\right)}^{\mathsf{CW}} \otimes_{\mathsf{pr}} \bigotimes_{\mathsf{v}\in S_f}{}' V_{\mathsf{v}}^{K_{n,\mathsf{v}}},$$

Using [Vog08], Lem. 9.9.(3) and our Prop. 8.11, for every $n \in \mathbb{N}$ we get an isomorphism of G_∞-representations

$$V^{K_n} \cong \overline{\bigotimes_{\mathsf{in}}}_{\mathsf{v}\in S_\infty} V_{\mathsf{v}} \otimes_{\mathsf{in}} \bigotimes_{\mathsf{v}\in S_f}{}' V_{\mathsf{v}}^{K_{n,\mathsf{v}}}, \tag{12.19}$$

for the Casselman–Wallach representations $(\pi_{\mathsf{v}}, V_{\mathsf{v}}) := (\overline{\pi_{0,\mathsf{v}}}^{\mathsf{CW}}, \overline{V_{0,\mathsf{v}}}^{\mathsf{CW}})$ of the local archimedean group G_{v}. Taking the inductive limit we obtain bicontinuous G_∞-equivariant linear bijections

$$\begin{aligned}
V &\overset{10.21}{=} \varinjlim V^{K_n} \\
&\overset{(12.19)}{\cong} \varinjlim \left(\overline{\bigotimes_{\mathsf{in}}}_{\mathsf{v}\in S_\infty} V_{\mathsf{v}} \otimes_{\mathsf{in}} \bigotimes_{\mathsf{v}\in S_f}{}' V_{\mathsf{v}}^{K_{n,\mathsf{v}}}\right) \\
&\overset{\text{Thm.8.12}}{\cong} \varinjlim \overline{\bigotimes_{\mathsf{in}}}_{\mathsf{v}\in S_\infty} V_{\mathsf{v}} \otimes_{\mathsf{in}} \varinjlim \bigotimes_{\mathsf{v}\in S_f}{}' V_{\mathsf{v}}^{K_{n,\mathsf{v}}} \\
&\overset{\text{Lem.8.2}}{\cong} \overline{\bigotimes_{\mathsf{in}}}_{\mathsf{v}\in S_\infty} V_{\mathsf{v}} \otimes_{\mathsf{in}} \varinjlim \bigotimes_{\mathsf{v}\in S_f}{}' V_{\mathsf{v}}^{K_{n,\mathsf{v}}} \\
&\overset{(12.15)}{\cong} \overline{\bigotimes_{\mathsf{in}}}_{\mathsf{v}\in S_\infty} V_{\mathsf{v}} \otimes_{\mathsf{in}} \varinjlim \left(\bigotimes_{\mathsf{v}\in S_f}{}' V_{\mathsf{v}}\right)^{K_n},
\end{aligned}$$

where the vector space $\varinjlim\left(\bigotimes'_{\mathsf{v}\in S_f} V_{\mathsf{v}}\right)^{K_n} \cong \varinjlim \bigotimes'_{\mathsf{v}\in S_f} V_{\mathsf{v}}^{K_{n,\mathsf{v}}}$ carries the finest locally convex topology by Cor. 8.3 and our above observation that $\bigotimes'_{\mathsf{v}\in S_f} V_{\mathsf{v}}^{K_{n,\mathsf{v}}}$ is finite-dimensional for every $n \in \mathbb{N}$. Clearly, $\varinjlim\left(\bigotimes'_{\mathsf{v}\in S_f} V_{\mathsf{v}}\right)^{K_n} = \bigotimes'_{\mathsf{v}\in S_f} V_{\mathsf{v}}$ as vector spaces, hence, if we equip

$\bigotimes'_{v\in S_f} V_v$ with the finest locally convex topology, we obtain a bicontinuous G_∞-equivariant linear bijection

$$V \cong \overline{\bigotimes_{\text{in}}}_{v\in S_\infty} V_v \otimes_{\text{in}} \bigotimes_{v\in S_f}' V_v.$$

As V is complete by assumption, so must be $\overline{\bigotimes_{\text{in}}}_{v\in S_\infty} V_v \otimes_{\text{in}} \bigotimes'_{v\in S_f} V_v$, whence we may finally replace it by $\overline{\bigotimes_{\text{in}}}_{v\in S_\infty} V_v \,\overline{\otimes}_{\text{in}} \bigotimes'_{v\in S_f} V_v$ without changing the rest of the assertion, i.e., there is a bicontinuous G_∞-equivariant linear bijection

$$V \cong \overline{\bigotimes_{\text{in}}}_{v\in S_\infty} V_v \,\overline{\otimes}_{\text{in}} \bigotimes_{v\in S_f}' V_v. \tag{12.20}$$

Since the actions of G_∞ and $G(\mathbb{A}_f)$ commute, we may finish the proof by showing that the map in (12.20) is also $G(\mathbb{A}_f)$-equivariant. Invoking [Fla79], Thm. 3 once more, we firstly observe that (12.20) is $G(\mathbb{A}_f)$-equivariant on the dense subspaces of K_∞-finite vectors. Hence, the theorem follows from the bicontinuity of (12.20), the continuity of the linear operators $\pi(g_f) : V \to V$, $g_f \in G(\mathbb{A}_f)$, and a simple combination of Lem. 8.2, Thm. 1.8, Lem. 1.15 and Prop. 8.11, which guarantees that $G(\mathbb{A}_f)$ operates continuously on the right hand side of (12.20) as well. □

Chapter 13

Cuspidality and Square-integrability

13.1 Unitary representations on spaces defined by square-integrable functions

In the theory of (smooth-)automorphic forms, square-integrability is best defined on spaces of finite (co-)volume: They share the advantage, that the constant functions are square-integrable on them, which is in practice a very desirable property. In this regard, we shall restrict to spaces of functions now, which are well-defined on the quotient $[G] = A_G^{\mathbb{R}} G(F)\backslash G(\mathbb{A})$, cf. Prop. 9.11.

Consider the space of measurable functions $f : G(\mathbb{A}) \to \mathbb{C}$, which are left $A_G^{\mathbb{R}} G(F)$-invariant and square-integrable on $[G]$, i.e.,

$$\int_{[G]} |f(g)|^2 \; \mathrm{d}g < \infty.$$

As usual, we shall call two such functions equivalent, if they are equal off a subset of $G(\mathbb{A})$ of measure zero. Let us denote by $L^2([G])$ the vector space of all such equivalence classes $[f]$, endowed with the Hermitian form

$$\langle [f_1], [f_2] \rangle := \int_{[G]} f_1(g)\overline{f_2(g)} \; \mathrm{d}g \,. \tag{13.1}$$

Exercise 13.2. Show that $\langle [f_1], [f_2] \rangle$ is indeed a well-defined Hermitian form on $L^2([G])$. *Hint:* For convergence, recall the Hölder inequality. For positive-definiteness firstly recall that $G(\mathbb{A}) = G_\infty \times G(\mathbb{A}_f)$ is countable at infinity and then combine [BouI04], IV, §5, Sect. 2, Cor. 3 with [BouII04], VII, §2, Prop. 6.

It is a general fact that with this Hermitian form $L^2([G])$ becomes a Hilbert space, cf. [Rud70], Thm. 3.11, and that right-translation of functions R induces a unitary representation of $G(\mathbb{A})$ on $L^2([G])$ (for a proof, see, for instance, the argument on pp. 135–136 in [Kni-Li06], which transfers verbatim to the setup at hand).

Exercise 13.3. Show that $L^2([G])$ is separable, i.e., contains a countable dense subset. *Hint*: Use that $G(\mathbb{A})$ is second countable.

Analogously, we may consider the space of measurable functions $f : G(\mathbb{A}) \to \mathbb{C}$, which are left $A_G^{\mathbb{R}} G(F)$-invariant and integrable on $[G]$, i.e.,

$$\int_{[G]} |f(g)| \, dg < \infty.$$

Again, we will call two such functions equivalent, if they are equal off a subset of $G(\mathbb{A})$ of measure zero and we let $L^1([G])$ be the vector space of all equivalence classes $[f]$ of such integrable functions, equipped with the taxi-norm

$$\|[f]\|_1 := \int_{[G]} |f(g)| \, dg \, .$$

Then, $L^1([G])$ is a Banach space, cf. [Rud70], Thm. 3.11. Recalling the Hölder inequality, we have

$$\int_{[G]} |f(g)| \, dg \leq \left(\int_{[G]} |f(g)|^2 \, dg \right)^{1/2} \cdot \left(\int_{[G]} dg \right)^{1/2},$$

whence the identity map provides a continuous embedding of the Banach space $L^2([G])$ into the Banach space $L^1([G])$. (Recall once more that the volume of $[G]$ is finite.)

Definition 13.4. A measurable, left $G(F)$-invariant function $f : G(\mathbb{A}) \to \mathbb{C}$ is called *cuspidal*, if for all proper standard parabolic F-subgroups $\mathbf{P} = \mathbf{L} \cdot \mathbf{N}$ of $\mathbf{G}$,

$$f_{\mathbf{P}}(g) := \int_{N(F)\backslash N(\mathbb{A})} f(ng) \, dn = 0 \tag{13.5}$$

for all $g \in G(\mathbb{A})$ off a zero set.

In Def. 13.4, dn denotes the unique Haar measure giving the quotient $N(F)\backslash N(\mathbb{A})$ total volume 1. This is indeed well-defined, as $N(F)\backslash N(\mathbb{A})$ is compact, cf. [GGPS69], Thm. 3 in §6.1. Also, observe that the notion of cuspidality is independent of our actual choice of standard parabolic F-subgroups $\mathbf{P}$ by [Bor91], Prop. 21.12 (Here, we use that f is supposed to be left $G(F)$-invariant). We let $L^p_{cusp}([G])$ be the subspace of $L^p([G])$, $p = 1, 2$, consisting of all classes $[f]$, which are representable by cuspidal functions and we equip this space with the respective subspace topology. We get the following.

Lemma 13.6. *$(R, L^2_{cusp}([G]))$ is a unitary $G(\mathbb{A})$-subrepresentation of $(R, L^2([G]))$.*

Proof. Obviously, $L^2_{cusp}([G])$ is stable under $R(g)$ for all $g \in G(\mathbb{A})$. It hence remains to show that $L^2_{cusp}([G])$ is closed in $L^2([G])$ with the subspace topology. Recalling our observation that the identity map gives rise to a continuous linear injection $L^2([G]) \hookrightarrow L^1([G])$, it is enough to show that $L^1_{cusp}([G])$ is closed in $L^1([G])$. We adapt the idea of the proof of [Bor97], Prop. 8.2, and define

$$\lambda_{\mathbf{P},\kappa}([f]) := \int_{N(F)\backslash G(\mathbb{A})} f(g) \cdot \kappa(g) \; dg$$

for $[f] \in L^1([G])$, $\mathbf{P} = \mathbf{L} \cdot \mathbf{N}$ a standard parabolic F-subgroup of $\mathbf{G}$ and $\kappa \in C_c^\infty(N(\mathbb{A})\backslash G(\mathbb{A}))$, i.e., $\kappa : G(\mathbb{A}) \to \mathbb{C}$ is a smooth function, which is left $N(\mathbb{A})$-invariant and has compact support modulo $N(\mathbb{A})$. Invoking Fubini's theorem, cf. [Rud70], Thm. 7.8, we get

$$\begin{aligned}\lambda_{\mathbf{P},\kappa}([f]) &= \int_{N(\mathbb{A})\backslash G(\mathbb{A})} \kappa(g) \int_{N(F)\backslash N(\mathbb{A})} f(ng) \; dn \, dg \\ &= \int_{N(\mathbb{A})\backslash G(\mathbb{A})} \kappa(g) \cdot f_{\mathbf{P}}(g) \; dg\,.\end{aligned}$$

This implies that a class $[f] \in L^1([G])$ is cuspidal, if and only if $\lambda_{\mathbf{P},\kappa}([f]) = 0$ for all proper standard parabolic F-subgroups $\mathbf{P} = \mathbf{L} \cdot \mathbf{N}$ of $\mathbf{G}$ and for all $\kappa \in C_c^\infty(N(\mathbb{A})\backslash G(\mathbb{A}))$, i.e., shortly said,

$$L^1_{cusp}([G]) = \bigcap_{\mathbf{P},\kappa} \ker(\lambda_{\mathbf{P},\kappa}).$$

It therefore suffices to show that $\lambda_{\mathbf{P},\kappa} : L^1([G]) \to \mathbb{C}$ is continuous for all $\mathbf{P}$ and κ as above. To this end, let us write $N(\mathbb{A}) \cdot C$ for the support of κ, where C is a compact subset of $G(\mathbb{A})$. (This is always possible.) Since $N(F)$ is cocompact in $N(\mathbb{A})$, we may rewrite this as $N(F) \cdot B$, with B a compact subset of $G(\mathbb{A})$. Therefore,

$$|\lambda_{\mathbf{P},\kappa}([f])| \leq \int_B |f(g)| \cdot |\kappa(g)| \; dg \leq \sup_{b \in B} |\kappa(b)| \cdot \int_B |f(g)| \; dg\,,$$

where obviously $\sup_{b\in B} |\kappa(b)|$ is finite, cf. Lem. 10.4. We may cover B with finitely many compact sets D_1, ..., D_k, such that the natural projection $G(\mathbb{A}) \twoheadrightarrow [G]$ is a homeomorphism of D_i onto its image, whence, finally,

$$\begin{aligned}|\lambda_{\mathbf{P},\kappa}([f])| &\leq \sup_{b\in B} |\kappa(b)| \cdot \int_B |f(g)| \; dg \\ &\leq \sup_{b\in B} |\kappa(b)| \cdot k \cdot \int_{[G]} |f(g)| \; dg \\ &\leq \sup_{b\in B} |\kappa(b)| \cdot k \cdot \|[f]\|_1.\end{aligned}$$

This shows that $\lambda_{\mathbf{P},\kappa} : L^1([G]) \to \mathbb{C}$ is continuous for all $\mathbf{P}$ and κ, whence $L^1_{cusp}([G])$ is closed in $L^1([G])$ and hence so is $L^2_{cusp}([G])$ in $L^2([G])$. □

The space $L^2_{cusp}([G])$ is also called the *cuspidal spectrum of* $G(\mathbb{A})$. The following fundamental result – originally due to Godement, [God66], and also Gel'fand–Graev–Piatetski-Shapiro, [GGPS69] – justifies this notion:

Theorem 13.7. *As a $G(\mathbb{A})$-representation $(R, L^2_{cusp}([G]))$ decomposes as a countable direct Hilbert sum of irreducible subrepresentations, each of which appearing with finite multiplicity, i.e., there exists a countable family of irreducible subrepresentations $\mathcal{H}_i$, indexed by $i \in I$, such that summation of classes induces an isomorphism of $G(\mathbb{A})$-representations*

$$L^2_{cusp}([G]) \cong \widehat{\bigoplus_{i\in I}} \mathcal{H}_i, \tag{13.8}$$

and $\mathcal{H}_i \cong \mathcal{H}_j$ only for finitely many $i \neq j$.

Proof. We give a proof by references. If $A^{\mathbb{R}}_G$ is trivial, then (applying restriction of scalars $\mathrm{Res}_{F/\mathbb{Q}}$ to $\mathbf{G}$) the result follows from [God66], §7, and [Wol07], Thm. 4.5.3. Alternatively, the reader may consult [GGPS69], §7.2 and [Bor97], Lem. 16.1 and show by him-/herself that there is indeed a Dirac sequence of non-negative Schwartz-Burhat functions $\alpha_n : G(\mathbb{A}) \to \mathbb{C}$, in the sense of [GGPS69], §3.5, as needed in order to apply [Bor97], Lem. 16.1 (a task, which is actually not too difficult). If $A^{\mathbb{R}}_G$ is not trivial, then, working with $G(\mathbb{A})^{(1)}$ instead of $G(\mathbb{A})$ in [God66], one obtains by literally the same argument as presented *ibidem*, that the convolution operators $R(f)$ are compact on $L^2_{cusp}([G])$ for all $f \in C_c(G(\mathbb{A})^{(1)})$. Recalling from Prop. 9.11 that $G(\mathbb{A}) = A^{\mathbb{R}}_G \times G(\mathbb{A})^{(1)}$ and that $A^{\mathbb{R}}_G$ acts trivially on $L^2_{cusp}([G])$, the result follows again from [Wol07], Thm. 4.5.3. □

Definition 13.9. Let $L^2_{dis}([G])$ be the maximal $G(\mathbb{A})$-subrepresentation of $L^2([G])$, which is isomorphic to a direct Hilbert sum $\widehat{\bigoplus}_{j\in J} \mathcal{H}_j$ of irreducible subrepresentations $\mathcal{H}_j$ of $L^2([G])$, such that $\mathcal{H}_j \cong \mathcal{H}_k$ only for finitely many $j \neq k$.

The space $L^2_{dis}([G])$ is also called the *discrete spectrum of* $G(\mathbb{A})$. We observe that $L^2([G])$ being separable, cf. Exc. 13.3, the indexing set J must necessarily be countable. Moreover, as a direct corollary of Lem. 13.6 and Thm. 13.7, $L^2_{cusp}([G])$ is a subrepresentation of $L^2_{dis}([G])$.

Lemma 13.10. *The discrete spectrum $L^2_{dis}([G])$ equals the smallest $G(\mathbb{A})$-subrepresentation of $(R, L^2([G]))$, which contains all irreducible $G(\mathbb{A})$-subrepresentations of $(R, L^2([G]))$.*

Proof. The proof is very simple and left as an exercise. *Hint*: Use the unitarity of $(R, L^2([G]))$ and recall Warning 1.14 and Lem. 1.12. □

As a direct consequence of Lem. 13.10, unless $\mathbf{G}$ has no proper parabolic F-subgroups at all (in which case the defining condition (13.5) of being cuspidal becomes vacuous, whence $L^2_{cusp}([G]) = L^2_{dis}([G]) = L^2([G])$), the cuspidal spectrum $L^2_{cusp}([G])$ is strictly smaller than the discrete spectrum $L^2_{dis}([G])$, since the latter space contains the one-dimensional (and hence irreducible) $G(\mathbb{A})$-subrepresentation of $L^2([G])$ on the space of constant functions $c : G(\mathbb{A}) \to \mathbb{C}$, $g \mapsto c(g) = c$, which are obviously not cuspidal (unless in the trivial case $c = 0$). We define $L^2_{res}([G])$ to be the natural orthogonal complement of $L^2_{cusp}([G])$ in $L^2_{dis}([G])$ and call $L^2_{res}([G])$ the *residual spectrum of* $G(\mathbb{A})$. As it is obvious from Def. 13.9 and Thm. 13.7,

$$L^2_{res}([G]) \cong \widehat{\bigoplus_{r \in J \setminus I}} \mathcal{H}_r$$

as a $G(\mathbb{A})$-representation.

Remark 13.11. There are *versions* of the cuspidal, residual and the discrete spectrum, which we shall shortly discuss, without going into the very details. Firstly, one may consider the right-regular action of $G(\mathbb{A})$ on the Hilbert space $L^2(Z_G(\mathbb{A})G(F)\backslash G(\mathbb{A}))$, consisting of all equivalence classes of measurable functions $f : G(\mathbb{A}) \to \mathbb{C}$, which are left $Z_G(\mathbb{A})G(F)$-invariant and square-integrable on the quotient $Z_G(\mathbb{A})G(F)\backslash G(\mathbb{A})$ (and equivalence being defined as before). Here, $Z_G(\mathbb{A})$ denotes the center of $G(\mathbb{A})$. Then, all results from §13.1 above hold verbatim, replacing the quotient $[G]$ with the smaller quotient $Z_G(\mathbb{A})G(F)\backslash G(\mathbb{A})$. Secondly, another variation is to introduce characters: Let $\omega : Z_G(\mathbb{A}) \to S^1 \subset \mathbb{C}^\times$ be a continuous group homomorphism. We denote by $L^2(Z_G(\mathbb{A})G(F)\backslash G(\mathbb{A}), \omega)$, the space of all equivalence classes of measurable left-$G(F)$-equivariant functions $f : G(\mathbb{A}) \to \mathbb{C}$, which satisfy

(i) $f(zg) = \omega(z) \cdot f(g), \quad \forall z \in Z_G(\mathbb{A}), g \in G(\mathbb{A})$.
(ii) $\int_{Z_G(\mathbb{A})G(F)\backslash G(\mathbb{A})} |f(g)|^2 \, dg < \infty$.

Then, again, all results from §13.1 above hold verbatim, replacing $L^2([G])$ with $L^2(Z_G(\mathbb{A})G(F)\backslash G(\mathbb{A}), \omega)$.

13.2 $(\mathfrak{g}_\infty, K_\infty, G(\mathbb{A}_f))$-modules on spaces defined by square-integrable functions

We now consider the space $\mathcal{A}_{(2)}([G])$ of all automorphic forms φ, cf. Def. 11.21, which are constant on $A_G^{\mathbb{R}}$ and square-integrable on $[G]$, i.e., such that $[\varphi]$ is a well-defined element of $L^2([G])$. As usual, we shall denote by a lower "$(\mathcal{Z}(\mathfrak{g}_\infty))$" the space of $\mathcal{Z}(\mathfrak{g}_\infty)$-finite elements. The main goal of this subsection is to prove the following theorem.

Theorem 13.12. *Acted upon by right-translation, the space $\mathcal{A}_{(2)}([G])$ is a $(\mathfrak{g}_\infty, K_\infty, G(\mathbb{A}_f))$-module. It coincides with the $(\mathfrak{g}_\infty, K_\infty, G(\mathbb{A}_f))$-module on the space of all continuous representatives of the classes in $L^2([G])^{\infty_\mathbb{A}}_{(K_\infty,\mathcal{Z}(\mathfrak{g}_\infty))}$, which is furthermore equal to the $(\mathfrak{g}_\infty, K_\infty, G(\mathbb{A}_f))$-module on the space of all continuous representatives of the classes in $L^2_{dis}([G])^{\infty_\mathbb{A}}_{(K_\infty,\mathcal{Z}(\mathfrak{g}_\infty))}$.*

Proof. We proceed in several steps.
Step 1: Let us first fix a map ϑ which selects in each equivalence class $[f]$ of measurable functions in sight its unique continuous representative, if such a function exists at all, and is set to 0 else. We will show that ϑ restricts to an isomorphism of vector spaces

$$\vartheta : L^2([G])^{\infty_\mathbb{A}}_{(K_\infty,\mathcal{Z}(\mathfrak{g}_\infty))} \xrightarrow{\sim} \mathcal{A}_{(2)}([G]). \tag{13.13}$$

To this end, let $[f] \in L^2([G])^{\infty_\mathbb{A}}_{(K_\infty,\mathcal{Z}(\mathfrak{g}_\infty))}$. As $[f]$ is right K_∞-finite and $\mathcal{Z}(\mathfrak{g}_\infty)$-finite by assumption, there must be a representative f of $[f]$ with formally the same properties. Here, observe that since $f|_{G_\infty}$ is locally integrable, the notion of $\mathcal{Z}(\mathfrak{g}_\infty)$-finiteness for representatives of $[f]$ still makes sense, see [Bor97], remark on p. 22, for the recipe. Hence, amplifying the argument presented in [Bor07], §3.1 to our setup at hand shows that $f|_{G_\infty}$ is annihilated by an elliptic differential operator. Therefore, [LBer-Sch71], Thm. 1 in §4.2, implies that $f|_{G_\infty}$ is smooth. As the actions of K_∞ and $\mathcal{Z}(\mathfrak{g}_\infty)$ commute with the action of $G(\mathbb{A}_f)$, the same applies to $f(\_\_ \cdot g_f)$ for all $g_f \in G(\mathbb{A}_f)$, so f satisfies condition (1) of Def. 10.3. Next, take an open compact subgroup K_n in our fixed sequence of neighborhoods $(K_n)_{n\in\mathbb{N}}$ such that $[f] \in L^2([G])^{K_n} \subset L^2([G])^{\infty_\mathbb{A}}$, i.e., such that $[f] = [f(\_\_ \cdot k_f)]$ for all $k_f \in K_n$. Using (11.4), we obtain an isomorphism of vector spaces

$$L^2([G])^{K_n} \xrightarrow{\sim} \bigoplus_{c\in C_n} L^2(A_G^{\mathbb{R}}\Gamma_{c,K_n}\backslash G_\infty), \tag{13.14}$$

$$[\phi] \mapsto ([\phi(\_\_ \cdot c)])_{c\in C_n}$$

where each space on the right hand side has the obvious meaning as the space of all classes of left $A_G^{\mathbb{R}}\Gamma_{c,K_n}$-invariant measurable functions $G_\infty \to \mathbb{C}$, which are square-integrable on $A_G^{\mathbb{R}}\Gamma_{c,K_n}\backslash G_\infty$. Inserting $[f]$ from above, we get $[f(\_\_ \cdot c)] = [f(\_\_ \cdot k_f c)]$ for all $k_f \in K_n$ and $c \in C_n$. However, as $f(\_\_ \cdot g_f) : G_\infty \to \mathbb{C}$ is smooth and hence continuous for all $g_f \in G(\mathbb{A}_f)$, necessarily $f(\_\_ \cdot c) = f(\_\_ \cdot k_f c)$ as functions on G_∞ for all $k_f \in K_n$ and $c \in C_n$. Therefore, $f = f(\_\_ \cdot k_f)$ as functions on $G(\mathbb{A})$ for all $k_f \in K_n$. Hence, f also satisfies condition (2) of Def. 10.3, whence f is smooth, i.e., in $C^\infty(G(\mathbb{A}))$. Applying Lem. 10.4 shows that ϑ defines an injection $L^2([G])^{\infty_{\mathbb{A}}}_{(K_\infty,\mathcal{Z}(\mathfrak{g}_\infty))} \hookrightarrow C^\infty(G(\mathbb{A}))$, whose image obviously consists of left $A_G^{\mathbb{R}}G(F)$-invariant, right K_∞-finite and $\mathcal{Z}(\mathfrak{g}_\infty)$-finite functions φ. Of course, the latter property may be equivalently reformulated by saying that φ is annihilated by an ideal $\mathcal{J}$ of $\mathcal{Z}(\mathfrak{g}_\infty)$ of finite codimension. Next, amplifying the argument underlying [Bor-Jac79], 4.3.(ii), shows that $\vartheta([f])$ is in fact of moderate growth for $[f] \in L^2([G])^{\infty_{\mathbb{A}}}_{(K_\infty,\mathcal{Z}(\mathfrak{g}_\infty))}$. Hence, in summary, and recalling that each K_n is of finite index in $K_{\mathbb{A}_f}$, our assignment ϑ defines an injection of vector spaces $L^2([G])^{\infty_{\mathbb{A}}}_{(K_\infty,\mathcal{Z}(\mathfrak{g}_\infty))} \hookrightarrow \mathcal{A}_{(2)}([G])$.

Step 2: In order to see that (13.13) is indeed a bijection, we still need to show that the class $[\varphi] \in L^2([G])$ defined by a $\varphi \in \mathcal{A}_{(2)}([G])$ is a globally smooth vector for $(R, L^2([G]))$, i.e., $[\varphi] \in L^2([G])^{\infty_{\mathbb{A}}}$. Indeed, φ being right K_n-invariant by some K_n in our sequence of neighborhoods $(K_n)_{n\in\mathbb{N}}$, it suffices to show that $[\varphi] \in L^2([G])^{\infty_{\mathbb{R}}}$, where clearly we viewed the Hilbert space $L^2([G])$ as a representation of G_∞ under right-translation R_∞. To this end, we reconsider (13.14). With a little bit of work it can be shown that this linear bijection is in fact an isomorphism of G_∞-representations, the only non-trivial part being that it may be normalized to become an isometry of Hilbert spaces, which follows from [Mui-Žun20], Lem. 4, and whose details (including providing the obvious definition for the Hermitian form on the spaces $L^2(A_G^{\mathbb{R}}\Gamma_{c,K_n}\backslash G_\infty)$ and defining the Hilbert space structure on the direct sum following the explanations of §1.2.2, in particular, Warning 1.14) we leave as an exercise to the reader. Taking this for granted, it follows from Exc. 2.22 and the finiteness of C_n that (13.14) restricts to an isomorphism of smooth G_∞-representations

$$(L^2([G])^{K_n})^{\infty_{\mathbb{R}}} \xrightarrow{\sim} \bigoplus_{c\in C_n} L^2(A_G^{\mathbb{R}}\Gamma_{c,K_n}\backslash G_\infty)^{\infty_{\mathbb{R}}} \tag{13.15}$$

and so it suffices to show that for $\varphi \in \mathcal{A}_{(2)}([G])$ the classes $[\varphi(\_\_ \cdot c)] \in L^2(A_G^{\mathbb{R}}\Gamma_{c,K_n}\backslash G_\infty)$, $c \in C_n$, are indeed smooth vectors for the right-regular action of G_∞. However, as each $\varphi(\_\_ \cdot c)$ is smooth, right K_∞-finite and $\mathcal{Z}(\mathfrak{g}_\infty)$-finite, this follows from combining Harish-Chandra's theorem [HCh66], Thm. 1 in §8 (as extended to groups with finitely many connected components), with Dixmier–Malliavin's theorem [Dix-Mal78], Thm. 3.13. Therefore, (13.13) is indeed an isomorphism of vector spaces.

Step 3: It follows from Lem. 10.43 and the fact that $(R, L^2([G]))$ is a representation of $G(\mathbb{A})$ that $L^2([G])^{\infty_\mathbb{A}}_{(K_\infty)}$ is a $(\mathfrak{g}_\infty, K_\infty, G(\mathbb{A}_f))$-module. Since the action of $\mathcal{Z}(\mathfrak{g}_\infty)$ commutes with the actions of $\mathfrak{g}_\infty$, K_∞ and $G(\mathbb{A}_f)$, hence so is also $L^2([G])^{\infty_\mathbb{A}}_{(K_\infty,\mathcal{Z}(\mathfrak{g}_\infty))}$. As the linear bijection in (13.13) is clearly equivariant under right-translation, this shows that $\mathcal{A}_{(2)}([G])$ is a $(\mathfrak{g}_\infty, K_\infty, G(\mathbb{A}_f))$-module, which is equal to $\vartheta(L^2([G])^{\infty_\mathbb{A}}_{(K_\infty,\mathcal{Z}(\mathfrak{g}_\infty))})$.

Step 4: We are hence left to show that

$$\vartheta(L^2([G])^{\infty_\mathbb{A}}_{(K_\infty,\mathcal{Z}(\mathfrak{g}_\infty))}) = \vartheta(L^2_{dis}([G])^{\infty_\mathbb{A}}_{(K_\infty,\mathcal{Z}(\mathfrak{g}_\infty))}), \tag{13.16}$$

for which it is obviously sufficient to prove that

$$\vartheta(L^2([G])^{\infty_\mathbb{A}}_{(K_\infty,\mathcal{Z}(\mathfrak{g}_\infty))}) \subseteq \vartheta(L^2_{dis}([G])^{\infty_\mathbb{A}}_{(K_\infty,\mathcal{Z}(\mathfrak{g}_\infty))}).$$

So, let $0 \neq \varphi \in \mathcal{A}_{(2)}([G]) = \vartheta(L^2([G])^{\infty_\mathbb{A}}_{(K_\infty,\mathcal{Z}(\mathfrak{g}_\infty))})$. According to Lem. 13.10, it is enough to show that the $G(\mathbb{A})$-representation $V_{[\varphi]} := \mathrm{Cl}_{L^2([G])}(\langle R(G(\mathbb{A}))[\varphi]\rangle)$ is the finite direct Hilbert sum of irreducible $G(\mathbb{A})$-representations.

We firstly observe that $V_{[\varphi]}$ is an admissible $G(\mathbb{A})$-representation. Indeed, in order to see this, following Lem. 10.34, we have to show that for all open compact subgroups K_f of $G(\mathbb{A}_f)$ and all irreducible representations (ρ_∞, W_∞) of K_∞, the isotypic component $V_{[\varphi]}^{K_f}(\rho_\infty)$ is finite-dimensional. Let K_n be the biggest open compact subgroup in our sequence of neighborhoods $(K_n)_{n\in\mathbb{N}}$, such that $\varphi \in \mathcal{A}_{(2)}([G])^{K_n}$. A short moment of thought shows that it is enough to prove that $V_{[\varphi]}^{K_m}(\rho_\infty)$ is finite-dimensional for all K_m, $m \geq n$ and all irreducible representations (ρ_∞, W_∞) of K_∞. To this end, fix an arbitrary $m \geq n$ and observe that, recalling (10.8), the composition of ϑ^{-1} with (13.15) maps φ onto the tuple $([\varphi(\_\_ \cdot c)])_{c\in C_m}$, for which each component $[\varphi(\_\_ \cdot c)]$ is a right K_∞-finite and $\mathcal{Z}(\mathfrak{g}_\infty)$-finite vector in the G_∞-representation $L^2(A_G^\mathbb{R}\Gamma_{c,K_m}\backslash G_\infty)^{\infty_\mathbb{R}}$. Since the latter representation is smooth, cf. Lem. 2.18, the $(\mathfrak{g}_\infty, K_\infty)$-module $V_{[\varphi(\_\_\cdot c)],0}$ inside $L^2(A_G^\mathbb{R}\Gamma_{c,K_m}\backslash G_\infty)^{\infty_\mathbb{R}}$ spanned by $[\varphi(\_\_ \cdot c)]$ is admissible for each $c \in C_m$ by [Bor-Jac79], Lem. 2.1.

Step 5: We now claim that the topological closure of $V_{[\varphi(\_\_\cdot c)],0}$ inside $L^2(A_G^{\mathbb{R}}\Gamma_{c,K_m}\backslash G_\infty)^{\infty_{\mathbb{R}}}$ is identical to the (smooth, cf. Cor. 2.21 and Lem. 2.18 again) G_∞-subrepresentation

$$V_{[\varphi(\_\_\cdot c)]} := \mathrm{Cl}_{L^2(A_G^{\mathbb{R}}\Gamma_{c,K_m}\backslash G_\infty)^{\infty_{\mathbb{R}}}}(\langle R(G_\infty)[\varphi(\_\_\cdot c)]\rangle)$$

of $L^2(A_G^{\mathbb{R}}\Gamma_{c,K_m}\backslash G_\infty)^{\infty_{\mathbb{R}}}$ generated by the class $[\varphi(\_\_\cdot c)]$, i.e., we claim that

$$\mathrm{Cl}_{L^2(A_G^{\mathbb{R}}\Gamma_{c,K_m}\backslash G_\infty)^{\infty_{\mathbb{R}}}}(V_{[\varphi(\_\_\cdot c)],0}) = V_{[\varphi(\_\_\cdot c)]}. \tag{13.17}$$

In order to show (13.17), we imitate the proof of Prop. 12.7: By construction, $\mathrm{Cl}_{L^2(A_G^{\mathbb{R}}\Gamma_{c,K_m}\backslash G_\infty)^{\infty_{\mathbb{R}}}}(V_{[\varphi(\_\_\cdot c)],0}) \subseteq V_{[\varphi(\_\_\cdot c)]}$. Suppose that the inequality is strict. Then, by the Hahn-Banach theorem, cf. [Jar81], Sect. 7.2, Cor. 2.(a), there must be a non-trivial continuous linear functional $u : V_{[\varphi(\_\_\cdot c)]} \to \mathbb{C}$ which is trivial on $\mathrm{Cl}_{L^2(A_G^{\mathbb{R}}\Gamma_{c,K_m}\backslash G_\infty)^{\infty_{\mathbb{R}}}}(V_{[\varphi(\_\_\cdot c)],0})$. For an arbitrary but fixed $[\phi] \in V_{[\varphi(\_\_\cdot c)],0}$, consider the map

$$u_{c,[\phi]} : G_\infty \to \mathbb{C}$$

$$g \mapsto u(R_\infty(g)[\phi]).$$

As obviously, $u_{c,[\phi]} = u \circ c_{[\phi]}$ and $[\phi]$ is a smooth vector for R_∞, the orbit map $c_{[\phi]}$ is a smooth function $G_\infty \to V_{[\varphi(\_\_\cdot c)]}$. Hence, as moreover $u : V_{[\varphi(\_\_\cdot c)]} \to \mathbb{C}$ is continuous and so bounded, u maps smooth curves to smooth curves by [Kri-Mic97], Cor. 2.11, and therefore is a smooth function by Cor. 3.14 *ibidem*. Hence, so are the compositions $u_{c,[\phi]}$ for all $[\phi] \in V_{[\varphi(\_\_\cdot c)],0}$.

We now claim that the functions $u_{c,[\phi]}$ are also $\mathcal{Z}(\mathfrak{g}_\infty)$-finite. In order to see this, observe that by the just verified smoothness of $u_{c,[\phi]}$, for every $X \in \mathfrak{g}_\infty$ and $g \in G_\infty$, we have the (well-defined) equalities

$$(X \cdot u_{c,[\phi]})(g) = \frac{\mathrm{d}}{\mathrm{d}t}\bigg|_{t=0} u(R_\infty(g\exp(tX))[\phi]) = u(R_\infty(g)R_\infty(X)[\phi]),$$

and obviously these extend linearly to all $D \in \mathcal{U}(\mathfrak{g}_\infty)$, i.e., we finally obtain

$$(D \cdot u_{c,[\phi]})(g) = u(R_\infty(g)R_\infty(D)[\phi]) \tag{13.18}$$

for all $D \in \mathcal{U}(\mathfrak{g}_\infty)$ and $g \in G_\infty$. Thus, fixing any ideal $\mathcal{J} \lhd \mathcal{Z}(\mathfrak{g}_\infty)$ of finite codimension such that $\mathcal{J} \cdot [\phi] = 0$ – which must exist, since $[\varphi]$ (and so all classes in $V_{[\varphi(\_\_\cdot c)],0}$) is $\mathcal{Z}(\mathfrak{g}_\infty)$-finite by assumption – (13.18) shows that also $\mathcal{J} \cdot u_{c,[\phi]} = 0$, hence $u_{c,[\phi]}$ is indeed $\mathcal{Z}(\mathfrak{g}_\infty)$-finite as asserted.

As obviously $u_{c,[\phi]}$ is also right K_∞-finite by the right K_∞-finiteness of $[\varphi]$, [LBer-Sch71], Appendix to §4, implies that $u_{c,[\phi]}|_{G^\circ_\infty} : G^\circ_\infty \to \mathbb{C}$ is real analytic. As a consequence, we may suppose that $u_{c,[\phi]}|_{G^\circ_\infty}$ is represented by a multi-variable Taylor series around the identity $id_\infty \in G^\circ_\infty$. However, as

$$(D \cdot u_{c,[\phi]})(id_\infty) \overset{(13.18)}{=} u(R_\infty(D)[\phi]) = 0,$$

for all $D \in \mathcal{U}(\mathfrak{g}_\infty)$ (as $R_\infty(D)[\phi] \in V_{[\varphi(\_\_\cdot c)],0}$, on which u is supposed to be identically zero), all partial derivatives of $u_{c,[\phi]}|_{G^\circ_\infty}$ vanish at $id_\infty \in G^\circ_\infty$. But this shows that the above Taylor series of $u_{c,[\phi]}|_{G^\circ_\infty}$ is constantly zero, whence so is $u_{c,[\phi]}|_{G^\circ_\infty}$. Hence, as $[\phi] \in V_{[\varphi(\_\_\cdot c)],0}$ was arbitrary, our continuous linear form u must be zero on

$$\begin{aligned}
&\mathrm{Cl}_{L^2(A^{\mathbb{R}}_G\Gamma_{c,K_m}\backslash G_\infty)^{\infty_\mathbb{R}}}\left(\langle R_\infty(G^\circ_\infty)V_{[\varphi(__\cdot c)],0}\rangle\right)\\
&= \mathrm{Cl}_{L^2(A^{\mathbb{R}}_G\Gamma_{c,K_m}\backslash G_\infty)^{\infty_\mathbb{R}}}\left(\langle R_\infty(G^\circ_\infty)R_\infty(K_\infty)V_{[\varphi(__\cdot c)],0}\rangle\right)\\
&= \mathrm{Cl}_{L^2(A^{\mathbb{R}}_G\Gamma_{c,K_m}\backslash G_\infty)^{\infty_\mathbb{R}}}\left(\langle R_\infty(G_\infty)V_{[\varphi(__\cdot c)],0}\rangle\right)\\
&\supseteq \mathrm{Cl}_{L^2(A^{\mathbb{R}}_G\Gamma_{c,K_m}\backslash G_\infty)^{\infty_\mathbb{R}}}(\langle R_\infty(G_\infty)[\varphi(__\cdot c)]\rangle)\\
&= V_{[\varphi(__\cdot c)]},
\end{aligned}$$

where the first equality follows from the fact that $V_{[\varphi(\_\_\cdot c)],0}$ is K_∞-invariant and the second equality holds by [Kna02], Prop. 7.19.(b). However, this implies that u is identically zero – a contradiction. Hence, (13.17) is true.

Step 6: Knowing (13.17), we may now mimic even another strategy, which we have already used earlier. Indeed, replacing $E^{K_n}_{\rho_\infty}$ by E_{ρ_∞}, defined in (2.32), and arguing verbatim as in the proof of Prop. 10.48, one gets that

$$V_{[\varphi(\_\_\cdot c)],0}(\rho_\infty) = V_{[\varphi(\_\_\cdot c)]}(\rho_\infty)$$

for all irreducible representations (ρ_∞, W_∞) of K_∞ and $c \in C_m$, whence, $V_{[\varphi(\_\_\cdot c)],0}$ being admissible, $V_{[\varphi(\_\_\cdot c)]}(\rho_\infty)$ is finite-dimensional for all such c and ρ_∞. Since C_m is finite, $\bigoplus_{c\in C_m} V_{[\varphi(\_\_\cdot c)]}(\rho_\infty)$ is finite-dimensional for all irreducible representations (ρ_∞, W_∞) of K_∞ and therefore so is its image under the composition of isomorphisms of vector spaces (13.15) and ϑ. However, by equivariance, the latter image is nothing but $V^{K_m}_{[\varphi]}(\rho_\infty)$, whence $V_{[\varphi]}$ is admissible.

Step 7: The right K_∞-finiteness of $[\varphi]$ implies that there must be a finite selection of irreducible representations $(\rho_{\infty,i}, W_{\infty,i})$, $1 \leq i \leq k$, of K_∞

such that $[\varphi] \in V^{[\varphi]} := V_{[\varphi]}^{K_n}(\rho_{\infty,1}) \oplus ... \oplus V_{[\varphi]}^{K_n}(\rho_{\infty,k})$. As $V_{[\varphi]}$ is admissible, $V^{[\varphi]}$ is hence finite-dimensional as well. We claim that for each non-trivial subrepresentation U of $V_{[\varphi]}$, the intersection $U \cap V_{[\varphi]}$ is non-trivial. Indeed, $U \cap V^{[\varphi]} = U_{[\varphi]}^{K_n}(\rho_{\infty,1}) \oplus ... \oplus U_{[\varphi]}^{K_n}(\rho_{\infty,k})$, so, if the latter space were equal to $\{0\}$, then $V^{[\varphi]}$ would be contained in the orthogonal complement $U^\perp$ of U in $V_{[\varphi]}$, which is a $G(\mathbb{A})$-subrepresentation of $V_{[\varphi]}$ by the unitarity of the latter.[1] But then $[\varphi]$ would be in $U^\perp$, whence, the latter space being closed and $G(\mathbb{A})$-stable, all of $V_{[\varphi]}$ would be in $U^\perp$, which is a proper subspace of $V_{[\varphi]}$. Hence, we arrived at a contradiction. Consequently, $U \cap V_{[\varphi]}$ is non-trivial for all non-trivial subrepresentation U of $V_{[\varphi]}$.

Now, take U such that $\dim_{\mathbb{C}}(U \cap V^{[\varphi]})$ is minimal (and by what we have just observed necessarily non-zero). Then, U is irreducible and $V_{[\varphi]} = U \widehat{\oplus} U^\perp$ as $G(\mathbb{A})$-subrepresentations: The latter assertion being clear, we only argue the irreducibility of U: If there were a proper $G(\mathbb{A})$-subrepresentation U' of U, then $\dim_{\mathbb{C}}(U' \cap V^{[\varphi]}) < \dim_{\mathbb{C}}(U \cap V^{[\varphi]})$, contradicting the minimality of $\dim_{\mathbb{C}}(U \cap V^{[\varphi]})$. Indeed, if $\dim_{\mathbb{C}}(U' \cap V^{[\varphi]}) = \dim_{\mathbb{C}}(U \cap V^{[\varphi]})$, then necessarily $\dim_{\mathbb{C}}((U')^\perp \cap V^{[\varphi]}) = 0$, where $(U')^\perp$ denotes the orthogonal complement of U' in U. As the latter is a non-trivial $G(\mathbb{A})$-representation of U (and hence of $V_{[\varphi]}$) under the right-regular action, this contradicted the fact that $V_{[\varphi]}$ has non-trivial intersection with all of its non-trivial subrepresentations. Therefore, U must be irreducible.

Now, iterating this process by replacing $V_{[\varphi]}$ by $U^\perp$ we end up with a finite direct Hilbert sum decomposition

$$V_{[\varphi]} = \widehat{\bigoplus_{1 \le i \le d}} U_i$$

of $V_{[\varphi]}$ into irreducible $G(\mathbb{A})$-representations. Therefore,

$$\vartheta(L^2([G])^{\infty_{\mathbb{A}}}_{(K_\infty, \mathcal{Z}(\mathfrak{g}_\infty))}) \subseteq \vartheta(L^2_{dis}([G])^{\infty_{\mathbb{A}}}_{(K_\infty, \mathcal{Z}(\mathfrak{g}_\infty))}),$$

which finally shows (13.16). This completes the proof of Thm. 13.12. □

Remark 13.19. Slightly less precisely said, but more elegantly summarized, the very assertion of Thm. 13.12 amounts to the dictum that the

[1] Indeed, recall that for $v \in U$, $w \in U^\perp$ and $g \in G(\mathbb{A})$ we have

$$\langle v, R(g)w\rangle = \langle R(g)R(g)^{-1}v, R(g)w\rangle = \langle R(g^{-1})v, w\rangle = 0,$$

as $R(g^{-1})v \in U$, U being a $G(\mathbb{A})$-subrepresentation of $V_{[\varphi]}$. So, $R(g)w \in U^\perp$.

space of square-integrable automorphic forms $\mathcal{A}_{(2)}([G])$ allows an isomorphism of $(\mathfrak{g}_\infty, K_\infty, G(\mathbb{A}_f))$-modules

$$\mathcal{A}_{(2)}([G]) \cong L^2_{dis}([G])^{\infty_\mathbb{A}}_{(K_\infty, \mathcal{Z}(\mathfrak{g}_\infty))} = L^2([G])^{\infty_\mathbb{A}}_{(K_\infty, \mathcal{Z}(\mathfrak{g}_\infty))},$$

or, even less precisely summarized, that a square-integrable automorphic form necessarily belongs to the discrete spectrum and cannot show up in the *continuous spectrum*, i.e., in the orthogonal complement of $L^2_{dis}([G])$ in $L^2([G])$.

We let $\mathcal{A}_{cusp}([G])$ be the subspace of $\mathcal{A}_{(2)}([G])$, which consists of cuspidal automorphic forms, cf. Def. 13.4. As a consequence of Thm. 13.12 and Thm. 13.7 we obtain the following.

Corollary 13.20. *Resuming the notation from Thm. 13.7 and Def. 13.9, the map ϑ and summation of functions induce the following decompositions as $(\mathfrak{g}_\infty, K_\infty, G(\mathbb{A}_f))$-modules:*

$$\mathcal{A}_{cusp}([G]) \cong \bigoplus_{i\in I} (\mathcal{H}_i^{\infty_\mathbb{A}})_{(K_\infty)} \quad \text{and} \quad \mathcal{A}_{(2)}([G]) \cong \bigoplus_{j\in J} (\mathcal{H}_j^{\infty_\mathbb{A}})_{(K_\infty)}.$$

Each summand $(\mathcal{H}_j^{\infty_\mathbb{A}})_{(K_\infty)}$ is irreducible and admissible.

Proof. In order to establish both direct sum decompositions, it suffices to prove that $\mathcal{A}_{(2)}([G]) \cong \bigoplus_{j\in J}(\mathcal{H}_j^{\infty_\mathbb{A}})_{(K_\infty)}$. To this end, let us first show that $\bigoplus_{j\in J}(\mathcal{H}_j^{\infty_\mathbb{A}})_{(K_\infty)}$ injects via ϑ to a $(\mathfrak{g}_\infty, K_\infty, G(\mathbb{A}_f))$-submodule of $\mathcal{A}_{(2)}([G])$: Indeed, recall from (10.46), that for each (irreducible unitary) summand $\mathcal{H}_j$ and each open compact subgroup K_n in our fixed sequence $(K_n)_{n\in\mathbb{N}}$, there is an isomorphism of $(\mathfrak{g}_\infty, K_\infty)$-modules

$$(\mathcal{H}_j^{K_n})_{(K_\infty)} \cong (\mathcal{H}_{j,\infty})_{(K_\infty)} \otimes (\mathcal{H}_{j,f})^{K_n},$$

where obviously $\mathcal{H}_{j,\infty}$ took the role of V_∞ and likewise $\mathcal{H}_{j,f}$ took the role of V_f. It hence follows from [Vog81], Prop. 0.3.19.(a), that every $D \in \mathcal{Z}(\mathfrak{g}_\infty)$ acts by multiplication by a scalar on $(\mathcal{H}_j^{K_n})_{(K_\infty)}$ and therefore, as $n \in \mathbb{N}$ was arbitrary, also on $(\mathcal{H}_j^{\infty_\mathbb{A}})_{(K_\infty)}$. Consequently, each element of the algebraic direct sum $\bigoplus_{j\in J}(\mathcal{H}_j^{\infty_\mathbb{A}})_{(K_\infty)}$ is $\mathcal{Z}(\mathfrak{g}_\infty)$-finite. As such an element is obviously also right K_∞-finite and smooth, reconsidering the isomorphism of $(\mathfrak{g}_\infty, K_\infty, G(\mathbb{A}_f))$-modules (13.13) together with Lem. 10.43 shows that ϑ maps $\bigoplus_{j\in J}(\mathcal{H}_j^{\infty_\mathbb{A}})_{(K_\infty)}$ onto a $(\mathfrak{g}_\infty, K_\infty, G(\mathbb{A}_f))$-submodule of $\mathcal{A}_{(2)}([G])$.

In order to prove that ϑ and summation of functions induce an isomorphism $\mathcal{A}_{(2)}([G]) \cong \bigoplus_{j\in J}(\mathcal{H}_j^{\infty_\mathbb{A}})_{(K_\infty)}$ of $(\mathfrak{g}_\infty, K_\infty, G(\mathbb{A}_f))$-modules, we are left to show exhaustion of $\mathcal{A}_{(2)}([G])$ by $\vartheta(\bigoplus_{j\in J}(\mathcal{H}_j^{\infty_\mathbb{A}})_{(K_\infty)})$. To this end, let $0 \neq \varphi \in \mathcal{A}_{(2)}([G])$. We have seen in the proof of Thm. 13.12 that the $G(\mathbb{A})$-subrepresentation $V_{[\varphi]} = \mathrm{Cl}_{L^2([G])}(\langle R(G(\mathbb{A}))[\varphi]\rangle)$ of $L^2([G])$ is the direct Hilbert sum of finitely many irreducible $G(\mathbb{A})$-subrepresentations of $L^2([G])$

$$V_{[\varphi]} \cong \widehat{\bigoplus_{1\leq i\leq d}} U_i$$

and hence

$$(V_{[\varphi]})_{(K_\infty)} \cong \bigoplus_{1\leq i\leq d} (U_i)_{(K_\infty)}$$

as vector spaces. Consequently, we may write $[\varphi] = \sum_{i=1}^d [\varphi_i]$, where in suggestive notation $[\varphi_i] \in (U_i)_{(K_\infty)}$. Clearly, each class $[\varphi_i] \in L^2_{dis}([G]) \cong \widehat{\bigoplus}_{j\in J} \mathcal{H}_j$ and so a representative φ_i has an expansion into a Fourier-series: $\varphi_i = \sum_k \varphi_{i,k}$, with $[\varphi_{i,k}] \in \mathcal{H}_k$. We observe that if $[\varphi_{i,k}] \neq 0$, then necessarily $\mathcal{H}_k \cong U_i$. But since each summand U_i has finite multiplicity in $L^2_{dis}([G])$, this implies that the sums $\varphi_i = \sum_k \varphi_{i,k}$ are finite and so $[\varphi_i] \in \bigoplus_{j\in J}(\mathcal{H}_j^{\infty_\mathbb{A}})_{(K_\infty)}$ for $1 \leq i \leq d$. However, this shows that $[\varphi] \in \bigoplus_{j\in J}(\mathcal{H}_j^{\infty_\mathbb{A}})_{(K_\infty)}$ and so $\varphi \in \vartheta(\bigoplus_{j\in J}(\mathcal{H}_j^{\infty_\mathbb{A}})_{(K_\infty)})$, which is just what we had to show. Hence, ϑ and summation of functions induce an isomorphism $\mathcal{A}_{(2)}([G]) \cong \bigoplus_{j\in J}(\mathcal{H}_j^{\infty_\mathbb{A}})_{(K_\infty)}$ of $(\mathfrak{g}_\infty, K_\infty, G(\mathbb{A}_f))$-modules.

Irreducibility of $(\mathcal{H}_j^{\infty_\mathbb{A}})_{(K_\infty)}$ is clear from Prop. 10.44, while admissibility follows from Thm. 10.37 and Lem. 10.43. This completes the proof. □

Now, let $\mathcal{J}$ be an arbitrary but fixed ideal of finite codimension in $\mathcal{Z}(\mathfrak{g}_\infty)$. We let $\mathcal{A}_\mathcal{J}([G])$ be the space of all automorphic forms in $\mathcal{A}_\mathcal{J}(G)$, which are constant on $A_G^\mathbb{R}$ and we set

$$\mathcal{A}_{cusp,\mathcal{J}}([G]) := \mathcal{A}_{cusp}([G]) \cap \mathcal{A}_\mathcal{J}([G]) \tag{13.21}$$

and

$$\mathcal{A}_{(2),\mathcal{J}}([G]) := \mathcal{A}_{(2)}([G]) \cap \mathcal{A}_\mathcal{J}([G]).$$

Then, Cor. 13.20 implies that there are uniquely determined subsets $I_\mathcal{J} \subseteq I$ and $J_\mathcal{J} \subseteq J$ such that as $(\mathfrak{g}_\infty, K_\infty, G(\mathbb{A}_f))$-modules:

$$\mathcal{A}_{cusp,\mathcal{J}}([G]) \cong \bigoplus_{i\in I_\mathcal{J}} (\mathcal{H}_i^{\infty_\mathbb{A}})_{(K_\infty)} \quad \text{and} \quad \mathcal{A}_{(2),\mathcal{J}}([G]) \cong \bigoplus_{j\in J_\mathcal{J}} (\mathcal{H}_j^{\infty_\mathbb{A}})_{(K_\infty)}. \tag{13.22}$$

Definition 13.23. An automorphic representation (π_0, V_0) is called *cuspidal* (respectively, *square-integrable*), if $V_0 \subseteq \mathcal{A}_{cusp,\mathcal{J}}([G])$ (respectively, $V_0 \subseteq \mathcal{A}_{(2),\mathcal{J}}([G])$) for some ideal $\mathcal{J} \lhd \mathcal{Z}(\mathfrak{g}_\infty)$ of finite codimension.

Cor. 13.20 hence shows that the irreducible cuspidal (respectively, square-integrable) automorphic representations are precisely those isomorphic to a $(\mathfrak{g}_\infty, K_\infty, G(\mathbb{A}_f))$-module $(\mathcal{H}_i^{\infty_\mathbb{A}})_{(K_\infty)}$, $i \in I_\mathcal{J}$ (respectively, $(\mathcal{H}_j^{\infty_\mathbb{A}})_{(K_\infty)}$, $j \in J_\mathcal{J}$).

As a final ingredient before we turn to smooth-automorphic forms again, we define

$$L^2_{cusp,\mathcal{J}}([G]) \tag{13.24}$$

to be the inverse images of $\widehat{\bigoplus}_{i\in I_\mathcal{J}} \mathcal{H}_i$ under the isomorphism (13.8) in Thm. 13.7. As it is obvious from the definitions, $L^2_{cusp,\mathcal{J}}([G])$ is a unitary representation of $G(\mathbb{A})$ under right-translation.

Chapter 14

Parabolic Support

14.1 Restricting to smooth-automorphic forms on $[G]$

As we would like to build a bridge[1] between the results of the previous chapter and smooth-automorphic forms, we will now introduce $\mathcal{A}^{\infty}_{\mathcal{J}}([G])$, which we define to be the subspace of all smooth-automorphic forms in $\mathcal{A}^{\infty}_{\mathcal{J}}(G)$, which are constant on $A^{\mathbb{R}}_{G}$. The following lemma, which is taken from [Gro-Žun23], will be fundamental:

Lemma 14.1. *The subspace topology on $\mathcal{A}^{\infty}_{\mathcal{J}}([G])$ coming from $\mathcal{A}^{\infty}_{\mathcal{J}}(G)$ is identical with the LF-topology given by $\varinjlim \mathcal{A}^{\infty}_{d}([G])^{K_n,\mathcal{J}^n}$.*

Proof. Let us, in this proof, for simplicity equip $\mathcal{A}^{\infty}_{\mathcal{J}}([G])$ with the semi-normed topology given by $\varinjlim \mathcal{A}^{\infty}_{d}([G])^{K_n,\mathcal{J}^n}$. We first observe that the spaces $\mathcal{A}^{\infty}_{d}([G])^{K_n,\mathcal{J}^n}$ are indeed Fréchet spaces by Prop. 11.9 and the argument used in the proof of the second part of Prop. 11.5, whence $\mathcal{A}^{\infty}_{\mathcal{J}}([G])$ is an LF-space. We now show that this topology on $\mathcal{A}^{\infty}_{\mathcal{J}}([G])$ is identical with the subspace topology inherited from $\mathcal{A}^{\infty}_{\mathcal{J}}(G)$.

As we have just observed, $\mathcal{A}^{\infty}_{d}([G])^{K_n,\mathcal{J}^n}$ is a closed topological subspace of $\mathcal{A}^{\infty}_{d}(G)^{K_n,\mathcal{J}^n}$ for every $n \in \mathbb{N}$, hence the natural inclusion map provides a continuous embedding of strict inductive limits

$$\mathcal{A}^{\infty}_{\mathcal{J}}([G]) = \varinjlim \mathcal{A}^{\infty}_{d}([G])^{K_n,\mathcal{J}^n} \hookrightarrow \varinjlim \mathcal{A}^{\infty}_{d}(G)^{K_n,\mathcal{J}^n} = \mathcal{A}^{\infty}_{\mathcal{J}}(G),$$

[1] This shall be compared to the following lines of R. Wagner:

Zur Burg führt die Brücke,
leicht, doch fest eurem Fuß:
beschreitet kühn ihren schrecklosen Pfad!

cf. Lem. 8.2. We may hence finish the proof by showing that there exists a continuous linear projection $\mathcal{A}^\infty_{\mathcal{J}}(G) \to \varinjlim \mathcal{A}^\infty_d([G])^{K_n,\mathcal{J}^n}$. We recall that this is not automatically satisfied, as one may not simply apply the Open Mapping Theorem to the image of $\mathcal{A}^\infty_{\mathcal{J}}([G])$ in $\mathcal{A}^\infty_{\mathcal{J}}(G)$, cf. Warning 8.6.

Let us denote by $\mathrm{d}a := \mathrm{d}g\,|_{A^{\mathbb{R}}_G}$ the Haar measure on $A^{\mathbb{R}}_G$ given by restricting $\mathrm{d}g$. It follows from [Wal94], Lem. 2.2 that there exists an $m \in \mathbb{Z}_{>0}$, such that

$$\int_{A^{\mathbb{R}}_G} \|a\|^{-m}\,\mathrm{d}a < \infty. \tag{14.2}$$

We put $m_0 := d + m \in \mathbb{Z}_{>0}$ and $A_0 := \int_{A^{\mathbb{R}}_G} \|a\|^{-m_0}\,\mathrm{d}a$, which is a well-defined positive real number, since $m_0 \geq m$ (and $\|a\| \geq 1$ for all $a \in A^{\mathbb{R}}_G$). Next, for $\varphi \in \mathcal{A}^\infty_{\mathcal{J}}(G)$, define a new function $P(\varphi) : G(\mathbb{A}) \to \mathbb{C}$ by

$$P(\varphi)(ag) := A_0^{-1} \int_{A^{\mathbb{R}}_G} \varphi(ag)\|a\|^{-m_0}\,\mathrm{d}a\,, \qquad a \in A^{\mathbb{R}}_G,\ g \in G(\mathbb{A})^{(1)}.$$

We claim that the assignment $\varphi \mapsto P(\varphi)$ gives rise to a well-defined continuous linear projection $P : \mathcal{A}^\infty_{\mathcal{J}}(G) \twoheadrightarrow \mathcal{A}^\infty_{\mathcal{J}}([G])$. Indeed, using (9.8) and [Beu21], Prop. A.1.1(i), one directly checks that for every $n \in \mathbb{N}$ and $\varphi \in \mathcal{A}^\infty_d(G)^{K_n,\mathcal{J}^n}$, $P(\varphi)$ is a well-defined, left $G(F)A^{\mathbb{R}}_G$-invariant, right K_n-invariant, smooth function $G(\mathbb{A}) \to \mathbb{C}$. Moreover, $P(\varphi)$ satisfies

$$D \cdot P(\varphi) = P(D_{\mathfrak{g}^{(1)}_\infty}\varphi), \qquad D \in \mathcal{U}(\mathfrak{g}_\infty), \tag{14.3}$$

Here, $G^{(1)}_\infty := G_\infty \cap G(\mathbb{A})^{(1)}$, and $D_{\mathfrak{g}^{(1)}_\infty}$ denotes the projection of D in $\mathcal{U}(\mathfrak{g}^{(1)}_\infty)$ with respect to the direct sum decomposition $\mathcal{U}(\mathfrak{g}_\infty) = \langle \mathfrak{a}_G \rangle_{\mathcal{U}(\mathfrak{g}_\infty)} \oplus \mathcal{U}(\mathfrak{g}^{(1)}_\infty)$. Obviously, the latter decomposition restricts to $\mathcal{Z}(\mathfrak{g}_\infty) = \langle \mathfrak{a}_G \rangle_{\mathcal{Z}(\mathfrak{g}_\infty)} \oplus \mathcal{Z}(\mathfrak{g}^{(1)}_\infty)$. Now, recall that $\mathrm{Lie}(A^{\mathbb{R}}_G) = \mathfrak{a}_G$. Since obviously $\mathcal{A}^\infty_{\mathcal{J}}([G]) = \mathcal{A}^\infty_{\mathcal{J}+\langle \mathfrak{a}_G \rangle_{\mathcal{Z}(\mathfrak{g})}}([G])$, we may assume without loss of generality that $\mathcal{J} \supseteq \mathfrak{a}_G$. Doing so, we get the decomposition $\mathcal{J} = \langle \mathfrak{a}_G \rangle_{\mathcal{Z}(\mathfrak{g}_\infty)} \oplus \left(\mathcal{J} \cap \mathcal{Z}(\mathfrak{g}^{(1)}_\infty)\right)$, and hence for each $n \in \mathbb{N}$

$$\begin{aligned}\mathcal{J}^n \cdot P(\varphi) &\overset{(14.3)}{=} P\Big(\Big(\Big(\langle \mathfrak{a}_G \rangle_{\mathcal{Z}(\mathfrak{g}_\infty)} \oplus \left(\mathcal{J} \cap \mathcal{Z}(\mathfrak{g}^{(1)}_\infty)\right)\Big)^n\Big)_{\mathfrak{g}^{(1)}_\infty}\varphi\Big) \\ &= P\Big(\left(\mathcal{J} \cap \mathcal{Z}(\mathfrak{g}^{(1)}_\infty)\right)^n \varphi\Big),\end{aligned}$$

which vanishes for all powers $\mathcal{J}^n$ such that $\mathcal{J}^n \cdot \varphi = 0$. It hence remains to show that $P(\varphi)$ is of uniform moderate growth, whence indeed in $\mathcal{A}^\infty_{\mathcal{J}}([G])$, and that the resulting map $P : \mathcal{A}^\infty_{\mathcal{J}}(G) \to \mathcal{A}^\infty_{\mathcal{J}}([G])$ (which is then by obvious reasons a linear projection onto $\mathcal{A}^\infty_{\mathcal{J}}([G])$) is continuous. We will do both in one go.

Recalling [Mœ-Wal95], I.2.2(viii), and (9.8), we obtain an $M_0 \in \mathbb{R}_{>0}$ and a $t_0 \in \mathbb{Z}_{>0}$, such that

$$\|g\| \leq M_0 \|ag\|^{t_0} \tag{14.4}$$

for all $a \in A_G^{\mathbb{R}}$ and $g \in G(\mathbb{A})^{(1)}$. This yields

$$\begin{aligned}
p_{td,D}(P(\varphi)) &\overset{(14.3)}{=} \sup_{\substack{g \in G(\mathbb{A})^{(1)},\\ a' \in A_G^{\mathbb{R}}}} \left| \left(P(D_{\mathfrak{g}^{(1)}_\infty}\varphi) \right)(a'g) \right| \|a'g\|^{-t_0 d} \\
&\overset{(14.4)}{\leq} \sup_{g \in G(\mathbb{A})^{(1)}} \left(A_0^{-1} \int_{A_G^{\mathbb{R}}} \left| \left(D_{\mathfrak{g}^{(1)}_\infty}\varphi \right)(ag) \right| \|a\|^{-m_0} \, da \cdot M_0^d \|g\|^{-d} \right) \\
&\leq \frac{M_0^d}{A_0} \sup_{g \in G(\mathbb{A})^{(1)}} \int_{A_G^{\mathbb{R}}} p_{d,D_{\mathfrak{g}^{(1)}_\infty}}(\varphi) \, \|ag\|^d \, \|a\|^{-m_0} \, \|g\|^{-d} \, da \\
&\overset{(9.8)}{\leq} \underbrace{\frac{(M_0 C_0)^d}{A_0} \int_{A_G^{\mathbb{R}}} \|a\|^{-m} \, da}_{<\infty, (14.2)} \cdot p_{d,D_{\mathfrak{g}^{(1)}_\infty}}(\varphi),
\end{aligned}$$

for all $D \in \mathcal{U}(\mathfrak{g}_\infty)$ and $\varphi \in \mathcal{A}^\infty_d(G)^{K_n, \mathcal{J}^n}$.

Summing up, we have shown that for every $n \in \mathbb{N}$, P restricts to a well-defined, continuous linear operator $\mathcal{A}^\infty_d(G)^{K_n,\mathcal{J}^n} \to \mathcal{A}^\infty_{t_0 d}([G])^{K_n,\mathcal{J}^n}$. Passing to the inductive limit, it follows from Lem. 8.2 that P is a well-defined, continuous linear projection

$$P : \mathcal{A}^\infty_{\mathcal{J}}(G) = \varinjlim \mathcal{A}^\infty_d(G)^{K_n,\mathcal{J}^n} \twoheadrightarrow \varinjlim \mathcal{A}^\infty_{t_0 d}([G])^{K_n,\mathcal{J}^n} = \mathcal{A}^\infty_{\mathcal{J}}([G]),$$

where the last equality of seminormed spaces holds by the same argument, which led to Lem. 11.14. This shows the lemma. □

14.2 LF-compatible smooth-automorphic representations

As a consequence of Lem. 14.1, $(R, \mathcal{A}^\infty_{\mathcal{J}}([G]))$ is a smooth-automorphic subrepresentation of $(R, \mathcal{A}^\infty_{\mathcal{J}}(G))$. Indeed, it is also *LF-compatible*, according to the following definition.

Definition 14.5. A smooth-automorphic representation (π, V) of $G(\mathbb{A})$ is called *LF-compatible*, if $V = \varinjlim V^{K_n,\mathcal{J}^n}$ holds topologically. Here, we wrote $V^{K_n,\mathcal{J}^n}$ for the closed G_∞-invariant subspace of K_n-invariant elements of V, which are annihilated by $\mathcal{J}^n$.

We invoke that by Lem. 8.2 the LF-space $\mathcal{A}^\infty_\mathcal{J}(G) = \varinjlim \mathcal{A}^\infty_d(G)^{K_n,\mathcal{J}^n}$ induces the original Fréchet-topology on $\mathcal{A}^\infty_\mathcal{J}(G)^{K_n,\mathcal{J}^n} = \mathcal{A}^\infty_d(G)^{K_n,\mathcal{J}^n}$: Hence, a smooth-automorphic subrepresentation (π, V) of $G(\mathbb{A})$ is LF-compatible, if and only if $V = \varinjlim(V \cap \mathcal{A}^\infty_d(G)^{K_n,\mathcal{J}^n})$ holds topologically (which shall explain the appearance of the word "compatible" in "LF-compatible").

Example 14.6. A finitely generated smooth-automorphic subrepresentation is LF-compatible, since it is annihilated by a power of $\mathcal{J}$. In particular, every irreducible smooth-automorphic subrepresentation is LF-compatible.

The following general proposition is contained in [Gro-Žun23]:

Proposition 14.7. *Let V be an LF-compatible smooth-automorphic subrepresentation, and denote $V_0 := V_{(K_\infty)}$. Suppose that summation of functions induces an isomorphism*

$$V_0 \cong \bigoplus_{i \in I} V_{0,i} \tag{14.8}$$

of $(\mathfrak{g}_\infty, K_\infty, G(\mathbb{A}_f))$-modules for an arbitrary family of $(\mathfrak{g}_\infty, K_\infty, G(\mathbb{A}_f))$-submodules $V_{0,i} \subseteq V_0$, $i \in I$. Denote $V_i := \mathrm{Cl}_{\mathcal{A}^\infty_\mathcal{J}(G)}(V_{0,i})$. Then, summation of functions induces an isomorphism of $G(\mathbb{A})$-representations,

$$V \cong \bigoplus_{i \in I} V_i, \tag{14.9}$$

where each summand V_i is an LF-compatible smooth-automorphic subrepresentation.

Proof. Lem. 12.7 implies that V_i is a smooth-automorphic subrepresentation for every $i \in I$. Furthermore, we recall from Lem. 8.2 that $\mathrm{Cl}_{\mathcal{A}^\infty_\mathcal{J}(G)}(V_{0,i}^{K_n,\mathcal{J}^n}) = \mathrm{Cl}_{\mathcal{A}^\infty_d(G)^{K_n,\mathcal{J}^n}}(V_{0,i}^{K_n,\mathcal{J}^n})$, since $\mathcal{A}^\infty_\mathcal{J}(G)$ induces on $\mathcal{A}^\infty_d(G)^{K_n,\mathcal{J}^n}$ its original topology. Therefore, for every fixed $n \in \mathbb{N}$, the sum of the spaces $\varinjlim \mathrm{Cl}_{\mathcal{A}^\infty_\mathcal{J}(G)}(V_{0,i}^{K_n,\mathcal{J}^n})$ ranging over all $i \in I$ is direct, as it follows from Lem. 11.24. We will now show that summation of functions induces an isomorphism

$$V \cong \bigoplus_{i \in I} \varinjlim \mathrm{Cl}_{\mathcal{A}^\infty_\mathcal{J}(G)}(V_{0,i}^{K_n,\mathcal{J}^n}). \tag{14.10}$$

of seminormed spaces. To this end, we first observe that we have an equality of seminormed spaces

$$\bigoplus_{i\in I} \varinjlim \mathrm{Cl}_{\mathcal{A}^\infty_{\mathcal{J}}(G)}(V_{0,i}^{K_n,\mathcal{J}^n}) = \varinjlim \bigoplus_{i\in I} \mathrm{Cl}_{\mathcal{A}^\infty_{\mathcal{J}}(G)}(V_{0,i}^{K_n,\mathcal{J}^n}).$$

Indeed, the equality of vector spaces is obvious and the equality of topologies follows directly from the definitions of the locally convex topology on direct sums and on inductive limits: More precisely, recalling the definitions, it turns out that both topologies are the finest locally convex topology with respect to which the inclusion maps of the subspaces $\mathrm{Cl}_{\mathcal{A}^\infty_{\mathcal{J}}(G)}(V_{0,i}^{K_n,\mathcal{J}^n})$ are continuous, whence these topologies are identical. Hence, in order to prove (14.10), exploiting the LF-compatibility of V, it suffices to prove that for every $n \in \mathbb{N}$, summation of functions induces an isomorphism

$$V^{K_n,\mathcal{J}^n} \cong \bigoplus_{i\in I} \mathrm{Cl}_{\mathcal{A}^\infty_{\mathcal{J}}(G)}(V_{0,i}^{K_n,\mathcal{J}^n}). \tag{14.11}$$

We recall that since $\mathcal{A}^\infty_d(G)^{K_n,\mathcal{J}^n}$ is an admissible G_∞-representation, cf. Prop. 11.9, $\mathrm{Cl}_{\mathcal{A}^\infty_{\mathcal{J}}(G)}(V_{0,i}^{K_n,\mathcal{J}^n}) = \mathrm{Cl}_{\mathcal{A}^\infty_d(G)^{K_n,\mathcal{J}^n}}(V_{0,i}^{K_n,\mathcal{J}^n})$ is a G_∞-subrepresentation of $\mathcal{A}^\infty_d(G)^{K_n,\mathcal{J}^n}$ by Thm. 3.16. Hence, combining Lem. 3.27 with Prop. 11.9, $\mathrm{Cl}_{\mathcal{A}^\infty_{\mathcal{J}}(G)}(V_{0,i}^{K_n,\mathcal{J}^n})$ is even a Casselman–Wallach representation of G_∞, which by Thm. 3.16 satisfies

$$\left(\mathrm{Cl}_{\mathcal{A}^\infty_{\mathcal{J}}(G)}(V_{0,i}^{K_n,\mathcal{J}^n})\right)_{(K_\infty)} = V_{0,i}^{K_n,\mathcal{J}^n}. \tag{14.12}$$

We claim that for every given $n \in \mathbb{N}$, $V_{0,i}^{K_n,\mathcal{J}^n} \neq \{0\}$ only for finitely many $i \in I$. Indeed, by (14.8), summation of functions induces an isomorphism of $(\mathfrak{g}_\infty, K_\infty)$-modules $V_0^{K_n,\mathcal{J}^n} \cong \bigoplus_{i\in I} V_{0,i}^{K_n,\mathcal{J}^n}$, whence, in order to show that $V_{0,i}^{K_n,\mathcal{J}^n} \neq \{0\}$ only for finitely many $i \in I$, it is enough to prove that the $(\mathfrak{g}_\infty, K_\infty)$-module $V_0^{K_n,\mathcal{J}^n}$ is finitely generated. However, as $V_0^{K_n,\mathcal{J}^n}$ is obviously $\mathcal{Z}(\mathfrak{g}_\infty)$-finite and moreover admissible by [Bor-Jac79], §4.3.(i), $V_0^{K_n,\mathcal{J}^n}$ is finitely generated by Thm. 3.14. Hence, for each $n \in \mathbb{N}$ there is a finite subset $I_n \subseteq I$ such that

$$V_0^{K_n,\mathcal{J}^n} \cong \bigoplus_{i\in I_n} V_{0,i}^{K_n,\mathcal{J}^n}, \tag{14.13}$$

I follows that for each $n \in \mathbb{N}$,

$$\bigoplus_{i\in I} \mathrm{Cl}_{\mathcal{A}^\infty_{\mathcal{J}}(G)}(V_{0,i}^{K_n,\mathcal{J}^n}) = \bigoplus_{i\in I_n} \mathrm{Cl}_{\mathcal{A}^\infty_{\mathcal{J}}(G)}(V_{0,i}^{K_n,\mathcal{J}^n})$$

is a Casselman–Wallach representation of G_∞, whose $(\mathfrak{g}_\infty, K_\infty)$-module of K_∞-finite vectors satisfies

$$\begin{aligned}
\left(\bigoplus_{i\in I} \mathrm{Cl}_{\mathcal{A}^\infty_{\mathcal{J}}(G)}(V_{0,i}^{K_n,\mathcal{J}^n})\right)_{(K_\infty)} &= \left(\bigoplus_{i\in I_n} \mathrm{Cl}_{\mathcal{A}^\infty_{\mathcal{J}}(G)}(V_{0,i}^{K_n,\mathcal{J}^n})\right)_{(K_\infty)} \\
&= \bigoplus_{i\in I_n} \left(\mathrm{Cl}_{\mathcal{A}^\infty_{\mathcal{J}}(G)}(V_{0,i}^{K_n,\mathcal{J}^n})\right)_{(K_\infty)} \\
&\overset{(14.12)}{=} \bigoplus_{i\in I_n} V_{0,i}^{K_n,\mathcal{J}^n} \\
&\overset{(14.13)}{\cong} V_0^{K_n,\mathcal{J}^n},
\end{aligned}$$

where the last isomorphism is provided by summation of functions. However, Prop. 11.9 together with Lem. 3.27 implies that $V^{K_n,\mathcal{J}^n}$ is a Casselman–Wallach representation of G_∞ and obviously, as the actions of K_∞, K_n and $\mathcal{J}^n$ all commute, $(V^{K_n,\mathcal{J}^n})_{(K_\infty)} = (V_{(K_\infty)})^{K_n,\mathcal{J}^n} = V_0^{K_n,\mathcal{J}^n}$. It follows now from Thm. 3.26 that summation of functions induces an isomorphism of G_∞-representations, i.e., that (14.11) holds. Therefore, we have finally shown (14.10).

We claim that (14.10) implies (14.9), or, more precisely that for every $i \in I$,

$$V_i = \varinjlim \mathrm{Cl}_{\mathcal{A}^\infty_{\mathcal{J}}(G)}(V_{0,i}^{K_n,\mathcal{J}^n}) \tag{14.14}$$

as seminormed spaces. In fact, as by Lem. 8.2, $\varinjlim \mathrm{Cl}_{\mathcal{A}^\infty_{\mathcal{J}}(G)}(V_{0,i}^{K_n,\mathcal{J}^n})$ is complete and hence a closed subspace of V for each $i \in I$, and as by definition V_i is also a closed subspace of V, we only need to verify that (14.14) holds as sets. To this end, we observe that obviously $V_i = \mathrm{Cl}_{\mathcal{A}^\infty_{\mathcal{J}}(G)}(V_{0,i}) \supseteq \mathrm{Cl}_{\mathcal{A}^\infty_{\mathcal{J}}(G)}(V_{0,i}^{K_n,\mathcal{J}^n})$ for every $n \in \mathbb{N}$, whence

$$V_i \supseteq \bigcup_{n\in\mathbb{N}} \mathrm{Cl}_{\mathcal{A}^\infty_{\mathcal{J}}(G)}(V_{0,i}^{K_n,\mathcal{J}^n}) = \varinjlim \mathrm{Cl}_{\mathcal{A}^\infty_{\mathcal{J}}(G)}(V_{0,i}^{K_n,\mathcal{J}^n})$$

for every $i \in I$. Vice versa, $\varinjlim \mathrm{Cl}_{\mathcal{A}^\infty_{\mathcal{J}}(G)}(V_{0,i}^{K_n,\mathcal{J}^n})$ contains $V_{0,i}$. So, as $\varinjlim \mathrm{Cl}_{\mathcal{A}^\infty_{\mathcal{J}}(G)}(V_{0,i}^{K_n,\mathcal{J}^n})$ is closed, it also contains $\mathrm{Cl}_{\mathcal{A}^\infty_{\mathcal{J}}(G)}(V_{0,i}) = V_i$. This shows (14.14) and hence (14.9).

It remains to show that V_i is LF-compatible. In order to see this, we observe that (14.8) and (14.9) imply that $(V_i)_{(K_\infty)} = V_{0,i}$. It follows that

$\left(V_i^{K_n,\mathcal{J}^n}\right)_{(K_\infty)} = V_{0,i}^{K_n,\mathcal{J}^n}$, whence

$$V_i^{K_n,\mathcal{J}^n} \overset{\text{Lem. 3.8}}{=} \mathrm{Cl}_{V_i^{K_n,\mathcal{J}^n}}\left(\left(V_i^{K_n,\mathcal{J}^n}\right)_{(K_\infty)}\right)$$
$$= \mathrm{Cl}_{V_i^{K_n,\mathcal{J}^n}}(V_{0,i}^{K_n,\mathcal{J}^n})$$
$$= \mathrm{Cl}_{\mathcal{A}_{\mathcal{J}}^\infty(G)}(V_{0,i}^{K_n,\mathcal{J}^n}).$$

Therefore, V_i is LF-compatible by (14.14). □

14.3 The parabolic support of a smooth-automorphic form

We let $\mathcal{A}_{cusp}^\infty([G])$ be the vector-subspace of $\mathcal{A}^\infty(G)$, which consists of all cuspidal smooth-automorphic forms, cf. §11.5 and Def. 13.4. For our purposes, it is not important to specify a locally convex topology on it (although we easily could, cf. (11.44)). The following notion is of central importance in the theory of (smooth-)automorphic forms:

Definition 14.15. A function $f \in C_{umg}^\infty(G(F)\backslash G(\mathbb{A}))$ is said to be *negligible* along a standard parabolic F-subgroup $\mathbf{P}$ of $\mathbf{G}$ with Levi decomposition $\mathbf{P} = \mathbf{L}\cdot\mathbf{N}$, if for all $g \in G(\mathbb{A})$ and $\varphi \in \mathcal{A}_{cusp}^\infty([L])$,

$$\lambda_{\mathbf{P},g,\varphi}(f) := \int_{L(F)\backslash L(\mathbb{A})^{(1)}} f_{\mathbf{P}}(\ell g)\,\overline{\varphi(\ell)}\,\mathrm{d}\ell = 0. \tag{14.16}$$

For every standard parabolic F-subgroup $\mathbf{P}$, let $\{\mathbf{P}\}$ denote the associate class of $\mathbf{P}$, i.e., the set of standard parabolic F-subgroups $\mathbf{Q} = \mathbf{L_Q N_Q}$ of $\mathbf{G}$ such that $\mathbf{L_Q}$ is conjugate to $\mathbf{L_P}$ by an element of $\mathbf{G}(F)$. We define $\mathcal{A}_{\mathcal{J},\{\mathbf{P}\}}^\infty([G])$ to be the space of all smooth-automorphic forms $\varphi \in \mathcal{A}_{\mathcal{J}}^\infty([G])$, which are negligible along all standard parabolic F-subgroups $\mathbf{Q} \notin \{\mathbf{P}\}$[2] and equip this space with the subspace topology from $\mathcal{A}_{\mathcal{J}}^\infty(G)$ (or, which amounts to the same by Lem. 14.1, with the subspace topology coming from $\mathcal{A}_{\mathcal{J}}^\infty([G])$). As every $\varphi \in \mathcal{A}_{\mathcal{J}}^\infty([G])$ is left $G(F)$-invariant by definition, the spaces $\mathcal{A}_{\mathcal{J},\{\mathbf{P}\}}^\infty([G])$ are independent of our fixed choice of standard parabolic F-subgroups, cf. [Bor91], Prop. 21.12.

[2]One could alternatively – and more positively – have said that φ is *concentrated* in the associate class $\{\mathbf{P}\}$ of $\mathbf{P}$. We will, however, not make use of this notion and stick with the seemingly more common conventions. See Rem. III.2.6 in [Mœ-Wal95], and [Lan79II], Prop. 2, though.

We remark that the way we have defined negligibility is intrinsic to the space of smooth-automorphic forms. In the relevant literature, the reader will find a slightly different definition of negligibility, which – in absence of the space $\mathcal{A}^{\infty}_{cusp}([L])$ – refers to cuspidal automorphic forms, i.e., one assumes (14.16) only for all $\varphi \in \mathcal{A}_{cusp}([L])$, cf. [Osb-War81], p. 82, which refers to Langlands's book [Lan76]. See also [BLS96], §2.2, [Bor06], §6.7, or [Fra-Sch98], §1.1. We will show in the proof of Thm. 14.17 below, however, that our Def. 14.15 is equivalent to the definition using $\mathcal{A}_{cusp}([L])$.

The following theorem is the main result of this section, which will finally allow us to give the definition of the parabolic support of a smooth-automorphic form $\varphi \in \mathcal{A}^{\infty}_{\mathcal{J}}([G])$ (see [Gro-Žun23] for the original source).

Theorem 14.17. *For every standard parabolic F-subgroup $\mathbf{P}$, the space $\mathcal{A}^{\infty}_{\mathcal{J},\{\mathbf{P}\}}([G])$ is an LF-compatible smooth-automorphic subrepresentation and summation of functions induces an isomorphism of $G(\mathbb{A})$-representations,*

$$\mathcal{A}^{\infty}_{\mathcal{J}}([G]) \cong \bigoplus_{\{\mathbf{P}\}} \mathcal{A}^{\infty}_{\mathcal{J},\{\mathbf{P}\}}([G]).$$

Proof. We will proceed in several steps.
Step 1: Let $\mathbf{Q} = \mathbf{LN}$ be an arbitrary, but fixed standard parabolic F-subgroup of $\mathbf{G}$. We will now introduce the *global Schwartz space* attached to L. To this end, we recall once more our fixed sequence $(K_n)_{n\in\mathbb{N}}$ of open compact subgroups of $G(\mathbb{A}_f)$ and denote by $K_{L,n} := K_n \cap L(\mathbb{A}_f)$. For $f \in C^{\infty}([L])$, $d \in \mathbb{Z}$ and $D \in \mathcal{U}(\mathfrak{l}_{\infty})$, we write

$$q_{d,D}(f) := \sup_{\ell \in L(\mathbb{A})^{(1)}} \left(|(Df)(\ell)| \inf_{\gamma \in L(F)} \|\gamma\ell\|^{d} \right),$$

Of course, the expression $q_{d,D}(f)$ may not always be finite, so we define

$$\mathcal{S}([L])^{K_{L,n}} := \{f \in C^{\infty}([L])^{K_{L,n}} \mid q_{d,D}(f) < \infty \quad \forall d \in \mathbb{Z}, D \in \mathcal{U}(\mathfrak{l}_{\infty})\}.$$

We observe that the maps $q_{d,D}$ define seminorms on $\mathcal{S}([L])^{K_{L,n}}$ and that equipping $\mathcal{S}([L])^{K_{L,n}}$ with the locally convex topology generated by the $q_{d,D}$, $d \in \mathbb{Z}$, $D \in \mathcal{U}(\mathfrak{l}_{\infty})$, yields a Fréchet space, which follows along the lines leading to [Mui-Žun20], Lem. 6 and Lem. 7. As a consequence of this construction (or, more precisely from amplifying Lem. 10 in [Mui-Žun20]) we see that the inductive limit of the spaces $\mathcal{S}([L])^{K_{L,n}}$ is in fact strict and so we may define the global Schwartz space of L as the LF-space

$$\mathcal{S}([L]) := \varinjlim \mathcal{S}([L])^{K_{L,n}}.$$

Now, let C be any compact subset $C \subseteq N(\mathbb{A})$ such that $N(\mathbb{A}) = N(F)C$, cf. [GGPS69], Thm. 3 in §6.1. Moreover, we let $d \in \mathbb{Z}_{>0}$ be as in Lem. 11.14. Then, for all $\varphi \in \mathcal{A}^{\infty}_{\mathcal{J}}([G])$ and all $\ell \in L(\mathbb{A})^{(1)}$, $g \in G(\mathbb{A})$,

$$\begin{aligned} |\varphi_{\mathbf{Q}}(\ell g)| &\leq \int_C |\varphi(n\ell g)| \,\mathrm{d}n \\ &\leq p_{d,1}(\varphi) \int_C \|n\ell g\|^d \,\mathrm{d}n \\ &\overset{(9.8)}{\leq} C_0^{2d}\, p_{d,1}(\varphi)\, \|\ell\|^d\, \|g\|^d \int_C \|n\|^d \,\mathrm{d}n\,. \end{aligned} \tag{14.18}$$

Therefore, for all $\varphi \in \mathcal{A}^{\infty}_{\mathcal{J}}([G])$, $f \in \mathcal{S}([L])$, and all $\ell \in L(\mathbb{A})^{(1)}$, $g \in G(\mathbb{A})$,

$$\begin{aligned} \left| \int_{L(F)\backslash L(\mathbb{A})^{(1)}} \varphi_{\mathbf{Q}}(\ell g) f(\ell)\,\mathrm{d}\ell \right| &\leq \int_{L(F)\backslash L(\mathbb{A})^{(1)}} |\varphi_{\mathbf{Q}}(\ell g) f(\ell)| \,\mathrm{d}\ell \\ &\overset{(14.18)}{\leq} C_0^{2d}\, p_{d,1}(\varphi)\, \|g\|^d \int_C \|n\|^d \,\mathrm{d}n \\ &\quad \cdot \int_{L(F)\backslash L(\mathbb{A})^{(1)}} |f(\ell)| \inf_{\gamma \in L(F)} \|\gamma\ell\|^d \,\mathrm{d}\ell \\ &\leq \left(C_0^{2d}\, \|g\|^d \,\mathrm{vol}_{\mathrm{d}\ell}(L(F)\backslash L(\mathbb{A})^{(1)}) \int_C \|n\|^d \,\mathrm{d}n \right) \\ &\quad \cdot \;\; p_{d,1}(\varphi)\; q_{d,1}(f). \end{aligned}$$

The expression in brackets is finite by Prop. 9.11 and the compactness of C. Hence, this implies that for every $g \in G(\mathbb{A})$,

$$\lambda_{\mathbf{Q},g,\bullet} : \mathcal{A}^{\infty}_{\mathcal{J}}([G]) \times \mathcal{S}([L]) \to \mathbb{C}$$

$$(\varphi, f) \mapsto \lambda_{\mathbf{Q},g,f}(\varphi) := \int_{L(F)\backslash L(\mathbb{A})^{(1)}} \varphi_{\mathbf{Q}}(\ell g)\, \overline{f(\ell)} \,\mathrm{d}\ell$$

is a well-defined, linear and separately continuous map.

Step 2: We observe that by compactness of C, (14.18) implies that the linear map

$$\mathcal{A}^{\infty}_{\mathcal{J}}([G]) \to \mathbb{C}$$

$$\varphi \mapsto \varphi_{\mathbf{Q}}(g)$$

is continuous for all standard parabolic F-subgroups $\mathbf{Q} = \mathbf{LN}$ of $\mathbf{G}$ and all $g \in G(\mathbb{A})$. In particular, the space

$$\mathcal{A}^{\infty}_{cusp,\mathcal{J}}([G]) := \{\varphi \in \mathcal{A}^{\infty}_{\mathcal{J}}([G]) \mid \varphi \text{ is cuspidal}\}$$

being the intersection of the kernels of the maps $\varphi \mapsto \varphi_{\mathbf{Q}}(g)$, for $\mathbf{Q}$ running over the proper standard parabolic F-subgroups and g through the elements in $G(\mathbb{A})$, is closed in $\mathcal{A}^{\infty}_{\mathcal{J}}([G])$, if equipped with the subspace topology. In particular, for every $n \in \mathbb{N}$, so is $\mathcal{A}^{\infty}_{cusp,\mathcal{J}}([G])^{K_n,\mathcal{J}^n} = \mathcal{A}^{\infty}_{cusp,\mathcal{J}}([G]) \cap \mathcal{A}^{\infty}_{d}([G])^{K_n,\mathcal{J}^n}$, which is hence by Lem. 8.2, Prop. 11.9 (and Lem. 1.11) a Fréchet space. We shall now give a different description of it using the global Schwartz-space.

In fact, we will slightly broaden our point of view: As above, we let $\mathbf{Q} = \mathbf{LN}$ be any standard parabolic F-subgroup of $\mathbf{G}$, and we let $\mathcal{L}$ be any ideal of finite codimension in $\mathcal{Z}(\mathfrak{l}_\infty)$. Then, for every $n \in \mathbb{N}$ we claim that the space $\mathcal{S}_{cusp}([L])^{K_{L,n},\mathcal{L}^n}$ of cuspidal, $\mathcal{L}^n$-annihilated functions in $\mathcal{S}([L])^{K_{L,n}}$ is a Fréchet space and as such equal to $\mathcal{A}^{\infty}_{cusp,\mathcal{L}}([L])^{K_{L,n},\mathcal{L}^n}$.

Indeed, we firstly observe that for every proper standard parabolic F-subgroup $\mathbf{Q}' = \mathbf{L}'\mathbf{N}'$ of $\mathbf{L}$, $f \in \mathcal{S}([L])^{K_{L,n}}$, and $\ell \in L(\mathbb{A})$, we get

$$|f_{\mathbf{Q}'}(\ell)| \leq \int_{N'(F)\backslash N'(\mathbb{A})} |f(n'\ell)| \, dn' \leq q_{0,1}(f)$$

and moreover for all $d \in \mathbb{Z}$, $D \in \mathcal{U}(\mathfrak{l}_\infty)$, and $Y \in \mathcal{L}^n$,

$$q_{d,D}(Yf) = \sup_{\ell \in L(\mathbb{A})^1} \left(|(DYf)(\ell)| \inf_{\gamma \in L(F)} \|\gamma\ell\|^d \right) = q_{d,DY}(f).$$

This implies that the linear maps

$$\mathcal{S}([L])^{K_{L,n}} \to \mathbb{C}$$

$$f \mapsto f_{\mathbf{Q}'}(\ell)$$

as well as

$$\mathcal{S}([L])^{K_{L,n}} \to \mathcal{S}([L])^{K_{L,n}}$$

$$f \mapsto Yf$$

are continuous for all proper standard parabolic F-subgroups $\mathbf{Q}' = \mathbf{L}'\mathbf{N}'$ of $\mathbf{L}$, $\ell \in L(\mathbb{A})$ and $Y \in \mathcal{L}^n$. Thus, the joint intersection $\mathcal{S}_{cusp}([L])^{K_{L,n},\mathcal{L}^n}$ of all their kernels is a closed subspace of the Fréchet space $\mathcal{S}([L])^{K_{L,n}}$ and hence indeed a Fréchet space itself (by Lem. 1.11).

Next we observe that [Mœ-Wal95], Cor. I.2.11 and I.2.2.(vi), imply, that $\mathcal{A}^{\infty}_{cusp,\mathcal{L}}([L])^{K_{L,n},\mathcal{L}^n} = \mathcal{S}_{cusp}([L])^{K_{L,n},\mathcal{L}^n}$ as vectors spaces. As by definition of the locally convex topologies on $\mathcal{A}^{\infty}_{cusp,\mathcal{L}}([L])^{K_{L,n},\mathcal{L}^n}$ and on $\mathcal{S}_{cusp}([L])^{K_{L,n},\mathcal{L}^n}$, the identity map $\mathcal{S}_{cusp}([L])^{K_{L,n},\mathcal{L}^n} \to \mathcal{A}^{\infty}_{cusp,\mathcal{L}}([L])^{K_{L,n},\mathcal{L}^n}$ is continuous, the Open Mapping Theorem, cf. Thm. 1.7, finally shows that

$$\mathcal{A}^{\infty}_{cusp,\mathcal{L}}([L])^{K_{L,n},\mathcal{L}^n} = \mathcal{S}_{cusp}([L])^{K_{L,n},\mathcal{L}^n} \tag{14.19}$$

as Fréchet spaces.

Step 3: We will now prove that a smooth-automorphic form $\varphi \in \mathcal{A}^{\infty}_{\mathcal{J}}([G])$ is negligible along a standard parabolic F-subgroup $\mathbf{Q} = \mathbf{LN}$ of $\mathbf{G}$ if and only if $\lambda_{\mathbf{Q},g,\phi}(\varphi) = 0$ for all $g \in G(\mathbb{A})$ and all $\phi \in \mathcal{A}_{cusp}([L])$.

Since $\mathcal{A}_{cusp}([L]) \subseteq \mathcal{A}^{\infty}_{cusp}([L])$, this condition is clearly necessary. We now show that it is also sufficient: To this end, assume that $\lambda_{\mathbf{Q},g,\phi}(\varphi) = 0$ for all $g \in G(\mathbb{A})$ and all $\phi \in \mathcal{A}_{cusp}([L])$. In other words, we assume that

$$\mathcal{A}_{cusp}([L]) \subseteq \bigcap_{g \in G(\mathbb{A})} \ker \lambda_{\mathbf{Q},g,\bullet}(\varphi). \tag{14.20}$$

We observe that (14.19) directly implies that $\mathcal{A}_{cusp}([L]) \subseteq \mathcal{S}([L])$. Hence, it makes sense to consider $\mathrm{Cl}_{\mathcal{S}([L])}(\mathcal{A}_{cusp}([L]))$. In fact, as $\lambda_{\mathbf{Q},g,\bullet}(\varphi) : \mathcal{S}([L]) \to \mathbb{C}$ is continuous, as shown in Step 1 above, our standing assumption (14.20) implies

$$\mathrm{Cl}_{\mathcal{S}([L])}(\mathcal{A}_{cusp}([L])) \subseteq \bigcap_{g \in G(\mathbb{A})} \ker \lambda_{\mathbf{Q},g,\bullet}(\varphi).$$

It is therefore enough to show that

$$\mathcal{A}^{\infty}_{cusp}([L]) \subseteq \mathrm{Cl}_{\mathcal{S}([L])}(\mathcal{A}_{cusp}([L])). \tag{14.21}$$

However, the following unions of sets being running over all $n \in \mathbb{N}$ and ideals $\mathcal{L}$ of finite codimension in $\mathcal{Z}(\mathfrak{l}_\infty)$

$$\begin{aligned}
\mathcal{A}^{\infty}_{cusp}([L]) &= \bigcup_{n,\mathcal{L}} \mathcal{A}^{\infty}_{cusp,\mathcal{L}}([L])^{K_{L,n},\mathcal{L}^n} \\
&= \bigcup_{n,\mathcal{L}} \mathrm{Cl}_{\mathcal{A}^{\infty}_{cusp,\mathcal{L}}([L])^{K_{L,n},\mathcal{L}^n}}(\mathcal{A}_{cusp,\mathcal{L}}([L])^{K_{L,n},\mathcal{L}^n}) \\
&\overset{(14.19)}{=} \bigcup_{n,\mathcal{L}} \mathrm{Cl}_{\mathcal{S}_{cusp}([L])^{K_{L,n},\mathcal{L}^n}}(\mathcal{A}_{cusp,\mathcal{L}}([L])^{K_{L,n},\mathcal{L}^n}) \\
&\subseteq \bigcup_{n,\mathcal{L}} \mathrm{Cl}_{\mathcal{S}([L])}(\mathcal{A}_{cusp,\mathcal{L}}([L])^{K_{L,n},\mathcal{L}^n}) \\
&\subseteq \mathrm{Cl}_{\mathcal{S}([L])}\left(\bigcup_{n,\mathcal{L}} \mathcal{A}_{cusp,\mathcal{L}}([L])^{K_{L,n},\mathcal{L}^n}\right) \\
&= \mathrm{Cl}_{\mathcal{S}([L])}(\mathcal{A}_{cusp}([L])).
\end{aligned}$$

Therefore, (14.21) is true and so finally a smooth-automorphic form $\varphi \in \mathcal{A}^\infty_{\mathcal{J}}([G])$ is negligible along a standard parabolic F-subgroup $\mathbf{Q} = \mathbf{LN}$ of $\mathbf{G}$ if and only if $\lambda_{\mathbf{Q},g,\phi}(\varphi) = 0$ for all $g \in G(\mathbb{A})$ and all $\phi \in \mathcal{A}_{cusp}([L])$.

Step 4: Taking the union over all $d \in \mathbb{Z}_{>0}$, Prop. 11.5 implies that the map $\varphi \mapsto (\varphi(\_ \cdot c))_{c \in C_n}$ induces an isomorphism of vector spaces

$$C^\infty_{umg}(G(F)\backslash G(\mathbb{A})) \cong \bigcup_{n\in\mathbb{N}} \bigoplus_{c\in C_n} C^\infty_{umg}(\Gamma_{c,K_n}\backslash G_\infty). \tag{14.22}$$

We apply [BLS96], Thm. 2.4 (or, alternatively, [Cas89II], Thm. 1.16 and Cor. 4.7) to $C^\infty_{umg}(\Gamma_{c,K_n}\backslash G_\infty)$ (and shortly invoke [Bor91], Prop. 21.12) and obtain for each $n \in \mathbb{N}$ and $c \in C_n$ Langlands's algebraic direct sum decomposition

$$C^\infty_{umg}(\Gamma_{c,K_n}\backslash G_\infty) \cong \bigoplus_{\{\mathbf{P}\}} C^\infty_{umg,\mathbf{P}}(\Gamma_{c,K_n}\backslash G_\infty), \tag{14.23}$$

where the spaces $C^\infty_{umg,\mathbf{P}}(\Gamma_{c,K_n}\backslash G_\infty)$ were denoted $V_{\Gamma_{c,K_n}}(\mathcal{P})$ in [BLS96], §2.2. We do not repeat their definition here, but only note that recalling [Bor-Jac79], 4.4.(3) and applying the inverse of the isomorphism in (14.22), (14.23) implies that summation of functions yields an isomorphims of vector spaces

$$C^\infty_{umg}(G(F)\backslash G(\mathbb{A})) \cong \bigoplus_{\{\mathbf{P}\}} C^\infty_{umg,\mathbf{P}}(G(F)\backslash G(\mathbb{A})), \tag{14.24}$$

where $C^\infty_{umg,\mathbf{P}}(G(F)\backslash G(\mathbb{A}))$ denotes the space of all functions $f \in C^\infty_{umg}(G(F)\backslash G(\mathbb{A}))$, that satisfy (14.16) for all standard parabolic F-subgroups $\mathbf{Q} \notin \{\mathbf{P}\}$, $g \in G(\mathbb{A})$ and $\varphi \in \mathcal{A}_{cusp}([L])$. Since the spaces $C^\infty_{umg,\mathbf{P}}(G(F)\backslash G(\mathbb{A}))$ are invariant under the action of $G(\mathbb{A})$ and $\mathcal{U}(\mathfrak{g}_\infty)$ by right-translation, (14.24) together with our Step 3 above implies that summation of functions induces a $G(\mathbb{A})$-equivariant decomposition of vector spaces

$$\mathcal{A}^\infty_{\mathcal{J}}([G]) \cong \bigoplus_{\{\mathbf{P}\}} \mathcal{A}^\infty_{\mathcal{J},\{\mathbf{P}\}}([G]). \tag{14.25}$$

We recall from Step 1 above that the linear functional $\lambda_{Q,g,\varphi} : \mathcal{A}^\infty_{\mathcal{J}}([G]) \to \mathbb{C}$ is continuous for every standard parabolic F-subgroup $\mathbf{Q} = \mathbf{LN}$ of $\mathbf{G}$, $g \in G(\mathbb{A})$ and $\varphi \in \mathcal{S}([L])$, hence, in particular for every $\varphi \in \mathcal{A}^\infty_{cusp}([L]) \subseteq \mathcal{S}([L])$. Therefore, the joint intersection of their kernels $\mathcal{A}^\infty_{\mathcal{J},\{\mathbf{P}\}}([G])$ is closed in $\mathcal{A}^\infty_{\mathcal{J}}([G])$. Together with Lem. 14.1 this shows that $\mathcal{A}^\infty_{\mathcal{J},\{\mathbf{P}\}}([G])$ is a smooth-automorphic subrepresentation. Hence,

$\mathcal{A}_{\mathcal{J},\{\mathbf{P}\}}([G]) := \mathcal{A}^{\infty}_{\mathcal{J},\{\mathbf{P}\}}([G])_{(K_\infty)}$ is a $(\mathfrak{g}_\infty, K_\infty, G(\mathbb{A}_f))$-module by Lem. 10.43, Thm. 11.17 and Prop. 10.23. Therefore, the linear $G(\mathbb{A})$-equivariant bijection (14.25) implies that summation of functions induces an isomorphism

$$\mathcal{A}_{\mathcal{J}}([G]) \cong \bigoplus_{\{P\}} \mathcal{A}_{\mathcal{J},\{\mathbf{P}\}}([G]) \tag{14.26}$$

of $(\mathfrak{g}_\infty, K_\infty, G(\mathbb{A}_f))$-modules. See also [Fra-Sch98], 1.1. It hence follows from our Lem. 14.1 and Prop. 14.7 that summation of functions induces an isomorphism of $G(\mathbb{A})$-representations

$$\mathcal{A}^{\infty}_{\mathcal{J}}([G]) \cong \bigoplus_{\{\mathbf{P}\}} \mathrm{Cl}_{\mathcal{A}^{\infty}_{\mathcal{J}}(G)}(\mathcal{A}_{\mathcal{J},\{\mathbf{P}\}}([G])), \tag{14.27}$$

where each summand $\mathrm{Cl}_{\mathcal{A}^{\infty}_{\mathcal{J}}(G)}(\mathcal{A}_{\mathcal{J},\{\mathbf{P}\}}([G]))$ is an LF-compatible smooth-automorphic subrepresentation. We are hence left to prove that for every $\{\mathbf{P}\}$, $\mathrm{Cl}_{\mathcal{A}^{\infty}_{\mathcal{J}}(G)}(\mathcal{A}_{\mathcal{J},\{\mathbf{P}\}}([G])) = \mathcal{A}^{\infty}_{\mathcal{J},\{\mathbf{P}\}}([G])$. By (14.27) and (14.25) it suffices to prove that $\mathrm{Cl}_{\mathcal{A}^{\infty}_{\mathcal{J}}(G)}(\mathcal{A}_{\mathcal{J},\{\mathbf{P}\}}([G])) \subseteq \mathcal{A}^{\infty}_{\mathcal{J},\{\mathbf{P}\}}([G])$. However, as we have just observed, $\mathcal{A}^{\infty}_{\mathcal{J},\{\mathbf{P}\}}([G])$ is closed in $\mathcal{A}^{\infty}_{\mathcal{J}}([G])$, whence by Lem. 14.1 in $\mathcal{A}^{\infty}_{\mathcal{J}}(G)$. As obviously $\mathcal{A}_{\mathcal{J},\{\mathbf{P}\}}([G]) \subseteq \mathcal{A}^{\infty}_{\mathcal{J},\{\mathbf{P}\}}([G])$, this shows that also $\mathrm{Cl}_{\mathcal{A}^{\infty}_{\mathcal{J}}(G)}(\mathcal{A}_{\mathcal{J},\{\mathbf{P}\}}([G])) \subseteq \mathcal{A}^{\infty}_{\mathcal{J},\{\mathbf{P}\}}([G])$ and hence the claim of Thm. 14.17. □

We are finally ready to give the definition of the parabolic support of a smooth-automorphic form.

Definition 14.28. For $\varphi \in \mathcal{A}^{\infty}_{\mathcal{J}}([G])$, write $\varphi = \sum_{\{\mathbf{P}\}} \varphi_{\{\mathbf{P}\}}$ according to Thm. 14.17. The summand $\varphi_{\{\mathbf{P}\}}$ is called the *parabolic support of* φ *along* $\{\mathbf{P}\}$.

Exercise 14.29. Show that the non-zero constant functions $\phi \equiv c$ in $\mathcal{A}^{\infty}_{\mathcal{J}}([G])$ have non-trivial parabolic support only along the associate class of the minimal parabolic $\{\mathbf{P}_0\}$. Conclude that the trivial representation $(\mathbf{1}, \mathbb{C})$ of $G(\mathbb{A})$ embeds as a smooth-automorphic subrepresentation into $\mathcal{A}^{\infty}_{\mathcal{J},\{\mathbf{P}_0\}}([G])$ and into no other summand $\mathcal{A}^{\infty}_{\mathcal{J},\{\mathbf{P}\}}([G])$, $\{\mathbf{P}\} \neq \{\mathbf{P}_0\}$. *Hint:* Recall §9.2, Fubini's theorem and the defining condition of cuspidality, in order to show that

$$\int_{L(F)\backslash L(\mathbb{A})^{(1)}} \overline{\varphi(\ell)}\, \mathrm{d}\ell = 0$$

for all non-minimal standard parabolic F-subgroups $\mathbf{P} = \mathbf{LN}$ of $\mathbf{G}$ and all $\varphi \in \mathcal{A}^{\infty}_{cusp}([L])$.

Chapter 15

Cuspidal Support

15.1 Decomposing spaces of cuspidal smooth-automorphic forms

We start off with the following observation.

Lemma 15.1. $\mathcal{A}^{\infty}_{\mathcal{J},\{\mathbf{G}\}}([G]) = \mathcal{A}^{\infty}_{cusp,\mathcal{J}}([G])$.

Proof. This is easy to see, once the reader recalls the transitivity of the constant term and the twofold fact that $\mathbf{L}_0$ has no proper parabolic F-subgroup (whence every smooth-automorphic form of $L_0(\mathbb{A})$ is cuspidal), whereas $\mathbf{G}$ has no proper unipotent radical (whence $\varphi_{\mathbf{G}} = \varphi$). We refer to [Bor07], §6.5 and §10.2, and leave the details to the reader. □

Corollary 15.2. *The space of all cuspidal smooth-automorphic forms* $\mathcal{A}^{\infty}_{cusp,\mathcal{J}}([G])$ *in* $\mathcal{A}^{\infty}_{\mathcal{J}}([G])$ *is an LF-compatible smooth-automorphic sub-representation.*

Proof. This is a direct consequence of Lem. 15.1 and Thm. 14.17. □

Remark 15.3. We remark again that this is far from being clear by general abstract considerations, cf. Warn. 8.6.

Definition 15.4. A smooth-automorphic representation (π, V) of $G(\mathbb{A})$ is called *cuspidal*, if $V \subseteq \mathcal{A}^{\infty}_{cusp,\mathcal{J}}([G])$ for some ideal $\mathcal{J} \lhd \mathcal{Z}(\mathfrak{g}_\infty)$ of finite codimension.

Let us now recall the space $L^2_{cusp,\mathcal{J}}([G])$ from (13.24), which is a unitary representation of $G(\mathbb{A})$. As it follows from (13.13), the map ϑ induces an isomorphism

$$\vartheta : L^2_{cusp,\mathcal{J}}([G])^{\infty_{\mathbb{A}}}_{(K_\infty,\mathcal{Z}(\mathfrak{g}_\infty))} \cong \mathcal{A}_{cusp,\mathcal{J}}([G]) \tag{15.5}$$

of $(\mathfrak{g}_\infty, K_\infty, G(\mathbb{A}_f))$-modules, which in fact extends to a $G(\mathbb{A})$-equivariant, linear bijection

$$\vartheta : L^2_{cusp,\mathcal{J}}([G])^{\infty_{\mathbb{A}}}_{(\mathcal{Z}(\mathfrak{g}_\infty))} \xrightarrow{\sim} \mathcal{A}^\infty_{cusp,\mathcal{J}}([G]). \tag{15.6}$$

Indeed, (15.6) this is a consequence of a combination of Lem. 15.1, our Prop. 11.5, Step 3 in the proof of Thm. 14.17 and [Bor-Jac79], 4.4.(3), [Bor06], (6.8.4), and (13.15).

A priori, it is by no means clear, that the linear bijection (15.6) is compatible with the direct Hilbert sum decomposition of $L^2_{cusp,\mathcal{J}}([G])$, cf. Thm. 13.7 and (13.24), i.e., it is unclear whether or not the LF-spaces $\mathcal{H}_i^{\infty_{\mathbb{A}}}$ of globally smooth vectors in the irreducible subrepresentations $\mathcal{H}_i$, $i \in I_{\mathcal{J}}$, identify with irreducible subrepresentations of the LF-space $\mathcal{A}^\infty_{cusp,\mathcal{J}}([G])$. However, we will show now the following.

Theorem 15.7. *The map ϑ, cf.* (15.5), *and summation of functions induce an isomorphism of $G(\mathbb{A})$-representations,*

$$\mathcal{A}^\infty_{cusp,\mathcal{J}}([G]) \cong \bigoplus_{i\in I_{\mathcal{J}}} \mathcal{H}_i^{\infty_{\mathbb{A}}}, \tag{15.8}$$

each summand being an irreducible (and hence) LF-compatible, smooth-automorphic subrepresentation. This characterizes the equivalence classes of irreducible cuspidal smooth-automorphic subrepresentations of $\mathcal{A}^\infty_{\mathcal{J}}([G])$ as being represented by the LF-spaces $\mathcal{H}_i^{\infty_{\mathbb{A}}}$ of globally smooth vectors in the unitary Hilbert space representations $\mathcal{H}_i$, $i \in I_{\mathcal{J}}$.

Proof. For $i \in I_{\mathcal{J}}$, set $H_i := \vartheta((\mathcal{H}_i^{\infty_{\mathbb{A}}})_{(K_\infty)})$. As $\mathcal{A}^\infty_{cusp,\mathcal{J}}([G])$ is LF-compatible by Cor. 15.2, our Cor. 13.20 (and its proof) implies that we may use Prop. 14.7, and hence obtain that summation of functions induces an isomorphism of $G(\mathbb{A})$-representations

$$\mathcal{A}^\infty_{cusp,\mathcal{J}}([G]) \cong \bigoplus_{i\in I_{\mathcal{J}}} \mathrm{Cl}_{\mathcal{A}^\infty_{\mathcal{J}}(G)}(H_i),$$

each summand being an LF-compatible smooth-automorphic subrepresentation. Recall from Cor. 13.20 that for each $i \in I_{\mathcal{J}}$, the automorphic subrepresentation H_i is irreducible. Therefore, $\mathrm{Cl}_{\mathcal{A}^\infty_{\mathcal{J}}(G)}(H_i)$ is in fact an irreducible smooth-automorphic subrepresentation by Thm. 12.10 and so we may finish the proof showing that for every $i \in I_{\mathcal{J}}$ the map ϑ defines an isomorphism $\mathcal{H}_i^{\infty_{\mathbb{A}}} \cong \mathrm{Cl}_{\mathcal{A}^\infty_{\mathcal{J}}(G)}(H_i)$ of $G(\mathbb{A})$-representations.

In order to do so, observe that $\mathrm{Cl}_{\mathcal{A}^\infty_{\mathcal{J}}(G)}(H_i)$ is a Casselman–Wallach representation of $G(\mathbb{A})$, which has H_i as its underlying $(\mathfrak{g}_\infty, K_\infty, G(\mathbb{A}_f))$-module, cf. Prop. 12.2 and Thm. 12.10.

On the other hand, we have seen in (10.45), that $\mathcal{H}_i \cong \mathcal{H}_{i,\infty}\widehat{\otimes}\mathcal{H}_{i,f}$, as $G(\mathbb{A})$-representation, with $(\mathcal{H}_{i,f})^{K_n}$ being finite-dimensional for each open compact subgroup K_n in our fixed sequence $(K_n)_{n\in\mathbb{N}}$. Hence, as representations of G_∞,

$$\begin{aligned}(\mathcal{H}_i^{\infty_\mathbb{R}})^{K_n} &\cong ((\mathcal{H}_{i,\infty}\widehat{\otimes}\mathcal{H}_{i,f})^{\infty_\mathbb{R}})^{K_n} \\ &= ((\mathcal{H}_{i,\infty}\widehat{\otimes}\mathcal{H}_{i,f})^{K_n})^{\infty_\mathbb{R}} \\ &= (\mathcal{H}_{i,\infty} \otimes \mathcal{H}_{i,f}^{K_n})^{\infty_\mathbb{R}} \\ &= (\mathcal{H}_{i,\infty})^{\infty_\mathbb{R}} \otimes \mathcal{H}_{i,f}^{K_n}.\end{aligned} \tag{15.9}$$

Indeed, the second line is (10.8), the third line follows in by the same argument leading to [Bor-Wal00], X.6.2.(2), whereas the last line is a simple combination of Lem. 1.16, Warning 1.14, Exc. 2.22 and Lem. 1.15. Since $\mathcal{H}_{i,\infty}$ is an irreducible unitary representation of G_∞ by Thm. 3.37, $\mathcal{H}_{i,\infty}^{\infty_\mathbb{R}}$ is a Casselman–Wallach representation of G_∞ by Lem. 2.18 (smoothness), Thm. 3.33 and Lem. 2.30 (admissibility), Lem. 3.24 (moderate growth) and Thm. 3.16 ($(\mathcal{H}_{i,\infty})_{(K_\infty)} = (\mathcal{H}_{i,\infty}^{\infty_\mathbb{R}})_{(K_\infty)}$ is finitely generated). By finite-dimensionality of $(\mathcal{H}_{i,f})^{K_n}$, hence so is $(\mathcal{H}_{i,\infty})^{\infty_\mathbb{R}} \otimes \mathcal{H}_{i,f}^{K_n} \cong (\mathcal{H}_i^{\infty_\mathbb{R}})^{K_n}$, and since $n \in \mathbb{N}$ was arbitrary, Lem. 10.19 implies that $\mathcal{H}_i^{\infty_\mathbb{A}}$ is a Casselman–Wallach representation of $G(\mathbb{A})$ according to Def. 10.50.

As by construction ϑ defines an isomorphism of $(\mathfrak{g}_\infty, K_\infty, G(\mathbb{A}_f))$-modules

$$\vartheta_0 : (\mathcal{H}_i^{\infty_\mathbb{A}})_{(K_\infty)} \xrightarrow{\sim} H_i = \mathrm{Cl}_{\mathcal{A}^\infty_{\mathcal{J}}(G)}(H_i)_{(K_\infty)},$$

Prop. 10.51 (and its proof) imply that ϑ_0 must extend to an isomorphism of $G(\mathbb{A})$-representations

$$\bar{\vartheta}_0 : \mathcal{H}_i^{\infty_\mathbb{A}} \xrightarrow{\sim} \mathrm{Cl}_{\mathcal{A}^\infty_{\mathcal{J}}(G)}(H_i).$$

It hence remains to show that this extension is equal to our original map ϑ, i.e., that $\bar{\vartheta}_0([f]) = \vartheta([f])$ for all $[f] \in \mathcal{H}_i^{\infty_\mathbb{A}}$. To this end, if $[f] \in \mathcal{H}_i^{\infty_\mathbb{A}}$, then $[f] \in (\mathcal{H}_i^{\infty_\mathbb{R}})^{K_n}$ for some $n \in \mathbb{N}$, and so Lem. 3.8 implies that there exists a sequence $([f_m])_{m\in\mathbb{N}}$, of elements $[f_m] \in (\mathcal{H}_i^{\infty_\mathbb{R}})^{K_n}_{(K_\infty)}$, which converges to $[f]$ in $(\mathcal{H}_i^{\infty_\mathbb{R}})^{K_n}$. By construction of the locally convex topology on $\mathcal{H}_i^{\infty_\mathbb{A}}$, cf. §8.1.1, this sequence also converges in $\mathcal{H}_i^{\infty_\mathbb{A}}$ and therefore also in $\mathcal{H}_i \subseteq L^2([G])$, see Lem. 10.19. Thus, replacing $([f_m])_{m\in\mathbb{N}}$ by a suitable subsequence, if necessary, and choosing any representatives h_m for $[f_m]$ and h for $[f]$, the corresponding sequence $(h_m)_{m\in\mathbb{N}}$ of functions must converge pointwise on $G(\mathbb{A})$ to h off a set of measure zero. However, by the continuity of $\bar{\vartheta}_0$, the sequence $\vartheta([f_m]) = \bar{\vartheta}_0([f_m])$ of smooth-automorphic forms converges to $\bar{\vartheta}_0([f])$ in $\mathcal{A}^\infty_{\mathcal{J}}([G])$, and so it also converges pointwise everywhere on $G(\mathbb{A})$ to $\bar{\vartheta}_0([f])$, cf. Exc. 11.16. Hence, $\bar{\vartheta}_0([f]) = h$ off a set of measure zero on $G(\mathbb{A})$, which, together with the continuity of $\bar{\vartheta}_0([f])$, cf. Lem. 10.4 and the fact that (15.6) is a linear bijection (whence $[h] = [f]$ must have a continuous representative) implies that $\bar{\vartheta}_0([f]) = \vartheta([h]) = \vartheta([f])$. This proves the claim. □

To round up our considerations, we will conclude this section with the following lemma.

Lemma 15.10. *For any two irreducible direct summands $\mathcal{H}$ and $\mathcal{H}'$ in the decomposition $L^2_{cusp,\mathcal{J}}([G]) \cong \widehat{\bigoplus}_{i\in I_{\mathcal{J}}} \mathcal{H}_i$, the following assertions are equivalent:*

(1) $\mathcal{H} \cong \mathcal{H}'$ as $G(\mathbb{A})$-representations,
(2) $\mathcal{H}^{\infty_\mathbb{A}} \cong \mathcal{H}'^{\infty_\mathbb{A}}$ as $G(\mathbb{A})$-representations,
(3) $\mathcal{H}^{\infty_\mathbb{A}}_{(K_\infty)} \cong \mathcal{H}'^{\infty_\mathbb{A}}_{(K_\infty)}$ as $(\mathfrak{g}_\infty, K_\infty, G(\mathbb{A}_f))$-modules.

Consequently, the finite multiplicities of $\mathcal{H}$ in $L^2_{cusp,\mathcal{J}}([G])$, of $\mathcal{H}^{\infty_\mathbb{A}}$ in $\mathcal{A}^\infty_{cusp,\mathcal{J}}([G])$ and of $\mathcal{H}^{\infty_\mathbb{A}}_{(K_\infty)}$ in $\mathcal{A}_{cusp,\mathcal{J}}([G])$ are all the same.

Proof. (1) ⇒ (2) ⇒ (3) is obviously trivial. That (3) ⇒ (1) may be seen as follows: Recall that in (15.9) we have seen that for each open compact subgroup K_n in our fixed sequence $(K_n)_{n\in\mathbb{N}}$, we have an isomorphism of G_∞-representations

$$(\mathcal{H}^{\infty_\mathbb{R}})^{K_n} \cong (\mathcal{H}_\infty)^{\infty_\mathbb{R}} \otimes \mathcal{H}_f^{K_n},$$

where $\mathcal{H}_f^{K_n}$ was finite-dimensional. In particular, there is an isomorphism of $(\mathfrak{g}_\infty, K_\infty)$-modules,

$$(\mathcal{H}^{\infty_\mathbb{R}})^{K_n}_{(K_\infty)} \cong (\mathcal{H}_\infty)^{\infty_\mathbb{R}}_{(K_\infty)} \otimes \mathcal{H}_f^{K_n},$$

which, according to Thm. 3.37, Cor. 3.34 and Thm. 3.16 simplifies to

$$(\mathcal{H}^{\infty_\mathbb{R}})^{K_n}_{(K_\infty)} \cong (\mathcal{H}_\infty)_{(K_\infty)} \otimes \mathcal{H}_f^{K_n}.$$

Hence, taking the union over all $n \in \mathbb{N}$ and reconsidering the idea of the proof of Lem. 4.35, we obtain that $\mathcal{H}^{\infty_\mathbb{A}}_{(K_\infty)}$ equals the space $\mathcal{H}_{(K_\mathbb{A})}$ of right $K_\mathbb{A}$-finite vectors in $\mathcal{H}$. Moreover, by Cor. 13.20, $\mathcal{H}^{\infty_\mathbb{A}}_{(K_\infty)}$ is irreducible and admissible. Of course, the same considerations also apply to $\mathcal{H}'$.

Invoking [Fla79], Thm. 3, we hence get for each $\mathsf{v} \in S_\infty$ irreducible $(\mathfrak{g}_\mathsf{v}, K_\mathsf{v})$-modules $(\pi_{0,\mathsf{v}}, V_{0,\mathsf{v}})$ and $(\pi'_{0,\mathsf{v}}, V'_{0,\mathsf{v}})$, and for each $\mathsf{v} \in S_f$ irreducible admissible G_v-representations $(\pi_\mathsf{v}, V_\mathsf{v})$ and $(\pi'_\mathsf{v}, V'_\mathsf{v})$, which are unramified outside a finite set $S_0 \subset S_f$, such that for every $(v^\circ_\mathsf{v})_{\mathsf{v}\in S_f\setminus S_0} \in \prod_{\mathsf{v}\in S_f\setminus S_0}(V_\mathsf{v}^{K_\mathsf{v}} \setminus \{0\})$ and, likewise, for every $(v'^\circ_\mathsf{v})_{\mathsf{v}\in S_f\setminus S_0} \in \prod_{\mathsf{v}\in S_f\setminus S_0}(V'^{K_\mathsf{v}}_\mathsf{v} \setminus \{0\})$, we obtain isomorphisms of $(\mathfrak{g}_\infty, K_\infty, G(\mathbb{A}_f))$-modules

$$\mathcal{H}_{(K_\mathbb{A})} = \mathcal{H}^{\infty_\mathbb{A}}_{(K_\infty)} \cong \bigotimes_{\mathsf{v}\in S_\infty} V_{0,\mathsf{v}} \otimes \bigotimes_{\mathsf{v}\in S_f}{}' V_\mathsf{v}.$$

and

$$\mathcal{H}'_{(K_\mathbb{A})} = \mathcal{H}'^{\infty_\mathbb{A}}_{(K_\infty)} \cong \bigotimes_{\mathsf{v}\in S_\infty} V'_{0,\mathsf{v}} \otimes \bigotimes_{\mathsf{v}\in S_f}{}' V'_\mathsf{v}.$$

As moreover all these $(\mathfrak{g}_\mathsf{v}, K_\mathsf{v})$-modules and G_v-representations are unique up to isomorphism, we must have

$$V_{0,\mathsf{v}} \underset{(\mathfrak{g}_\mathsf{v},K_\mathsf{v})}{\cong} V'_{0,\mathsf{v}} \quad \text{and} \quad V_\mathsf{v} \underset{G_\mathsf{v}}{\cong} V'_\mathsf{v} \tag{15.11}$$

for all $\mathsf{v} \in S_\infty$, respectively, for all $\mathsf{v} \in S_f$.

Now recall from (10.38) that we have decompositions of the irreducible unitary $G(\mathbb{A})$-representations $\mathcal{H}$ and $\mathcal{H}'$ into restricted Hilbert space tensor products

$$\mathcal{H} \cong \widehat{\bigotimes}'_{\mathsf{v}\in S} \mathcal{H}_\mathsf{v} \quad \text{and} \quad \mathcal{H}' \cong \widehat{\bigotimes}'_{\mathsf{v}\in S} \mathcal{H}'_\mathsf{v}. \tag{15.12}$$

Then, using Cor. 3.34 and Thm. 3.16 for the archimedean places and Lem. 4.35 for the non-archimedean places, it is not difficult to argue that (by construction) we must have isomorphisms

$$V_{0,\mathsf{v}} \underset{(\mathfrak{g}_\mathsf{v},K_\mathsf{v})}{\cong} (\mathcal{H}_\mathsf{v})_{(K_\mathsf{v})} \quad \text{and} \quad V_\mathsf{v} \underset{G_\mathsf{v}}{\cong} (\mathcal{H}_\mathsf{v})_{(K_\mathsf{v})}$$

for all $\mathsf{v} \in S_\infty$, respectively, for all $\mathsf{v} \in S_f$, and likewise for $\mathcal{H}'$. (We remark that this is the simplest part of [Fla79], Thm. 4, namely, Thm. 4.(3): Recall that we have verified that $\mathcal{H}_{(K_\mathbb{A})} = \mathcal{H}^{\infty_\mathbb{A}}_{(K_\infty)}$ and $\mathcal{H}'_{(K_\mathbb{A})} = \mathcal{H}'^{\infty_\mathbb{A}}_{(K_\infty)}$.)

Consequently, (15.11) shows that $(\mathcal{H}_\mathsf{v})_{(K_\mathsf{v})} \cong (\mathcal{H}'_\mathsf{v})_{(K_\mathsf{v})}$ as $(\mathfrak{g}_\mathsf{v}, K_\mathsf{v})$-modules, if $\mathsf{v} \in S_\infty$, respectively, as G_v-representations, if $\mathsf{v} \in S_f$. Therefore, as all $\mathcal{H}_\mathsf{v}$ and $\mathcal{H}'_\mathsf{v}$ are topologically irreducible unitary representations of G_v, our Thm. 3.40 together with Lem. 4.35 and Thm. 3.35 imply the isomorphy of the $G(\mathbb{A})$-representations (cf. Rem. 10.39)

$$\widehat{\bigotimes}'_{\mathsf{v}\in S} \mathcal{H}_\mathsf{v} \cong \widehat{\bigotimes}'_{\mathsf{v}\in S} \mathcal{H}'_\mathsf{v}.$$

As these are no other than $\mathcal{H}$ and $\mathcal{H}'$, cf. (15.12), the claim follows. □

15.2 Associate classes of cuspidal smooth-automorphic representations

We now want to generalize the considerations of §15.1 above to all associate classes $\{\mathbf{P}\}$ of standard parabolic F-subgroups of $\mathbf{G}$.

Let $\mathbf{P} = \mathbf{L} \cdot \mathbf{N}$ be a standard parabolic F-subgroup of $\mathbf{G}$ and let $\mathbf{A_P}$ be the fixed maximal F-split central torus of $\mathbf{L}$. By our conventions, cf. §9.2, $\mathbf{A_P} \subseteq \mathbf{A_{P_0}}$. We recall the spaces $\mathfrak{a}_\mathbf{P} = \mathrm{Hom}(X_F(\mathbf{L}), \mathbb{R})$ from *ibidem* and observe that $\mathfrak{a}_{\mathbf{P}_0,\mathbb{C}} := \mathfrak{a}_{\mathbf{P}_0} \otimes_\mathbb{R} \mathbb{C}$ (and hence each complexified space $\mathfrak{a}_{\mathbf{P},\mathbb{C}} := \mathfrak{a}_\mathbf{P} \otimes_\mathbb{R} \mathbb{C}$) can be extended to a Cartan subalgebra $\mathfrak{h}$ of $\mathfrak{g}_{\infty,\mathbb{C}}$, which we assume to have fixed. Then, this Cartan subalgebra $\mathfrak{h}$ is also a Cartan subalgebra of $\mathfrak{l}_{\infty,\mathbb{C}}$ (Exercise!) and so we obtain inclusions of Weyl groups

$$W(\mathfrak{g}_{\infty,\mathbb{C}}, \mathfrak{h}) \supseteq W(\mathfrak{l}_{\infty,\mathbb{C}}, \mathfrak{h}),$$

cf. [Wal89], §0.2.3. Hence, for each $\mathbf{P}$ as above, the Harish-Chandra isomorphism γ, cf. [Wal89], Thm. 3.2.3, furnishes a natural embedding

$$\imath_\mathbf{P} : \mathcal{Z}(\mathfrak{g}_\infty) \xrightarrow[\gamma]{\sim} \mathcal{U}(\mathfrak{h})^{W(\mathfrak{g}_{\infty,\mathbb{C}},\mathfrak{h})} \subseteq \mathcal{U}(\mathfrak{h})^{W(\mathfrak{l}_{\infty,\mathbb{C}},\mathfrak{h})} \xrightarrow[\gamma^{-1}]{\sim} \mathcal{Z}(\mathfrak{l}_\infty),$$

Moreover, we observe that using formally the identical arguments, the considerations of §4.3.1 carry over to the situation at hand and we obtain spaces $\check{\mathfrak{a}}^\mathbf{G}_{\mathbf{P},\mathbb{C}}$, as well as a set of roots $\Delta(\mathbf{P}, \mathbf{A_P})$ and hence a notion of positivity on $\check{\mathfrak{a}}^\mathbf{G}_{\mathbf{P},\mathbb{C}}$ as in Def. 4.24 by applying verbatim the same definitions, only keeping in mind that the ground field F is now global.

Let $\mathcal{J} \lhd \mathcal{Z}(\mathfrak{g}_\infty)$ be an arbitrary, but fixed ideal of finite codimension, and let us consider pairs $([\widetilde{\pi}], \Lambda)$, where

(i) $[\widetilde{\pi}]$ is an equivalence class of an irreducible cuspidal smooth-automorphic subrepresentation $\widetilde{\pi}$ of $L(\mathbb{A})$ and
(ii) $\Lambda : A^\mathbb{R}_P \to \mathbb{C}^*$ is a character of real Lie groups, which is trivial on $A^\mathbb{R}_G$,

satisfying the following compatibility-hypothesis with $\mathcal{J}$: Let $\lambda_0 \in \check{\mathfrak{a}}^\mathbf{G}_{\mathbf{P},\mathbb{C}}$ be the derivative of Λ and consider the irreducible smooth-automorphic subrepresentation $\pi := e^{\langle\lambda_0, H_P(\cdot)\rangle} \cdot \widetilde{\pi}$ of $L(\mathbb{A})$. We remind the reader that π is given by right-translation on the the space $\{e^{\langle\lambda_0, H_P(\cdot)\rangle} \cdot \varphi \mid \varphi \in \widetilde{\pi}\}$ of smooth-automorphic forms of $L(\mathbb{A})$. Recall now that we have verified in (the proof of) Thm. 15.7 that $\widetilde{\pi}$ is a Casselman–Wallach representation of $L(\mathbb{A})$, whence so is π. According to Thm. 12.16 we may hence factor it as

$$\pi \cong \overline{\bigotimes_{\mathsf{in}}}_{\mathsf{v}\in S_\infty} \pi_\mathsf{v} \; \overline{\otimes}_{\mathsf{in}} \; \bigotimes_{\mathsf{v}\in S_f}{}' \pi_\mathsf{v}$$

and we set $\pi_\infty := \overline{\bigotimes_{\mathsf{v}\in S_\infty}}_{\mathsf{in}} \pi_{\mathsf{v}}$. Then, the underlying $(\mathfrak{l}_\infty, K_\infty \cap L_\infty)$-module $(\pi_\infty)_0 = (\pi_\infty)_{(K_\infty \cap L_\infty)}$ is irreducible by construction, cf. the proof of Thm. 12.16 and Thm. 3.26, whence [Vog81], Prop. 0.3.19.(a), implies that every $D \in \mathcal{Z}(\mathfrak{l}_\infty)$ acts by multiplication by a scalar, i.e., there is a homomorphism $\chi_{\pi_\infty} : \mathcal{Z}(\mathfrak{l}_\infty) \to \mathbb{C}$ such that $(\pi_\infty)_0(D) = \chi_{\pi_\infty}(D) \cdot \mathrm{id}$, for all $D \in \mathcal{Z}(\mathfrak{l}_\infty)$. Our compatibility-hypothesis of $([\widetilde{\pi}], \Lambda)$ with $\mathcal{J}$ can now be stated as the assumption that

$$\chi_{\pi_\infty}(\imath_{\mathbf{P}}(\mathcal{J})) = \{0\}. \tag{15.13}$$

In order to give the definition of an associate class of smooth-automorphic subrepresentations, represented by $([\widetilde{\pi}], \Lambda)$ as above, we need one more ingredient. Let us write $W_{\mathbf{L}}$ for the Weyl group of $\mathbf{L}$ with respect to $\mathbf{A}_{\mathbf{P}_0}$, i.e., for the Weyl group attached to the root system of all $\pm\alpha$, with $\alpha \in \Delta(\mathbf{P}_0, \mathbf{A}_{\mathbf{P}_0}) \setminus \Delta(\mathbf{P}, \mathbf{A}_{\mathbf{P}})$. As in [Mœ-Wal95], II.1.7, p. 89, we write $W(\mathbf{L})$ for the set of representatives w of $W_{\mathbf{G}}/W_{\mathbf{L}}$ of minimal length (as a word in the Weyl reflections attached to a choice of simple roots), with the property that $w\mathbf{L}w^{-1}$ is again the Levi subgroup of a standard parabolic F-subgroup of $\mathbf{G}$.

To sum up, given an ideal $\mathcal{J} \lhd \mathcal{Z}(\mathfrak{g}_\infty)$ and a pair $([\widetilde{\pi}], \Lambda)$ as above, we define the *associate class of irreducible cuspidal smooth-automorphic subrepresentations*, represented by $([\widetilde{\pi}], \Lambda)$ as the family

$$\sigma(\pi) := \{\sigma([\pi])_{\mathbf{Q}}\}_{\mathbf{Q}\in\{\mathbf{P}\}}$$

of (finite) sets of equivalence classes

$$\sigma([\pi])_{\mathbf{Q}} := \{[w \cdot \pi] \mid w \in W(\mathbf{L}_{\mathbf{P}}) \text{ such that } w\mathbf{L}_{\mathbf{P}}w^{-1} = \mathbf{L}_{\mathbf{Q}}\}$$

of smooth-automorphic subrepresentations $w \cdot \pi$ of $L_Q(\mathbb{A})$, where we have set $(w \cdot \pi)(\ell) := \pi(w^{-1}\ell w)$ for $\ell \in L_Q(\mathbb{A})$.

For a fixed, but arbitrary ideal $\mathcal{J} \lhd \mathcal{Z}(\mathfrak{g}_\infty)$ of finite codimension we denote by $\Phi_{\mathcal{J},\{\mathbf{P}\}}$ the family of all associate classes $\sigma(\pi)$, represented by a pair $([\widetilde{\pi}], \Lambda)$ as above. By construction, $\Phi_{\mathcal{J},\{\mathbf{P}\}}$ only depends on the ideal $\mathcal{J} \lhd \mathcal{Z}(\mathfrak{g}_\infty)$ and the associate class $\{\mathbf{P}\}$.

Exercise 15.14. Convince yourself that for $\mathbf{P} = \mathbf{G}$, the associate classes of irreducible cuspidal smooth-automorphic subrepresentation look like $\sigma(\pi) = \{\{[\widetilde{\pi}]\}\}$, where $\widetilde{\pi}$ is an irreducible cuspidal smooth-automorphic subrepresentation of $\mathcal{A}^\infty_{cusp,\mathcal{J}}([G])$. Hence, in this way, convince yourself that the notion of associate classes of irreducible cuspidal smooth-automorphic subrepresentations represents a generalization of the notion of equivalence classes of irreducible cuspidal smooth-automorphic subrepresentations.

15.2.1 *Review of Eisenstein series*

In view of Exc. 15.14, in order to obtain the desired generalization of the decomposition of $\mathcal{A}^{\infty}_{\mathcal{J},\{\mathbf{G}\}}([G])$, given by Lem. 15.1 and Thm. 15.7, for all $\mathcal{A}^{\infty}_{\mathcal{J},\{\mathbf{P}\}}([G])$, i.e., where $\{\mathbf{P}\}$ is now any associate class of standard parabolic F-subgroups of $\mathbf{G}$, we still need to attach appropriate spaces of smooth-automorphic forms on $G(\mathbb{A})$ to the associate classes $\sigma(\pi)$ of irreducible cuspidal smooth-automorphic subrepresentations of $L(\mathbb{A})$. To this end, we will have to consider certain automorphic forms on $G(\mathbb{A})$ called *Eisenstein series*, which we will now define and whose important properties we shall now recall in due brevity.

Let $\widetilde{\pi}$ be an irreducible cuspidal smooth-automorphic subrepresentation of $L(\mathbb{A})$, i.e., letting $\mathbf{L}$ take the role of $\mathbf{G}$ in Thm. 15.7, a smooth-automorphic subrepresentation of $L(\mathbb{A})$, which is isomorphic to the space of globally smooth vectors $\mathcal{H}_i^{\infty_{\mathbb{A}}}$ for some $i \in I_{\mathcal{L}}$ and some ideal $\mathcal{L} \lhd \mathcal{Z}(\mathfrak{l}_\infty)$ of finite codimension. The $\widetilde{\pi}$-isotypic component of $\mathcal{A}^{\infty}_{cusp,\mathcal{L}}([L])$ is by definition the (closure of the) image of the natural map defined by linearly extending

$$\mathrm{Hom}_{L(\mathbb{A})}(\widetilde{\pi}, \mathcal{A}^{\infty}_{cusp,\mathcal{L}}([L])) \otimes \widetilde{\pi} \longrightarrow \mathcal{A}^{\infty}_{cusp,\mathcal{L}}([L])$$

$$\Xi \otimes \widetilde{\varphi} \mapsto \Xi(\widetilde{\varphi}).$$

Here, we understand that $\mathrm{Hom}_{L(\mathbb{A})}(\widetilde{\pi}, \mathcal{A}^{\infty}_{cusp,\mathcal{L}}([L]))$ denotes the vector space of all continuous, linear $L(\mathbb{A})$-equivariant maps $\Xi : \widetilde{\pi} \to \mathcal{A}^{\infty}_{cusp,\mathcal{L}}([L])$. Again by Thm. 15.7, the $\widetilde{\pi}$-isotypic component of $\mathcal{A}^{\infty}_{cusp,\mathcal{L}}([L])$ must be isomorphic to the $L(\mathbb{A})$-representation

$$\bigoplus_{\substack{i \in I_{\mathcal{L}} \\ \mathcal{H}_i^{\infty_{\mathbb{A}}} \cong \widetilde{\pi}}} \mathcal{H}_i^{\infty_{\mathbb{A}}} \cong \bigoplus_{k=1}^{m(\widetilde{\pi})} \widetilde{\pi} =: \widetilde{\pi}^{m(\widetilde{\pi})}, \tag{15.15}$$

where $m(\widetilde{\pi})$ denotes the finite (see Lem. 15.10 and Thm. 13.7) multiplicity of $\widetilde{\pi}$ in $\mathcal{A}^{\infty}_{cusp,\mathcal{L}}([L])$, i.e., $m(\widetilde{\pi}) = \dim_{\mathbb{C}} \mathrm{Hom}_{L(\mathbb{A})}(\widetilde{\pi}, \mathcal{A}^{\infty}_{cusp,\mathcal{L}}([L]))$. Observe that $m(\widetilde{\pi})$ is independent of $\mathcal{L}$, once $\widetilde{\pi} \subseteq \mathcal{A}^{\infty}_{cusp,\mathcal{L}}([L])$.

We consider the vector space

$$\mathcal{I}^{G(\mathbb{A})}_{P(\mathbb{A})}(\widetilde{\pi})$$

of all smooth, left $L(F)N(\mathbb{A})A_P^{\mathbb{R}}$-invariant functions $f: G(\mathbb{A}) \to \mathbb{C}$, such that for every $g \in G(\mathbb{A})$ the function $\ell \mapsto f(\ell g)$ on $L(\mathbb{A})$ is contained in the $\widetilde{\pi}$-isotypic component of $\mathcal{A}^{\infty}_{cusp,\mathcal{L}}([L])$.

It is easy to see that $\mathcal{I}_{P(\mathbb{A})}^{G(\mathbb{A})}(\widetilde{\pi})$ is stable under the action of $G(\mathbb{A})$ by right-translation R of functions and it makes hence sense to speak of K_∞-finite elements of $\mathcal{I}_{P(\mathbb{A})}^{G(\mathbb{A})}(\widetilde{\pi})$, i.e., of those $f \in \mathcal{I}_{P(\mathbb{A})}^{G(\mathbb{A})}(\widetilde{\pi})$, for which the $\mathbb{C}$-linear span of $R(K_\infty)f$ is finite-dimensional. Let us write $\mathcal{I}_{P(\mathbb{A})}^{G(\mathbb{A})}(\widetilde{\pi})_{(K_\infty)}$ for the vector subspace of K_∞-finite elements of $\mathcal{I}_{P(\mathbb{A})}^{G(\mathbb{A})}(\widetilde{\pi})$.

For a function $f \in \mathcal{I}_{P(\mathbb{A})}^{G(\mathbb{A})}(\widetilde{\pi})$, $\lambda \in \check{\mathfrak{a}}_{\mathbf{P},\mathbb{C}}^{\mathbf{G}}$ and $g \in G(\mathbb{A})$ an *Eisenstein series* may be formally defined as

$$E_P(f,\lambda)(g) := \sum_{\gamma \in P(F)\backslash G(F)} f(\gamma g)\, e^{\langle \lambda + \rho_P, H_P(\gamma g)\rangle}.$$

Here, in complete analogy to §4.3.1,

$$\rho_P := \frac{1}{2} \sum_{\alpha \in \Delta(\mathbf{P},\mathbf{A_P})} m(\alpha)\cdot \alpha,$$

where $m(\alpha)$ denotes the multiplicity of the root α in the adjoint action of $\mathbf{P}$ on $\mathbf{N}$.

The reader should note that our definition of $E_P(f,\lambda)(g)$ – so far – has been completely formal. Indeed, even if $f \in \mathcal{I}_{P(\mathbb{A})}^{G(\mathbb{A})}(\widetilde{\pi})$ is fixed, the question of convergence of $E_P(f,\lambda)(g)$ as a function on $G(\mathbb{A}) \times \check{\mathfrak{a}}_{\mathbf{P},\mathbb{C}}^{\mathbf{G}}$ is an extremely subtle one, which is far from being fully understood as yet. It is known, however, that the following holds:

Proposition 15.16. *If $f \in \mathcal{I}_{P(\mathbb{A})}^{G(\mathbb{A})}(\widetilde{\pi})$ is an arbitrary, but fixed, K_∞-finite element, the attached Eisenstein series $E_P(f,\lambda)(g)$ converges as a function on $G(\mathbb{A}) \times \check{\mathfrak{a}}_{\mathbf{P},\mathbb{C}}^{\mathbf{G}}$ absolutely and uniformly on compact subsets of*

$$G(\mathbb{A}) \times \{\lambda \in \check{\mathfrak{a}}_{\mathbf{P},\mathbb{C}}^{\mathbf{G}} |\ \lambda - \rho_P > 0\}.$$

As a consequence, for every fixed $f \in \mathcal{I}_{P(\mathbb{A})}^{G(\mathbb{A})}(\widetilde{\pi})_{(K_\infty)}$ and every fixed $\lambda \in \check{\mathfrak{a}}_{\mathbf{P},\mathbb{C}}^{\mathbf{G}}$ with the property that $\lambda - \rho_P > 0$,

$$E_P(f,\lambda): G(\mathbb{A}) \to \mathbb{C}$$

is a well-defined, K_∞-finite automorphic form.

Proof. This is a consequence of (our Lem. 11.22 and) [Mœ-Wal95], Prop. II.1.5, once we convince ourselves that all $f \in \mathcal{I}_{P(\mathbb{A})}^{G(\mathbb{A})}(\widetilde{\pi})_{(K_\infty)}$ satisfy the assumptions made in the latter reference.

To this end, observe that since $K_\mathbb{A} = K_\infty \cdot K_{\mathbb{A}_f}$, is chosen to be a maximal compact subgroup of $G(\mathbb{A})$ in good position, cf. §9.2, $K_\mathbb{A} \cap L(\mathbb{A})$ is a maximal compact subgroup of $L(\mathbb{A})$ in good position. Let $K_{L,\infty} := K_\infty \cap L_\infty$ and $K_{L,\mathbb{A}_f} := K_{\mathbb{A}_f} \cap L(\mathbb{A}_f)$. It follows from Thm. 12.10 that $\widetilde{\pi}_{(K_{L,\infty})}$ is an irreducible cuspidal automorphic representation according to Def. 13.23.

Hence, observing the shift by multiplication by the function $e^{\langle \rho_P, H_P(\cdot)\rangle}$ in our definition of Eisenstein series above and in the definition in [Mœ-Wal95], II.1.5, in order to see that $f \in \mathcal{I}_{P(\mathbb{A})}^{G(\mathbb{A})}(\widetilde{\pi})_{(K_\infty)}$ satisfies the conditions of [Mœ-Wal95], Prop. II.1.5, it is enough to prove that

$$e^{\langle \rho_P, H_P(\cdot)\rangle} \cdot f \in A(N(\mathbb{A})L(F)\backslash G(\mathbb{A}))_{\widetilde{\pi}_{(K_{L,\infty})}}. \tag{15.17}$$

Here, partly imitating the notation of [Mœ-Wal95] – see in particular their §I.2.17 and then pp. 78–79, *ibidem* – $A(N(\mathbb{A})L(F)\backslash G(\mathbb{A}))$ stands for the vector space of all smooth, left $N(\mathbb{A})L(F)$-invariant functions $f_0 : G(\mathbb{A}) \to \mathbb{C}$, which are $K_\mathbb{A}$-finite (under translation from the right), $\mathcal{Z}(\mathfrak{g}_\infty)$-finite and of moderate growth. Whereas the space $A(N(\mathbb{A})L(F)\backslash G(\mathbb{A}))_{\widetilde{\pi}_{(K_{L,\infty})}}$ denotes the vector space of all $f_0 \in A(N(\mathbb{A})L(F)\backslash G(\mathbb{A}))$, for which for all $k \in K_\mathbb{A}$ the function $\ell \mapsto e^{-\langle \rho_P, H_P(\ell)\rangle} \cdot f_0(\ell k)$ on $L(\mathbb{A})$ is contained in the $\widetilde{\pi}_{(K_{L,\infty})}$-isotypic component of $\mathcal{A}_{cusp,\mathcal{L}}([L])$. Taking one more breath, we remind the reader that by Cor. 13.20 and Lem. 15.10 the latter $\widetilde{\pi}_{(K_{L,\infty})}$-isotypic component is isomorphic as a $(\mathfrak{l}_\infty, K_{L,\infty}, L(\mathbb{A}_f))$-module to the finite direct sum

$$\widetilde{\pi}_{(K_{L,\infty})}^{m(\widetilde{\pi})} = \bigoplus_{k=1}^{m(\widetilde{\pi})} \widetilde{\pi}_{(K_{L,\infty})},$$

which the reader may equally well take as its definition.

Summarizing, we seek to verify that each $f \in \mathcal{I}_{P(\mathbb{A})}^{G(\mathbb{A})}(\widetilde{\pi})_{(K_\infty)}$ satisfies

(1) $e^{\langle \rho_P, H_P(\cdot)\rangle} \cdot f \in A(N(\mathbb{A})L(F)\backslash G(\mathbb{A}))$ and
(2) $(R(k)f)|_{L(\mathbb{A})} \in \widetilde{\pi}_{(K_{L,\infty})}^{m(\widetilde{\pi})}$ for all $k \in K_\mathbb{A}$.

In order to obtain (1), we observe that it follows trivially from the assumptions on f that $e^{\langle \rho_P, H_P(\cdot)\rangle} \cdot f$ is a smooth, left $N(\mathbb{A})L(F)$-invariant function

$G(\mathbb{A}) \to \mathbb{C}$ (recall that $L(F)N(\mathbb{A}) = N(\mathbb{A})L(F)$), which is $K_{\mathbb{A}}$-finite under right-translation ($K_{\mathbb{A}_f}$-finiteness being a consequence of smoothness, cf. §10.2). By (our Lem. 11.22 and) the proof on pp. [Mœ-Wal95] 37–38, we may hence show (1), if we prove that the function $(R(k)f)|_{L(\mathbb{A})} : \ell \mapsto f(\ell k)$ on $L(\mathbb{A})$ is automorphic for all $k \in K_{\mathbb{A}}$. We recall that by definition of $\mathcal{I}_{P(\mathbb{A})}^{G(\mathbb{A})}(\widetilde{\pi})$, we have $(R(k)f)|_{L(\mathbb{A})} \in \widetilde{\pi}^{m(\widetilde{\pi})}$ for every $k \in K_{\mathbb{A}}$.

We may hence, cf. Def. 11.21, finish the proof of (1) as well as of (2), if we manage to show that $(R(k)f)|_{L(\mathbb{A})}$ is $K_{L,\infty}$-finite for every $k \in K_{\mathbb{A}}$. So, let $k = k_\infty \cdot k_f \in K_{\mathbb{A}}$ be arbitrary but fixed. Then, inserting into the definitions and recalling that the actions of $K_{\mathbb{A}_f}$ and K_∞ commute, we immediately obtain

$$R(K_{L,\infty})\big((R(k)f)|_{L(\mathbb{A})}\big) = R(k_f)\big(R(K_{L,\infty})\big((R(k_\infty)f)|_{L(\mathbb{A})}\big)\big).$$

However, since f is K_∞-finite by assumption, so is $R(k_\infty)f$ and hence the set of linear operators $R(K_{L,\infty})\big((R(k_\infty)f)|_{L(\mathbb{A})}\big)$ spans a finite-dimensional vector space. In particular, its image under the linear map $R(k_f)$ must be finite-dimensional. This shows (1) and (2) and hence the proposition, □

Remark 15.18. In Prop. 15.16, the cuspidality of $\widetilde{\pi}$ can be relaxed to assuming that all $\varphi \in \widetilde{\pi}$ define classes in $L^2_{dis}([L])$. This does not follow from [Mœ-Wal95], Prop. II.1.5, but the reader may find the result stated as (a consequence of) Lem. 4 in [Art79] or Lem. 6.1.6 in [Sha10]. As we are not going to make any use of this more general result, we refer to [HCh68], p. 29 or [Sha13] for the idea of the proof.

In the last assertion of Prop. 15.16, $\lambda \in \check{\mathfrak{a}}^{\mathbf{G}}_{\mathbf{P},\mathbb{C}}$ is fixed and $g \in G(\mathbb{A})$ is allowed to vary arbitrarily. It will be important to understand what happens, if the roles of λ and g are interchanged, i.e., if $g \in G(\mathbb{A})$ is fixed and $\lambda \in \check{\mathfrak{a}}^{\mathbf{G}}_{\mathbf{P},\mathbb{C}}$ is allowed to vary. The following theorem is one of the pillars on which the theory of automorphic forms is built.

Theorem 15.19. *For fixed $f \in \mathcal{I}_{P(\mathbb{A})}^{G(\mathbb{A})}(\widetilde{\pi})_{(K_\infty)}$ and $g \in G(\mathbb{A})$, the map*

$$\{\lambda \in \check{\mathfrak{a}}^{\mathbf{G}}_{\mathbf{P},\mathbb{C}} |\ \lambda - \rho_P > 0\} \longrightarrow \mathbb{C}$$

$$\lambda \mapsto E_P(f,\lambda)(g)$$

may be analytically continued to a meromorphic function on all of $\check{\mathfrak{a}}^{\mathbf{G}}_{\mathbf{P},\mathbb{C}}$, whose singularities are poles, which lie along affine hyperplanes of the form

$$R_{\alpha,t} := \{\xi \in \check{\mathfrak{a}}^{\mathbf{G}}_{\mathbf{P},\mathbb{C}} | \langle \xi, \check{\alpha} \rangle = t\}$$

for some $t \in \mathbb{C}$ and some root $\alpha \in \Delta(\mathbf{P}, \mathbf{A_P})$. This entails the assertion that for each fixed $f \in \mathcal{I}_{P(\mathbb{A})}^{G(\mathbb{A})}(\widetilde{\pi})_{(K_\infty)}$, $g \in G(\mathbb{A})$ and $\lambda_0 \in \check{\mathfrak{a}}_{\mathbf{P},\mathbb{C}}^{\mathbf{G}}$, there is a neighbourhood $U \subseteq \check{\mathfrak{a}}_{\mathbf{P},\mathbb{C}}^{\mathbf{G}}$ of λ_0 and an integer $k \geq 0$, such that the function

$$U \to \mathbb{C}$$

$$\lambda \mapsto q_{\lambda_0}(\lambda) E_P(f, \lambda)(g)$$

is holomorphic, where

$$q_{\lambda_0}(\lambda) := \prod_{\alpha \in \Delta(\mathbf{P}, \mathbf{A_P})} \langle \lambda - \lambda_0, \check{\alpha} \rangle^k.$$

Proof. This is (encapsulated in) one of the main theorems of [Mœ-Wal95] and [Lan76]. More precisely, recalling our verification of the points (1) and (2), which appear in the proof of Prop. 15.16 above, for every $f \in \mathcal{I}_{P(\mathbb{A})}^{G(\mathbb{A})}(\widetilde{\pi})_{(K_\infty)}$, we refer to [Mœ-Wal95], Thm. IV.1.8.(a) and Prop. IV.1.11.(a) for analytic continuation and the fact that the poles lie along affine hyperplanes of the form $R_{\alpha,t}$ (see also [Lan76], pp. 170–171). However, for a proof of the claim of Thm. 15.19 in exactly its above form, the reader may want to consult [Ber-Lap23], Thm. 2.3, together with their definition of holomorphy in §3.1, as combined with Cor. 8.6, *ibidem.* □

Remark 15.20. The assumption of irreducibility as well as of cuspidality of $\widetilde{\pi}$ in Thm. 15.19 is in some sense artificial. Indeed, Thm. 2.3 in [Ber-Lap23] shows Thm. 15.19 *mutatis mutandis* for general admissible smooth-automorphic subrepresentations of $L(\mathbb{A})$. Their technique uses a new uniqueness result for automorphic forms and Langlands's computation of the constant term of Eisenstein series (which may be found as Prop. II.1.7 in [Mœ-Wal95]). Our formulation of Thm. 15.19 is (to a certain extent) more in line with the literature before [Ber-Lap23]. See Rem. 15.18 above, however. This match is also fine contentwise, as for our purposes, Thm. 15.19 in its more classical formulation is just right what we are going to need.

15.3 The cuspidal support of a smooth-automorphic form

Now, let

$$S(\check{\mathfrak{a}}_{\mathbf{P},\mathbb{C}}^{\mathbf{G}}) := \bigoplus_{m \geq 0} \mathrm{Sym}^m(\check{\mathfrak{a}}_{\mathbf{P},\mathbb{C}}^{\mathbf{G}})$$

be the symmetric algebra of $\check{\mathfrak{a}}_{\mathbf{P},\mathbb{C}}^{\mathbf{G}}$, where $\mathrm{Sym}^m(\check{\mathfrak{a}}_{\mathbf{P},\mathbb{C}}^{\mathbf{G}})$ denotes the symmetric tensor product of m copies of $\check{\mathfrak{a}}_{\mathbf{P},\mathbb{C}}^{\mathbf{G}}$. We remark that since the Lie algebra $\check{\mathfrak{a}}^{\mathbf{G}}_{\mathbf{P},\mathbb{C}}$ is abelian, the symmetric algebra of $\check{\mathfrak{a}}^{\mathbf{G}}_{\mathbf{P},\mathbb{C}}$ equals the universal enveloping algebra $\mathcal{U}(\check{\mathfrak{a}}^{\mathbf{G}}_{\mathbf{P},\mathbb{C}})$, whence there would not have been much need to introduce a new notion. We decided, however, to follow the usual conventions in the literature, which give preference to $S(\check{\mathfrak{a}}^{\mathbf{G}}_{\mathbf{P},\mathbb{C}})$. It is easy to see, only using the just mentioned fact that $S(\check{\mathfrak{a}}^{\mathbf{G}}_{\mathbf{P},\mathbb{C}}) = \mathcal{U}(\check{\mathfrak{a}}^{\mathbf{G}}_{\mathbf{P},\mathbb{C}})$ and our (2.17), that $S(\check{\mathfrak{a}}^{\mathbf{G}}_{\mathbf{P},\mathbb{C}})$ identifies with the space of differential operators ∂ with constant coefficients on $\check{\mathfrak{a}}^{\mathbf{G}}_{\mathbf{P},\mathbb{C}}$.

Let now $\sigma(\pi) \in \Phi_{\mathcal{J},\{\mathbf{P}\}}$ be an associate class of irreducible cuspidal smooth-automorphic subrepresentations of $L(\mathbb{A})$, represented by $([\widetilde{\pi}], \Lambda)$ and let $\lambda_0 \in \check{\mathfrak{a}}^{\mathbf{G}}_{\mathbf{P},\mathbb{C}}$ be the derivative of Λ. Then, Thm. 15.19 implies that for all $g \in G(\mathbb{A})$ the function

$$\mathcal{I}^{G(\mathbb{A})}_{P(\mathbb{A})}(\widetilde{\pi})_{(K_\infty)} \times S(\check{\mathfrak{a}}^{\mathbf{G}}_{\mathbf{P},\mathbb{C}}) \longrightarrow \mathbb{C}$$

$$(f, \partial) \mapsto \partial(q_{\lambda_0}(\lambda) E_P(f, \lambda)(g))|_{\lambda=\lambda_0}$$

is well-defined. Recalling Def. 11.21 (and using our Lem. 11.22), it can be shown that for each $f \in \mathcal{I}^{G(\mathbb{A})}_{P(\mathbb{A})}(\widetilde{\pi})_{(K_\infty)}$ and $\partial \in S(\check{\mathfrak{a}}^{\mathbf{G}}_{\mathbf{P},\mathbb{C}})$ the resulting map

$$G(\mathbb{A}) \longrightarrow \mathbb{C}$$

$$g \mapsto \partial(q_{\lambda_0}(\lambda) E_P(f, \lambda)(g))|_{\lambda=\lambda_0}$$

is a K_∞-finite automorphic form in $\mathcal{A}_{\mathcal{J}}([G])$. (Exercise!) We hence obtain a well-defined linear map by linearly extending the assignment

$$\mathcal{I}^{G(\mathbb{A})}_{P(\mathbb{A})}(\widetilde{\pi})_{(K_\infty)} \otimes S(\check{\mathfrak{a}}^{\mathbf{G}}_{\mathbf{P},\mathbb{C}}) \longrightarrow \mathcal{A}_{\mathcal{J}}([G])$$

$$f \otimes \partial \mapsto \partial(q_{\lambda_0}(\lambda) E_P(f, \lambda)(\cdot))|_{\lambda=\lambda_0},$$

whose image we will denote by $\mathcal{A}_{\mathcal{J},\{\mathbf{P}\},\sigma(\pi)}([G])$. Although a deep theorem, we only remark here that the definition of $\mathcal{A}_{\mathcal{J},\{\mathbf{P}\},\sigma(\pi)}([G])$ is – as it is indicated by our choice of notation – independent of the representatives $\mathbf{P}$ and $([\widetilde{\pi}], \Lambda)$, thanks to the functional equations satisfied by the Eisenstein series considered, cf. [Mœ-Wal95], Thm. IV.1.10 or, even more explicitly, [Ber-Lap23], Thm. 2.3.

We set

$$\mathcal{A}^\infty_{\mathcal{J},\{\mathbf{P}\},\sigma(\pi)}([G]) := \mathrm{Cl}_{\mathcal{A}^\infty_{\mathcal{J}}([G])}(\mathcal{A}_{\mathcal{J},\{\mathbf{P}\},\sigma(\pi)}([G])).$$

The following is our last main result. We refer again to [Gro-Žun23] for the original source:

Theorem 15.21. *For every associate class* $\{\mathbf{P}\}$ *of standard parabolic* F*-subgroups of* $\mathbf{G}$ *and every* $\sigma(\pi) \in \Phi_{\mathcal{J},\{\mathbf{P}\}}$*, the space* $\mathcal{A}^{\infty}_{\mathcal{J},\{\mathbf{P}\},\sigma(\pi)}([G])$ *is an LF-compatible smooth-automorphic subrepresentation and summation of functions induces an isomorphism of* $G(\mathbb{A})$*-representations,*

$$\mathcal{A}^{\infty}_{\mathcal{J},\{\mathbf{P}\}}([G]) \cong \bigoplus_{\sigma(\pi)\in\Phi_{\mathcal{J},\{\mathbf{P}\}}} \mathcal{A}^{\infty}_{\mathcal{J},\{\mathbf{P}\},\sigma(\pi)}([G]).$$

Proof. We divide our argument into several steps. Also interesting in its own right, we will first introduce another view on our spaces $\mathcal{I}^{G(\mathbb{A})}_{P(\mathbb{A})}(\widetilde{\pi})$, by relating them more closely to (both, local and global) concepts of induced representations.

Step 1: Letting $\mathbf{L}$ take the role of $\mathbf{G}$ in Thm. 13.7, denote by $\mathcal{H}$ an irreducible summand in the direct sum decomposition (13.8) of $L^2_{cusp}([L])$. Let $\mathscr{H}$ be the $\mathcal{H}$-isotypic component of $L^2_{cusp}([L])$, i.e., by definition the (closure of the) image of the natural map defined by linearly extending

$$\mathrm{Hom}_{L(\mathbb{A})}(\mathcal{H}, L^2_{cusp}([L])) \otimes \mathcal{H} \longrightarrow L^2_{cusp}([L])$$

$$\Theta \otimes [\widetilde{f}] \mapsto \Theta([\widetilde{f}]).$$

Here, $\mathrm{Hom}_{L(\mathbb{A})}(\mathcal{H}, L^2_{cusp}([L]))$ denotes the vector space of all continuous, linear $L(\mathbb{A})$-equivariant maps $\Theta : \mathcal{H} \to L^2_{cusp}([L])$. By Thm. 13.7, $\mathscr{H}$ is isomorphic to $\mathcal{H}^{m(\mathcal{H})} := \widehat{\bigoplus}_{i=1}^{m(\mathcal{H})} \mathcal{H}$ as $L(\mathbb{A})$-representation, where

$$m(\mathcal{H}) := \dim_{\mathbb{C}} \mathrm{Hom}_{L(\mathbb{A})}(\mathcal{H}, L^2_{cusp}([L]))$$

denotes the finite (see again Thm. 13.7) multiplicity of $\mathcal{H}$ in $L^2_{cusp}([L])$.

We now consider the space $I(\mathcal{H})$ of all functions $f : G(\mathbb{A}) \to \mathscr{H}$ satisfying the following two conditions:

(i) f is continuous, i.e., $f \in C(G(\mathbb{A}), \mathscr{H})$,
(ii) for all $p = \ell\, n \in L(\mathbb{A})N(\mathbb{A}) = P(\mathbb{A})$ and all $g \in G(\mathbb{A})$, it holds that

$$f(pg) = e^{\langle \rho_P, H_P(\ell)\rangle} R(\ell) f(g).$$

Recall from §9.2 that $G(\mathbb{A}) = P(\mathbb{A}) \cdot K_{\mathbb{A}}$. We define a non-degenerate Hermitian from on $I(\mathcal{H})$ by setting

$$\langle f_1, f_2\rangle_{I(\mathcal{H})} := \int_{K_{\mathbb{A}}} \langle f_1(k), f_2(k)\rangle \, dk\,,$$

where $\langle f_1(k), f_2(k)\rangle$ denotes the Hermitian form on $L^2([L])$ defined in (13.1). The Hilbert space closure of $I(\mathcal{H})$ with respect to $\langle\cdot,\cdot\rangle_{I(\mathcal{H})}$ shall be denoted by $\widehat{I_{P(\mathbb{A})}^{G(\mathbb{A})}(\mathscr{H})}$.

Recalling [Wal89], Lem. 1.5.3, shows that $\widehat{I_{P(\mathbb{A})}^{G(\mathbb{A})}(\mathscr{H})}$ is naturally a unitary representation of $G(\mathbb{A})$, whose action is defined through right-translation of functions. We call it the *global L^2-induction* of $\mathscr{H}$.

Analogously, we introduce the space $I_{P(\mathbb{A})}^{G(\mathbb{A})}(\mathscr{H}^{\infty_\mathbb{A}})$ of all functions satisfying the following two conditions:

(i) f is smooth, i.e., $f \in C^\infty(G(\mathbb{A}), \mathscr{H}^{\infty_\mathbb{A}})$,
(ii) for all $p = \ell n \in L(\mathbb{A})N(\mathbb{A}) = P(\mathbb{A})$ and all $g \in G(\mathbb{A})$, it holds that

$$f(pg) = e^{\langle \rho_P, H_P(\ell)\rangle} R(\ell) f(g).$$

As for $\mathcal{I}_{P(\mathbb{A})}^{G(\mathbb{A})}(\widetilde{\pi})$ above, we refrain from giving $I_{P(\mathbb{A})}^{G(\mathbb{A})}(\mathscr{H}^{\infty_\mathbb{A}})$ a topology, but only note that $G(\mathbb{A})$ obviously acts linearly on it by right-translation of functions. It therefore makes sense to consider its subspace of right K_∞-finite elements $I_{P(\mathbb{A})}^{G(\mathbb{A})}(\mathscr{H}^{\infty_\mathbb{A}})_{(K_\infty)}$. We call $I_{P(\mathbb{A})}^{G(\mathbb{A})}(\mathscr{H}^{\infty_\mathbb{A}})$ the *globally smooth induction* of the smooth $G(\mathbb{A})$-representation $\mathscr{H}^{\infty_\mathbb{A}}$.

We assume from now on – as we may by Thm. 15.7 – that $\mathcal{H}^{\infty_\mathbb{A}} \cong \widetilde{\pi}$ as $L(\mathbb{A})$-representations. Moreover, we will suppose for simplicity, that $m(\mathcal{H}) = 1$: Indeed, the general case of any $m(\mathcal{H}) \in \mathbb{Z}_{>0}$ will be a trivial consequence of our considerations below and Lem. 15.10, showing that $m(\mathcal{H}) = m(\widetilde{\pi})$. (The reader will only have to insert finite direct sums at the right places.)

Step 2: We claim that the vector space $\widehat{I_{P(\mathbb{A})}^{G(\mathbb{A})}(\mathscr{H})}^{\infty_\mathbb{A}}_{(K_\infty)}$ is in a natural way isomorphic to $\mathcal{I}_{P(\mathbb{A})}^{G(\mathbb{A})}(\widetilde{\pi})_{(K_\infty)}$. We will argue by passing through the globally smoothly induced representation $I_{P(\mathbb{A})}^{G(\mathbb{A})}(\mathscr{H}^{\infty_\mathbb{A}})$: Indeed, recalling the decomposition (10.38) of $\mathcal{H}$ as $G(\mathbb{A})$-representation and the isomorphism of G_∞-representations (15.9), it is easy to verify (using (4.34) for the local induced representations $\widehat{I_{P_\mathsf{v}}^{G_\mathsf{v}}(\mathcal{H}_\mathsf{v})}$), that there is a $G(\mathbb{A})$-equivariant isomorphism of vector spaces

$$\widehat{I_{P(\mathbb{A})}^{G(\mathbb{A})}(\mathscr{H})}^{\infty_\mathbb{A}} \xrightarrow{\sim} I_{P(\mathbb{A})}^{G(\mathbb{A})}(\mathscr{H}^{\infty_\mathbb{A}}).$$

From this we obtain an isomorphism of $(\mathfrak{g}_\infty, K_\infty, G(\mathbb{A}_f))$-modules

$$\widehat{I_{P(\mathbb{A})}^{G(\mathbb{A})}(\mathscr{H})}^{\infty_{\mathbb{A}}}_{(K_\infty)} \xrightarrow{\sim} I_{P(\mathbb{A})}^{G(\mathbb{A})}(\mathscr{H}^{\infty_{\mathbb{A}}})_{(K_\infty)}, \tag{15.22}$$

which is more explicitly given by assigning to a class $[f]$ on the left hand side its unique smooth representative.

Next, we consider the map

$$\begin{aligned} I_{P(\mathbb{A})}^{G(\mathbb{A})}(\mathscr{H}^{\infty_{\mathbb{A}}})_{(K_\infty)} &\longrightarrow \mathcal{I}_{P(\mathbb{A})}^{G(\mathbb{A})}(\widetilde{\pi})_{(K_\infty)} \\ \varphi &\longmapsto e^{-\langle \rho_P, H_P(\_)\rangle} \cdot \varphi(\_)(id). \end{aligned} \tag{15.23}$$

One immediately checks that this map is a well-defined, linear injection, and we leave it as an exercise to the reader to prove its surjectivity. (*Hint*: For a given $f \in \mathcal{I}_{P(\mathbb{A})}^{G(\mathbb{A})}(\widetilde{\pi})_{(K_\infty)}$, consider the function $\varphi : g \mapsto e^{\langle \rho_P, H_P(g)\rangle}.(R(g)f)$.)

Composing the maps in (15.23) and (15.22) shows that $\widehat{I_{P(\mathbb{A})}^{G(\mathbb{A})}(\mathscr{H})}^{\infty_{\mathbb{A}}}_{(K_\infty)}$ is isomorphic to $\mathcal{I}_{P(\mathbb{A})}^{G(\mathbb{A})}(\widetilde{\pi})_{(K_\infty)}$ as claimed.

Step 3: We will, yet, need another interpretation of $\widehat{I_{P(\mathbb{A})}^{G(\mathbb{A})}(\mathscr{H})}^{\infty_{\mathbb{A}}}_{(K_\infty)}$. In fact, we shall convince ourselves that

$$\widehat{I_{P(\mathbb{A})}^{G(\mathbb{A})}(\mathscr{H})}^{\infty_{\mathbb{A}}}_{(K_\infty)} = \widehat{I_{P(\mathbb{A})}^{G(\mathbb{A})}(\mathscr{H})}_{(K_{\mathbb{A}})}, \tag{15.24}$$

where the right hand side equals the space of right $K_{\mathbb{A}}$-finite vectors in $\widehat{I_{P(\mathbb{A})}^{G(\mathbb{A})}(\mathscr{H})}$: To this end, we exploit an argument, which we have already used in mild variations a few times above. Firstly, the argument used in the proof of Lem. 4.35, shows that $\widehat{I_{P(\mathbb{A})}^{G(\mathbb{A})}(\mathscr{H})}^{\infty_{\mathbb{A}}}_{(K_\infty)}$ is equal to the space of $K_{\mathbb{A}}$-finite vectors in $\widehat{I_{P(\mathbb{A})}^{G(\mathbb{A})}(\mathscr{H})}^{\infty_{\mathbb{R}}}$:

$$\begin{aligned} \widehat{I_{P(\mathbb{A})}^{G(\mathbb{A})}(\mathscr{H})}^{\infty_{\mathbb{A}}}_{(K_\infty)} &= \widehat{I_{P(\mathbb{A})}^{G(\mathbb{A})}(\mathscr{H})}^{\infty_{\mathbb{R}}} \cap \widehat{I_{P(\mathbb{A})}^{G(\mathbb{A})}(\mathscr{H})}^{\infty_f} \cap \widehat{I_{P(\mathbb{A})}^{G(\mathbb{A})}(\mathscr{H})}_{(K_\infty)} \\ &= \widehat{I_{P(\mathbb{A})}^{G(\mathbb{A})}(\mathscr{H})}^{\infty_{\mathbb{R}}} \cap \widehat{I_{P(\mathbb{A})}^{G(\mathbb{A})}(\mathscr{H})}_{(K_{\mathbb{A}_f})} \cap \widehat{I_{P(\mathbb{A})}^{G(\mathbb{A})}(\mathscr{H})}_{(K_\infty)} \\ &= \widehat{I_{P(\mathbb{A})}^{G(\mathbb{A})}(\mathscr{H})}^{\infty_{\mathbb{R}}}_{(K_{\mathbb{A}})}. \end{aligned}$$

Now, we apply the isomorphism of G_∞-representations (15.9) to $\mathcal{H}$ and use the admissibility of the local G_∞-representation $\widehat{I_{P_\infty}^{G_\infty}(\mathcal{H}_\infty)}$, cf. Prop. 4.33, in the light of Thm. 3.16 to see that taking the G_∞-smooth vectors can actually be removed, i.e., that finally $\widehat{I^{G(\mathbb{A})}_{P(\mathbb{A})}(\mathscr{H})}^{\infty_{\mathbb{A}}}_{(K_\infty)} = \widehat{I^{G(\mathbb{A})}_{P(\mathbb{A})}(\mathscr{H})}_{(K_{\mathbb{A}})}$.

Step 4: Let $I_A(\mathscr{H})$ be the vector space of all classes of left $L(F)N(\mathbb{A})A^{\mathbb{R}}_P$-invariant, right $K_{\mathbb{A}}$-finite, measurable functions $f : G(\mathbb{A}) \to \mathbb{C}$, such that for every $g \in G(\mathbb{A})$ the class represented by the function $\ell \mapsto f(\ell g)$ on $L(\mathbb{A})$ defines an element of $\mathscr{H}$. Amplifying the arguments of Step 2 above shows that the map

$$\begin{aligned} \widehat{I^{G(\mathbb{A})}_{P(\mathbb{A})}(\mathscr{H})}_{(K_{\mathbb{A}})} &\longrightarrow I_A(\mathscr{H}) \\ [f] &\mapsto [e^{-\langle \rho_P, H_P(\_)\rangle} \cdot f(\_)(id)] \end{aligned} \tag{15.25}$$

is a well-defined, linear bijection.

Step 5: Invoking our §10.2, (and Weil's restriction of scalars from F to $\mathbb{Q}$, as well as the fact that $L(F)N(\mathbb{A}) = N(\mathbb{A})L(F)$), it follows from the explanations in [Art05], p. 34, that picking the unique smooth representative of a class in $I_A(\mathscr{H})$ provides an isomorphism of vector spaces

$$I_A(\mathscr{H}) \xrightarrow{\sim} W_{P,\mathcal{H}}, \tag{15.26}$$

where $W_{P,\mathcal{H}}$ denotes the space defined in [Fra-Sch98], §1.3.

Step 6: Let us, just in this step of the proof, denote by $\mathfrak{l}$ the inverse of the isomorphism in (15.23), by $\mathfrak{a}$ the inverse of the isomorphism in (15.22), by $\mathfrak{u}$ the isomorphism (i.e., equality) in (15.24), by $\mathfrak{q}$ the isomorphism in (15.25), and, finally, by $\mathfrak{e}$ the isomorphism in (15.26). Then, composing all of them, we not only obtain an isomorphism, but even *equality*

$$\mathfrak{equal} : I^{G(\mathbb{A})}_{P(\mathbb{A})}(\widetilde{\pi})_{(K_\infty)} = W_{P,\mathcal{H}},$$

where we recall once more that $W_{P,\mathcal{H}}$ stands for the space defined in [Fra-Sch98], §1.3.

Invoking our (13.24), (13.22), Thm. 15.7 and Lem. 15.10 and the fact that we have shown in "Step 3" of the proof of Thm. 14.17, that for smooth-automorphic forms $\varphi \in \mathcal{A}^\infty_{\mathcal{J}}([G])$ our definition of negligibility, cf. Def. 14.15, matches the one to be found in [Fra-Sch98], it hence follows from [Fra-Sch98], Thm. 1.4 and Rem. 3.4, *ibidem*, that summation of functions induces a direct sum decomposition of $(\mathfrak{g}_\infty, K_\infty, G(\mathbb{A}_f))$-modules

$$\mathcal{A}_{\mathcal{J},\{\mathbf{P}\}}([G]) \cong \bigoplus_{\sigma(\pi)\in\Phi_{\mathcal{J},\{\mathbf{P}\}}} \mathcal{A}_{\mathcal{J},\{\mathbf{P}\},\sigma(\pi)}([G]).$$

Therefore, Thm. 14.17, Prop. 14.7 and Lem. 14.1 finally imply the assertion of Thm. 15.21. □

Thm. 15.21 finally allows us to give the definition of the cuspidal support of a smooth-automorphic form:

Definition 15.27. For $\varphi \in \mathcal{A}^\infty_{\mathcal{J}}([G])$, write

$$\varphi = \sum_{\{\mathbf{P}\}} \sum_{\sigma(\pi)\in\Phi_{\mathcal{J},\{\mathbf{P}\}}} \varphi_{\{\mathbf{P}\},\sigma(\pi)}$$

according to Thm. 14.17 and Thm. 15.21. The summand $\varphi_{\{\mathbf{P}\},\sigma(\pi)}$ is called the *cuspidal support of* φ *along* $\sigma(\pi)$.

In view of this book's approach to automorphic representation theory, it would be desirable to have all statements and results in §15.2.1 – §15.3 formulated purely on the level of smooth-automorphic forms, i.e., without reference to K_∞-finiteness throughout. The reader may find first, highly interesting results in this direction in [Lap08] and [Wal19]. Further achievements shall be the subject of future research.

References

[Alc85] J. Alcántara-Bode, On Grothendieck's Problem of Topologies, *Publ. RIMS, Kyoto Univ.* **21** (1985) 801–805.

[Aristoteles, Metaphysics] Aristotle's Metaphysics, edited by W. D. Ross, Vol. II, Oxford University Press (1975).

[Aristoteles, Physics] Aristotle's Physics, edited by W. D. Ross, Oxford University Press (1936).

[Art05] J. Arthur, An introduction to the trace formula, in: *Harmonic Analysis, the Trace Formula and Shimura Varieties*, J. Arthur, D. Ellwood, R. Kottwitz, eds., *Clay Math. Proc.*, **4** (2005) 1–264.

[Art79] J. Arthur, Eisenstein series and the trace formula, in: *Automorphic Forms, Representations, and L-functions*, A. Borel, W. Casselman, eds., *Proc. Sympos. Pure Math.*, Vol. XXXIII, Part I, *Amer. Math. Soc.* (1979) 253–274.

[Aue04] A. Auel, *Une démonstration d'un théorème de Bernstein sur les représentations de quasi-carré-intégrable de GLn(F) où F est un corps local non archimédien, Diplôme d'Études Approfondies (DEA) Mathématiques Pures*, (thesis at the Université Paris-Sud XI, Orsay; advisor Guy Henniart) (2004).

[Bad-Ren10] I. A. Badulescu, D. Renard, Unitary dual of GL(n) at archimedean places and global Jacquet-Langlands correspondence, *Comp. Math.* **145** (2010) 1115–1164.

[Bar03] E. M. Baruch, A proof of Kirillov's conjecture, *Ann. Math.* **158** (2003) 207–252.

[Ber74] J. N. Bernstein, All reductive p-adic groups are tame, *Funct. Anal. Appl.* **8** (1974) 91–93.

[Ber84] J. N. Bernstein, P-invariant distributions on GL(N) and the classification of unitary representations of GL(N) (non-Archimedean case), in: *Lie group representations, II* (College Park, Md., 1982/1983), (1984) 50–102.

[Ber-Krö14] J. N. Bernstein, B. Krötz, Smooth Fréchet globalizations of Harish-Chandra modules, *Israel J. Math.* **199** (2014) 45–111.

[Ber-Lap23] J. Bernstein, E. Lapid, On the meromorphic continuation of Eisenstein series, *J. Amer. Math. Soc.* (2023), DOI: https://doi.org/10.1090/J.Amer.Math.Soc./1020.

[LBer-Sch71] L. Bers, M. Schechter, Elliptic Equations, in: *Partial Differential Equations, Applied Math.*, Vol. 3A, *Amer. Math. Soc.* (1971) 131–299.

[Beu21] R. Beuzart-Plessis, Comparison of local relative characters and the Ichino-Ikeda conjecture for unitary groups, *J. Inst. Math. Jussieu* **20** (2021) 1803–1854.

[Beu-Cha-Zyd21] R. Beuzart-Plessis, P.-H. Chaudouard, M. Zydor, The global Gan-Gross-Prasad Conjecture for unitary groups: The endoscopic case, *Publ. IHES* **135** (2021) 183–336.

[Bor07] A. Borel, Automorphic Forms on Reductive Groups, in: *Automorphic Forms and Applications, IAS/Park City Mathematical Series* 12, *Amer. Math. Soc.* (2007) 7–39.

[Bor97] A. Borel, *Automorphic Forms on* $SL_2(\mathbb{R})$, Cambridge Tracts Math. **130**, Cambridge University Press (1997).

[Bor69] A. Borel, *Introduction aux Groupes Arithmétiques*, Hermann (1969).

[Bor06] A. Borel, Introduction to the Cohomology of Arithmetic Groups, In: *Lie Groups and Automorphic Forms*, Amer. Math. Soc./IP Studies in Advanced Mathematics **37**, *Amer. Math. Soc.*/International Press (2006) 51–86.

[Bor91] A. Borel, *Linear Algebraic Groups*, Second Enlarged Edition, Graduate Texts in Math. 126, Springer (1991).

[Bor72] A. Borel, *Representations de Groupes Localement Compacts*, Lecture Notes in Mathematics, **276**, Springer (1972).

[Bor63] A. Borel, Some finiteness properties of adele groups over number fields, *Publ. IHES* **16** (1963) 5–30.

[Bor-Jac79] A. Borel, H. Jacquet, Automorphic forms and automorphic representations, in: *Automorphic Forms, Representations, and L-functions*, A. Borel, W. Casselman, eds., *Proc. Sympos. Pure Math.*, Vol. XXXIII, Part I, *Amer. Math. Soc.* (1979) 189–202.

[BLS96] A. Borel, J.-P. Labesse, J. Schwermer, On the cuspidal cohomology of Sarithmetic subgroups of reductive groups over number fields, *Compos. Math.* **102** (1996) 1–40.

[Bor-Ser73] A. Borel, J.-P. Serre, Corners and Arithmetic Groups, *Commentarii Math. Helv.* **48** (1973) 436–491.

[Bor-Wal00] A. Borel, N. Wallach, *Continuous cohomology, discrete subgroups and representations of reductive groups*, Ann. of Math. Studies **94**, Princeton Univ. Press (2000).

[BLR90] S. Bosch, W. Lütkebohmert, M. Raynaud, *Néron Models*, Ergebnisse der Mathematik und ihrer Grenzgebiete **21**, Springer (1990).

[BouI04] N. Bourbaki, *Integration I*, Springer (2004).

[BouII04] N. Bourbaki, *Integration II*, Springer (2004).

[Bou03] N. Bourbaki, *Topological Vector Spaces*, Chapters 1–5, Springer (2003).

[Bru-Tit72] F. Bruhat, J. Tits, Groupes réductifs sur un corps local: I. Données radicielles valuées, *Publ. IHES* **41** (1972) 5–251.

[Bru-Tit84] F. Bruhat, J. Tits, Groupes réductifs sur un corps local: II. Schémas en groupes. Existence d'une donnée radicielle valuée, *Publ. IHES* **60** (1984) 5–84.

[Bum98] D. Bump, *Automorphic forms and Representations*, Cambridges Studies in Advanced Mathematics **55**, Cambridge Univ. Press (1998).

[Bus-Hen06] C. Bushnell, G. Henniart, *The Local Langlands Conjecture for GL(2)*, Grundlehren der math. Wissenschaften **335**, Springer (2006).

[Car79] P. Cartier, Representations of p-adic groups: A survey, in: *Automorphic Forms, Representations, and L-functions*, A. Borel, W. Casselman, eds., *Proc. Sympos. Pure Math.*, Vol. XXXIII, Part I, *Amer. Math. Soc.* (1979) 111–155.

[Cas89I] W. Casselman, Canonical extensions of Harish-Chandra modules to representations of G, *Can. J. Math.* **41** (1989) 385–438.

[Cas22] W. Casselman, Continuous representations, unpublished notes, dated Sept. 9, 2022, available at https://personal.math.ubc.ca/~cass/research/pdf/Continuous.pdf.

[Cas89II] W. Casselman, Introduction to the Schwartz space of $\Gamma\backslash G$, *Can. J. Math.* **40** (1989) 285–320.

[Cas95] W. Casselman, Introduction to the theory of admissible representations of p-adic reductive groups, unpublished notes (1995), available at https://personal.math.ubc.ca/~cass/research/pdf/p-adic-book.pdf.

[Cas12] W. Casselman, Remarks on Macdonald's book on p-adic spherical functions, unpublished notes, dated August 25, 2012, available at https://personal.math.ubc.ca/~cass/research/pdf/Macdonald.pdf.

[CHM00] W. Casselman, H. Hecht, D. Miličić, Bruhat filtrations and Whittaker vectors for real groups, in: *The Mathematical Legacy of Harish-Chandra: A Celebration of Representation Theory and Harmonic Analysis*, R. S. Doran, V. S. Varadarajan, eds., *Proc. Sympos. Pure Math.*, Vol. LXVIII, *Amer. Math. Soc.* (2000) 151–190.

[Cas-Sha98] W. Casselman, F. Shahidi, On irreducibility of standard modules for generic representations, *Ann. Sci. ENS* **31** (1998) 561–589.

[Cog04] J. Cogdell, Lectures on L-functions, Converse Theorems, and Functoriality for GLn, in: *Lectures in Automorphic L-functions, Fields Institute Monographs* **20**, *Amer. Math. Soc.*/Fields Institute (2004) 1–96.

[Con14] B. Conrad, Reductive group schemes, in: *Autour des Schémas en Groupes, École d'été "Schémas en groupes"*, Group Schemes, A celebration of SGA3, Vol. I, *Société Math. France* (2014) 93–458.

[Corvallis] Automorphic Forms, Representations, and L-functions, A. Borel, W. Casselman, eds., Proc. Sympos. Pure Math., Vol. XXXIII, Part I & II, *Amer. Math. Soc.* (1979).

[Dei10] A. Deitmar, *Automorphic Forms*, Universitext, Springer (2010).

[DKV84] P. Deligne, D. Kazdhan, M.-F. Vignéras, Représentations des algèbres centrales simples p-adiques, in: *Représentations des groupes réductifs sur un corps local*, J. N. Bernstein, P. Deligne, D. Kazdhan, M.-F. Vignéras, eds., Travaux en cours, Hermann (1984) 33–117.

[Dem-Gro70] M. Demazure, A. Grothendieck, Schémas en groupes III, *Lecture Notes in Mathematics*, **153**, Springer (1970).

[Dix77] J. Dixmier, *C-Algebras*, North-Holland Publishing Company (1977).

[Dix57] J. Dixmier, Sur les représentations unitaires des groupes de Lie algébriques, *Ann. Inst. Fourier* **7** (1957) 315–328.

[Dix-Mal78] J. Dixmier, P. Malliavin, Factorisations de fonctions et de vecteurs indèfiniment différentiables, *Bull. Sci. Math.* **102** (1978) 307–330.

[EGHLSVY11] P. Etingof, O. Golberg, S. Hensel, T. Liu, A. Schwendner, D. Vaintrob, E. Yodovina, *Introduction to Representation Theory*, Student math. library **59**, *Amer. Math. Soc.* (2011).

[Fla79] D. Flath, Decomposition of representations into tensor products, in: *Automorphic Forms, Representations, and L-functions*, A. Borel, W. Casselman, eds., Proc. Sympos. Pure Math., Vol. XXXIII, Part I, *Amer. Math. Soc.* (1979) 179–183.

[Fra-Sch98] J. Franke, J. Schwermer, A decomposition of spaces of automorphic forms, and the Eisenstein cohomology of arithmetic groups, *Math. Ann.* **311** (1998) 765–790.

[Gar05] P. Garrett, *Factorization of unitary representations of adele groups*, unpublished notes, dated February 19, 2005, available at https://www-users.cse.umn.edu/~garrett/m/v/factoring_repns.pdf.

[Gar09] P. Garrett, *Proving Admissibility of Irreducible Unitaries*, unpublished notes, dated January 15, 2009, available at https://www-users.cse.umn.edu/~garrett/m/v/proving_admissibility.pdf.

[Gar14] P. Garrett, *Unitary representations of topological groups*, unpublished notes, dated July 28, 2014, available at https://www-users.cse.umn.edu/~garrett/m/v/unitary_of_top.pdf.

[GGPS69] I. M. Gel'fand, M. I. Graev, I. I. Piatetski-Shapiro, Generalized functions, Volume 6: Representation Theory and Automorphic Forms, *Amer. Math. Soc.* (1969).

[God66] R. Godement, *The Spectral Decomposition of Cusp-Forms*, in: *Algebraic Groups and Discontinuous Subgroups*, A. Borel, G. D. Mostow, eds., Proc. Sympos. Pure Math., Vol. IX, Amer. Math. Soc. (1966) 223–234.

[Gol-Hun11] D. Goldfeld, J. Hundley, *Automorphic Representations and LFunctions for the General Linear Group: Volume 1 & 2*, Cambridge Studies in Advanced Mathematics **129**, Cambridge Univ. Press (2011).

[Gro-Žun23] H. Grobner, S. Žunar, On the notion of the parabolic and the cuspidal support of smooth-automorphic forms and smooth-automorphic representations, preprint (2023), available at http://homepage.univie.ac.at/harald.grobner/papers/SmoothAutomorphicForms.pdf.

[AGro66] A. Grothendieck, Produits tensoriels topologiques et espaces nucléaires, *Memoires Amer. Math. Soc.* **16**, Amer. Math. Soc. (1966).

[AGro54] A. Grothendieck, Sur les espaces (F) et (DF), *Summa Brasil. Math.* **3** (1954) 57–123.

[AGro73] A. Grothendieck, Topological vector spaces, Notes on mathematics and its applications, Gordon and Breach (1973).

[Hai-Ros10] T. J. Haines, S. Rostami, The Satake isomorphism for special maximal parahoric Hecke algebras, *Representation Th.* **14** (2010) 264–284.

[HCh68] Harish-Chandra, Automorphic forms on semisimple Lie groups, Notes by J. G. M. Mars, *Lecture Notes in Mathematics*, **68**, Springer (1968).

[HCh66] Harish-Chandra, Discrete series for semisimple Lie groups. II. Explicit determination of the characters, *Acta Math.* **116** (1966) 1–111.

[HCh53] Harish-Chandra, Representations of a semisimple Lie group on a Banach space I, *Trans. Amer. Math. Soc.* **75** (1953) 185–243.

[HCh54] Harish-Chandra, Representations of a semisimple Lie groups II, *Trans. Amer. Math. Soc.* **76** (1954) 26–65.

[Hei-Opd13] V. Heiermann, E. Opdam, On the tempered L-functions conjecture, *Amer. J. Math.* **135** (2013) 777–799.

[Hel78] S. Helgason, *Differential Geometry, Lie Groups, and Symmetric Spaces*, Academic Press (1978).

[Hen07] G. Henniart, *Representations of Reductive Groups over Local Non-Archimedean Fields*, unpublished notes of the Summer School and Conference on Automorphic Forms and Shimura Varieties, 9–27 July 2007, (2007), available at https://indico.ictp.it/event/a06207/session/5/contribution/3/material/0/0.pdf.

[Jac77] H. Jacquet, Generic representations, in: *Non-Commutative Harmonic Analysis*, J. Carmona, M. Vergne, eds., *Lecture Notes in Mathematics*, **587**, Springer (1977) 91–101.

[Jac-Lan70] H. Jacquet, R. P. Langlands, Automorphic Forms on GL(2), *Lecture Notes in Mathematics*, **114**, Springer (1970).

[Jar81] H. Jarchow, *Locally Convex Spaces, Mathematische Leitfäden*, Teubner (1981).

[JSZ11] D. Jiang, B. Sun, C.-B. Zhu, Uniqueness of Ginzburg–Rallis Models: The Archimedean Case, *Trans. Amer. Math. Soc.* **363** (2011) 2763–2802.

[Joh76] R. A. Johnson, Representation of compact groups on topological vector spaces: Some remarks, *Proc. Amer. Math. Soc.* **61** (1976) 131–136.

[Kas-Rot70] M. J. Kascic Jr., B. Roth, A Closed Subspace of $\mathcal{D}(\Omega)$ which is not a LF-Space, *Proc. Amer. Math. Soc.* **24** (1970) 801–802.

[Key82] C.G. Keys, On the Decomposition of Reducible Principal Series Representations of p-adic Chevalley Groups, *Pacific J. Math.* **101** (1982) 351–388.

[Kha82] S. M. Khaleelulla, *Counterexamples in Topological Vector Spaces*, Lecture Notes in Mathematics, **936**, Springer (1982).

[Kna76] A. W. Knapp, Commutativity of intertwining operators II, *Bull. Amer. Math. Soc.* **82** (1976) 271–273.

[Kna82] A. W. Knapp, Commutativity of intertwining operators for semisimple groups, *Compos. Math.* **46** (1982) 33–84.

[Kna02] A. W. Knapp, *Lie Groups Beyond an Introduction*, Progress in Mathematics, **140**, Birkhäuser (2002).

[Kna94] A. W. Knapp, Local Langlands correspondence: The Archimedean case, in: *Motives, Proc. Sympos. Pure Math.*, Vol. LV, Part II, American Mathematical Society, (1994) 393–410.

[Kna86] A. W. Knapp, *Representation Theory of Semisimple Groups — An Overview Based on Examples*, Princeton Univ. Press (1986).

[Kna-Vog95] A. W. Knapp, D. A. Vogan Jr., *Cohomological Induction and Unitary Representations*, Princeton Mathematical Series **45**, Princeton Univ. Press (1995).

[Kna-ZucI82] A. W. Knapp, G. J. Zuckerman, Classification of irreducible tempered representations of semisimple groups, Part I, *Ann. Math.* **116** (1982) 389–455.

[Kna-ZucII82] A. W. Knapp, G. J. Zuckerman, Classification of irreducible tempered representations of semisimple groups, Part II, *Ann. Math.* **116** (1982) 457–501.

[Kna-Zuc80] A. W. Knapp, G. J. Zuckerman, Multiplicity one fails for p-adic unitary principal series, *Hiroshima Math. J.* **10** (1980) 295–309.

[Kni-Li06] A. Knightly, C. Li, *Traces of Hecke Operators*, Mathematical Surveys Monographs **133**, *American Mathematical Society* (2006).

[Kon03] T. Konno, A note on Langlands' classification and irreducibility of induced representations of p-adic groups, *Kyushu J. Math.* **57** (2003) 383–409.

[Kos78] B. Kostant, On Whittaker vectors and representation theory, *Invent. Math.* **48** (1978) 101–184.

[Köt79] G. Köthe, *Topological Vector Spaces II, Grundlehren der math. Wissenschaften* **237**, Springer (1979).

[Kri-Mic97] A. Kriegl, P. Michor, *The Convenient Setting of Global Analysis*, Mathematical Surveys Monographs **53**, *American Mathematical Society* (1997).

[SLan02] S. Lang, *Algebra, Graduate Texts in Math.* **211**, Springer (2002).

[Lan79I] R. P. Langlands, Automorphic representations, Shimura varieties, and motives. Ein Märchen in: *Automorphic Forms, Representations, and Lfunctions*, A. Borel, W. Casselman, eds., *Proc. Sympos. Pure Math.*, Vol. XXXIII, Part II, American Mathematical Society (1979) 205–246.

[Lan89] R. P. Langlands, On the classification of irreducible representations of real algebraic groups, in: *Representation Theory and Harmonic Analysis of Semisimple Lie groups*, *American Mathematical Society* (1989) 101–170.

[Lan76] R. P. Langlands, On the Functional Equations Satisified by Eisenstein Series, *Lecture Notes in Mathematics*, **544**, Springer (1976).

[Lan79II] R. P. Langlands, On the notion of an automorphic representation. A supplement to the preceding paper in: *Automorphic Forms, Representations, and L-functions*, A. Borel, W. Casselman, eds., *Proc. Sympos. Pure Math.*, Vol. XXXIII, Part I, American Mathematical Society (1979) 203–207.

[Lap08] E. Lapid, A Remark on Eisenstein Series, in: *Eisenstein Series and Applications, Progress in Mathematics*, **258**, Birkhäuser (2008) 239–249.

[Lap-Mín15] E. Lapid, A. Míguez, On parabolic induction on inner forms of the general linear group over a non-archimedean local field, *Select. Math.* **22** (2016) 2347–2400.

[Lar73] R. Larsen, *Functional Analysis. An Introduction*, Marcel Dekker Inc. (1973).

[McD71] I. G. Macdonald, Spherical functions on a group of p-adic type, Ramanujan Institute Lecture Notes, Ramanujan Inst./Univ. Madras (1971).

[GMac53] G. W. Mackey, Induced representations of locally compact groups II. The Frobenius reciprocity theorem, *Ann. Math.* **58** (1953) 193–221.

[MMac17] M. Maculan, Maximality of hyperspecial compact subgroups avoiding Bruhat-Tits theory, *Ann. Inst. Fourier* **67** (2017) 1–21.

[DMil77] D. Miličić, Asymptotic behavior of matrix coefficients of the discrete series, *Duke Math. J.* **44** (1977) 59–88.

[JMil17] J. S. Milne, *Algebraic Groups — The Theory of Group Schemes of Finite Type over a Field*, Cambridge Studies in Advanced Mathematics **170**, Cambridge Univ. Press. (2017).

[Mín11] A. Mínguez, Unramified Representations of Unitary Groups in: *Stabilization of the Trace Formula, Shimura Varieties, and Arithmetic Applications Volume 1: On the Stabilization of the Trace Formula*, eds. L. Clozel, M. Harris, J.-P. Labesse, B. C. Ngô, Int. Press (2011), 389–410.

[Mín-Séc13] A. Mínguez, V. Sécherre, Représentations banales de $\mathrm{GL}_m(\mathrm{D})$, *Compos. Math.* **149** (2013) 679–704.

[Mín-Séc14] A. Mínguez, V. Sécherre, Représentations lisses modulo ℓ de $\mathrm{GL}_m(\mathrm{D})$, *Duke Math. J.* **163** (2014) 795–887.

[Mir86] I. Mirković, *Classification of Irreducible Tempered Representations of Semisimple Groups*, PhD-dissertation, Univ. Utah, Salt Lake City (1986).

[Mœ-Wal95] C. Moeglin, J.-L. Waldspurger, *Spectral Decomposition and Eisenstein Series*, Cambridge Univ. Press (1995).

[Mui-Žun20] G. Muić, S. Žunar, On the Schwartz space $\mathcal{S}(G(k)\backslash G(\mathbb{A}))$, *Monatsh. Math.* **192** (2020) 677–720.

[Neu99] J. Neukrich, *Algebraic Number Theory*, Grundlehren der math. Wissenschaften **322**, Springer (1999).

[Oes84] J. Oesterlé, Nombres de Tamagawa et groupes unipotents en caractéristique p, *Invent. Math.* **78** (1984) 13–88.

[Osb-War81] M. S. Osborne, G.Warner, *The Theory of Eisenstein Systems, Pure and Applied Mathematics*, **99**, Academic Press (1981).

[Pey87] M. R. Peyrovian, Maximal Compact Normal Subgroups, *Proc. Amer. Math. Soc.* **99** (1987) 389–394.

[Pla-Rap94] V. P. Platonov, A. Rapinchuck, *Algebraic Groups and Number Theory, Pure and Applied Mathematics*, **139**, Academic Press (1994).

[Rei95] M. Reid, *Undergraduate Commutative Algebra, London Math. Soc. Student Texts* **29**, Cambridge Univ. Press (1995).

[Ren10] D. Renard, Représentations des groupes réductifs p-adiques, Cours Specialises: Collection Soc. Math. France **17**, American Mathematical Society (2010).

[Rod75] F. Rodier, *Modèles de Whittaker des représentations admissibles des groupes réductifs p-adiques quasi-déploés*, unpublished article, dated 1975 and available at http://iml.univ-mrs.fr/~odier/Whittaker.pdf.

[Rod82] F. Rodier, Représentations de $GL(n, k)$ où k est un corps p-adique, *Astérisque* **92–93** (1982) 201–218.

[Rud70] W. Rudin, *Real and Complex Analysis*, McGraw-Hill (1970).

[Séc09] V. Sécherre, Proof of the Tadić conjecture (U0) on the unitary dual of $\mathrm{GL}_m(D)$, *J. reine angew. Math.* **626** (2009) 187–203.

[Ser97] J.-P. Serre, *Galois Cohomology, Springer Monographs in Mathematics*, Springer (1997).

[Sha90] F. Shahidi, A proof of Langlands' conjecture on Plancherel measures; complementary series of p-adic groups, *Ann. Math.* **132** (1990) 273–330.

[Sha13] F. Shahidi, An overview of the theory of Eisenstein series, in: *p-adic Representations, Θ-Correspondence, and the Langlands-Shahidi Theory*, Y.

Yangbo, T. Ye, eds., Lecture Series of Modern Number Theory, Vol. 1, Science Press (2013) 108–133.

[Sha10] F. Shahidi, *Eisenstein series and Automorphic L-functions*, Colloquium publications **58**, American Mathematical Society (2010).

[Sha85] F. Shahidi, Local coefficients as Artin factors for real groups, *Duke Math. J.* **52** (1985) 973–1007.

[JSha74] J. A. Shalika, The multiplicity one theorem for GL_n, *Ann. Math.* **100** (1974) 171–193.

[Sil82] A. J. Silberger, Asymptotics and lntegrability properties for matrix coefficients of admissible representations of reductive p-adic groups, *J. Funct. Ana.* **45** (1982) 391–402.

[Sil79] A. J. Silberger, *Introduction to Harmonic Analysis on Reductive p-adic Groups*, Princeton Univ. Press/Univ. Tokyo Press (1979).

[Sil78] A. J. Silberger, The Langlands quotient theorem for p-adic groups, *Math. Ann.* **236** (1978) 95–104.

[Soe88] W. Soergel, An irreducible not admissible Banach representation of $SL(2, \mathbb{R})$, *Proc. Amer. Math. Soc.* **104** (1988) 1322–1324.

[Spe-Vog80] B. Speh, D. A. Vogan, Reducibility of generalized principal series representations, *Acta Math.* **145** (1980) 227–299.

[Spr09] T. A. Springer, *Linear Algebraic Groups*, Birkhäuser (2009).

[Spr79] T. A. Springer, Reductive Groups, in: *Automorphic Forms, Representations, and L-functions*, A. Borel, W. Casselman, eds., *Proc. Sympos. Pure Math.*, Vol. XXXIII, Part I, American Mathematical Society (1979) 3–27.

[Spr94] T. A. Springer, Reduction theory over global fields, *Proc. Indian Acad. Sci. (Math. Sci.)* **104** (1994) 207–216.

[Swi01] H. P. F. Swinnerton-Dyer, *A Brief Guide to Algebraic Number Theory*, London Math. Soc. Student Texts **50**, Cambridge Univ. Press (2001).

[Tad90] M. Tadic, Induced representations of $\mathrm{GL}(n, A)$ for p-adic division algebras A, *J. Reine Angew. Math.* **405** (1990), 48–77.

[Tad94] M. Tadic, Representations of classical p-adic groups, in: *Representations of Lie Groups and Quantum Groups*, Pitman Research Notes in Mathematics Series, **311**, Longman Scientific & Technical (1995) 129–204.

[Tit79] J. Tits, Reductive groups over local fields, in: *Automorphic Forms, Representations, and L-functions*, A. Borel, W. Casselman, eds., Proc. Sympos. Pure Math., Vol. XXXIII, Part I, American Mathematical Society (1979) 29–69.

[Tre70] F. Treves, *Topological Vectors Spaces, Distributions and Kernels*, 3rd reprinting, Academic Press (1970).

[Var77] V. S. Varadarajan, *Harmonic Analysis on Real Reductive Groups*, Lecture Notes in Mathematics, **576**, Springer (1977).

[Vog78] D. A. Vogan, Jr., Gelfand–Kiriliov dimension for Harish-Chandra modules, *Invent. Math.* **48** (1978) 75–98.

[Vog81] D. A. Vogan, Jr., Representations of real reductive Lie groups, *Progress in Mathematics*, **15**, Birkhäuser (1981).

[Vog86] D. A. Vogan Jr., The unitary dual of $GL(n)$ over an Archimedean field, *Invent. Math.* **83** (1986) 449–505.

[Vog08] D. A. Vogan, Jr., Unitary Representations and Complex Analysis, in: *Representation Theory and Complex Analysis, Lecture Notes in Mathematics*, **1931**, Springer (2008) 259–344.

[Wld03] J.-L. Waldspurger, La formule de Plancherel pour les groupes p-adiques, *J. Inst. Math. Jussieu* **2** (2003) 235–333.

[Wal94] N. R. Wallach, C^∞ vectors, in: *Representations of Lie Groups and Quantum Groups*, Pitman Research Notes Math. Soc. **311**, Longman Scientific & Technical (1994) 205–270.

[Wal89] N. R. Wallach, *Real Reductive Groups I*, Pure and Applied Math. **132**, Academic Press (1989).

[Wal92] N. R. Wallach, *Real Reductive Groups II*, Pure and Applied Math. **132** II, Academic Press (1992).

[Wal19] N. R. Wallach, *The Meromorphic Continuation of* C^∞ *Eisenstein Series*, preprint, DOI: 10.13140/RG.2.2.23874.63689 (2019).

[WarI72] G. Warner, *Harmonic Analysis on Semi-Simple Lie Groups I*, Grundlehren der math. Wissenschaften **188**, Springer (1972).

[WarII72] G. Warner, *Harmonic Analysis on Semi-Simple Lie Groups II*, Grundlehren der math. Wissenschaften **189**, Springer (1972).

[Was82] L. C. Washington, *Introduction to cyclotomic fields*, Graduate Texts in Math. **83**, Springer (1982).

[Wei74] A. Weil, *Basic Number Theory*, Springer (1974).

[Wol07] J. A. Wolf, *Harmonic Analysis on Commutative Spaces, Math. Surveys Monogr.* **142**, Amer. Math. Soc. (2007).

[Zel80] A. V. Zelevinski, Induced representations of reductive p-adic groups. II. On irreducible representations of GL(n), *Ann. sci. ENS* **13** (1980) 165–210.

Index

Printed in the United States
by Baker & Taylor Publisher Services